Studies of Brain Function, Vol. 12

Studies of Brain Function

Volumes already published in the series:

1 *W. Heiligenberg*
Principles of Electrolocation and Jamming Avoidance in Electric Fish

2 *W. Precht*
Neuronal Operations in the Vestibular System

3 *J. T. Enright*
The Timing of Sleep and Wakefulness

4 *H. Braak*
Architectonics of the Human Telencephalic Cortex

5 *H. Collewijn*
The Oculomotor System of the Rabbit and Its Plasticity

6 *M. Abeles*
Local Cortical Circuits

7 *G. Palm*
Neural Assemblies

8 *J. Hyvärinen*
The Parietal Cortex of Monkey and Man

9 *E. Zrenner*
Neurophysiological Aspects of Color Vision in Primates

10 *U. Bässler*
Neural Basis of Elementary Behavior in Stick Insects

11 *G. A. Orban*
Neuronal Operations in the Visual Cortex

12 *M. Heisenberg, R. Wolf*
Vision in Drosophila

M. Heisenberg R. Wolf

Vision in Drosophila

Genetics of Microbehavior

With 112 Figures

Springer-Verlag
Berlin Heidelberg New York Tokyo 1984

Prof. Dr. MARTIN HEISENBERG
REINHARD WOLF
Institut für Genetik und Mikrobiologie
Röntgenring 11
8700 Würzburg

ISBN-13: 978-3-642-69937-5 e-ISBN-13: 978-3-642-69936-8
DOI: 10.107/978-3-642-69936-8

Library of Congress Cataloging in Publication Data. Heisenberg, Martin, 1940– . Vision in Drosophila. (Studies of brain function; vol. 12). Bibliography: p. . Includes index. 1. Drosophila – Behavior. 2. Drosophila – Genetics. 3. Drosophila – Physiology. 4. Behavior genetics. 5. Insects – Behavior. 6. Insects – Genetics. 7. Insects – Physiology. I. Wolf, Reinhard, 1947– . II. Title. III. Series: Studies of brain function; vol. 12) QL537.D76H45 1984 595.77′1 84-14088

Softcover reprint of the hardcover 1st edition 1984

Media conversion: Daten- und Lichtsatz-Service, Würzburg

2131/3130-543210

Preface

This book is written for our friends in Germany and abroad to give an account of our main interests during the last years. We gratefully take the opportunity provided by Springer Verlag with the Series *Studies of Brain Function* to present some of our recent work on vision in *Drosophila* in a coherent form and in a broader context than publications in periodicals would permit. We want to picture the visual system of the fly fitting it into the natural history from which our own vision emerged. In this perspective genetics seems to be a useful approach.

Our study is a rather personal exercise. We have neither the expertise nor the ambition to cover the whole literature or to compile all the knowledge about vision in flies. We are not even in a position to present, with the appropriate weight and breadth, the pioneering contributions of those who established this field of research. Our account mirrors our own intersects. We are selecting and interpreting other peoples' work from this subjective point of view, although we try to present it in its own right. For several years we have been fascinated by the first glimpses of the "inner life" of an insect and we hope to convey this fascination to those who are not discouraged by the scientific procedure.

To a large extent the book reviews published work, but it also describes many new experiments which have not been published before. From the published work we largely omit details and present conclusions without the evidence supporting them. Our new experiments we elaborate more throughly and in some cases we even include experimental procedures in the figures or figure captions. We try to show how these experiments should be understood in a broader context. Views, reviews, and new results are only partially separated into different chapters. The reader will find "views" mainly in *Introduction*, *Synopsis* and the introductory paragraphs to some other chapters. Chapter 1 to 4, 10, and 13 are to a large extent summaries of other peoples' work; new data from our own laboratory are to be found mainly in Chapters 5 to 9, 11, and 12.

We thank all our friends and colleagues who contributed to this book in one way or another. Without their support the book would not have been written. V. Braitenberg encouraged us to start this enterprise. His interest in general problems of brain function was a constant source of inspiration. F. Huber, over the years, shared with us his enthusaism in the discussion of central brain function of insects. W. Reichardt tought us a basic understanding of his theoretical approach to fly vision. We regard this approach as a necessary step in behavioral analysis. K. G. Götz, E. Buchner and K. F. Fischbach repeatedly acted as consultants in the evaluation and design of experiments. Their expertise has been a permanent resource in our work and a comfort for our sleep. I. Meinertzhagen, L. Schneider and N. Strausfeld generously provided unpublished material. We consider it a special honor to include their original work in the book. V. Braitenberg, E. Buchner, D. Byers, P. Coombe, K. F. Fischbach, K. G. Götz, F. Huber and W. Reichardt kindly went to the trouble of reading parts or all of the book in a preliminary form. They corrected errors in details, erradicated all too biased opinions and helped in smoothing and polishing. D. Byers, in particular, went through the sections on *Initiation* and scrutinized every sentence. E. Buchner, K. F. Fischbach, K. G. Götz and W. Reichardt generously granted permission to use their figures. K. Leopold, K. Schulze, M. Stengl, S. Wagner, J. Winter and B. Wolf typed, with great patience, parts of the manuscript. D. Dodel and A. Ledermann helped in the preparation of drawings. Their friendly cooperation made work fun.

We dedicate the book to our wives who created the time for the book in our well-packed schedules.

Würzburg, Summer 1984

M. Heisenberg
R. Wolf

Contents

Introduction

Initiating Activity 1
Input–Output Analysis 3
Orientation and Orientedness 5
Neurogenetics 6

Part A: Eye, Brain, and Simple Behavior

1 The Compound Eye 10

The Distribution of Optical Axes 10
Neural Superposition 13
The Equator 16
Angular Sensitivity of Visual Elements 17
Screening Pigments 17
Electrical Responses to Light 18
Spectral Sensitivity 20
Polarization Sensitivity 22
Ocelli 23
Summary 23

2 Neuronal Architecture of the Visual System . . . 25

Retinotopic Maps 25
Visual Foci 26
Lamina ganglionaris 27
Medulla, Lobula, Lobula Plate 31

3 Motion Sensitivity Under Open Loop Conditions . 33

Orientation 33
Optomotor Yaw Torque and Turning Behavior . . 34
Optomotor Equilibrium 43
Optomotor Roll and Pitch Responses 44
Translatory Optomotor Responses 46
Optomotor Posture Control by Visual Motion . . 48
Visually Induced Start of Flight and Landing . . 49

4 Toward Correlating Structure and Function . . . 52

Retinal Subsystems 52
Color Vision 53
Polarization Sensitivity in Behavior 56
Dark Adaptation on Account of Acuity 58
Elementary Movement Detectors 59
^{3}H-Deoxy-Glucose 61
Lobula Plate Giant Neurons 63
Other Visual Pathways 74

Part B: The Behavioral Structure of the Visual System

5 Flying Straight 82

Artificial Closed Loop Conditions 82
Optomotor Balance 83
Closed Loop Gain of Freely Rotating Flies . . . 92

6 Endogenous Behavior in Yaw Torque Fluctuations 95

Saccades and Torque Spikes 95
Torque Spikes as Fixed Action Patterns 98
Reafference Control 100
Spontaneous Occurrence of Torque Spikes 104
Torque Spikes Rules 104
Flight Modes 107

7 Orientation Toward Objects 110

Fixation Behavior in *Musca*, a Theory 113
The Object Response in *Drosophila* 119
Special Role of Front-to-Back Motion 120
Range of Stimulus Parameters for the Position Function 123
Two Responses Contribute to Yaw Torque 125
Analysis of Individual Torque Traces 129
Mutants with Altered Orientation Behavior . . . 131
Fixation in Flight 134
Fixation in Walking Flies 141

8 Menotaxis 146

Small Angle Oscillations 147
Significance of Reafferent Stimuli 153
The Object Response During Menotaxis 154

9 Foreground–Background Experiments 158
Figure–Ground Discrimination in *Musca* 158
One Output–Multiple Input Systems 160
A Simple Model Generating Figure–Ground Discrimination 162
Figure–Ground Discrimination in *Drosophila* . . 163
Tracking 165
Interplay Between Object Response and Optomotor Balance 166
Dissociation into Object and Background 174
Interpretation of Object Response 176
Can the *Musca* Theory on Pattern-Induced Flight Orientation Be Applied to *Drosophila*? 178

10 Visual Control in Free Flight 180

11 Selective Attention 183
Displacement Experiments 184
Open Loop Oscillations 187

12 Plasticity of Visuo-Motor Coordination 194

13 Valuation 205
Hypotheses 206
Significance of Simple Visual Stimuli 209
Context Dependence 210
Visual Components of Courtship Behavior . . . 213
Conditioning 215

Synopsis

The Three Levels of Orientation 217
Parallel Processing 218
Actions and Responses 219
Voluntary Behavior 220
Orientedness Revisited 223

Appendix 1: List of Neurological Mutants 226

Appendix 2: Symbols, Dimensions, Abbreviations . . 230

References 232

Subject Index 243

Introduction

Vision is naively considered by many people as a process originating in an organism and directed toward the world. In early scientific illustrations rays were drawn from the eye to the object. Only in the last few centuries with the growing awareness of the nature of light has it become generally accepted that the relevant light travels from the object to the eye and that information flows from the world into the body. While this concept today is the uncontested basis of vision research, it tends to overemphasize the role of the viewing subject as that of a passive receiver. Particularly in comparative studies of small animals, stimulus–response chains are considered the hard core of visual behavior. Our experience as humans with vision as a continuous series of verifications or rejections of guesses is taken as a specialty of higher vertebrates.

Initiating Activity

Among the most important brain functions of which we know introspectively is the ability to initiate activity. We do not have to wait until a stimulus gets us going. We can ourselves start a train of events which then leads to a change in the situation. Initiation is a constituent of nearly all behavior. Choice, will, self, the pursuit of goals, and probably all mental activity such as imagination and thinking require this all-pervasive faculty of the brain. We hope to convince the reader in this book that initiation cannot have been added late in evolution as a result of growing complexity. Rather, it must be a basic trait of all brains, even of small ones. The natural history of voluntary behavior must have started at the very beginning of the evolution of brains.

We emphasize the self-sufficiency of brains in generating activity since we feel that although this idea was put forward long ago, it has not been fully appreciated by many people in the field and is still questioned by some of them. One of the reasons for scepticism has been that the evidence for truly endogenous activity has never been compelling. Animal observations in the natural environment cannot exclude the possibility that some uncontrolled stimulus might have caused a suddenly occuring behavior. In fact, in the tradition of empiricist philosophy the so-called behaviorists for a long time maintained that all behavior, including spontaneous behavior, could in principle be traced back to stimuli even if these were not apparent to the observer (e.g., Skinner, 1938). Later the so-called ethologist school, by studying social interactions among animals, showed that much of behavior consists of fixed action patterns, stereotyped motor programs so characteristic for all members of one species that each individual must have inherited a set of them via its genome. Thus behavior does not need to be shaped in detail by the interaction with the environment. A signal for which the animal possesses an – again – inborn releasing mechanism may touch off the whole ready-made behavioral sequence. Thus in this approach only few highly specific constellations in the environment are considered as "releasing stimuli".

To cope with the apparent spontaneity of behavior ethologists proposed the so-called "inner stimuli" (innere Reize; e.g., Lorenz 1965). Since at that time the importance of drives, moods, appetencies and rhythmically changing dispositions in behavior became more and more apparent, fluctuating inner states and their mutual interaction were thought to be equivalent to external signals in triggering behavior patterns. They were made responsible for initiating activity.

With this conjecture the ethologists designed a perfect parallel to the "behaviorist's" argument. Both regarded spontaneous activity as actually nonspontaneous, i.e., stimulated. Both invoked as stimuli an important well-established item of the behavioral repertoire: external stimuli in one case and internal states in the other. Finally, both parties, being unsatisfied with their solution, proposed that complex interactions of these stimuli account for the apparent spontaneity.

Prompted by some experiments which will be described in the second part of this book, we adopt a different view in dealing with the origin of spontaneity. We assume that in many instances a change in behavior is not caused by external stimuli or by a change in the motivational state of an animal. Fluctuating moods and drives, as well as sensory messages, are no doubt very important, but we assume that they are usually not sufficient to elicit behavior; they merely bias the animal in its choice of what to do next. We suggest that much of behavior is irreducibly the result of initiation. By this term we mean an internal process of the brain whose physical nature is unknown but which we assume to comprise the elements of variation and selection, trial, and the elimination of error familiar from the process of Darwinian evolution (Popper, 1976).

Our proposal bears some resemblance to the presently held view of many neurophysiologists, who venture the possibility that stochastic processes may be the basis of spontaneity. With the application of electrophysiological recording techniques to brains it became evident that endogenous neural activity accompanies all phases of behavior, even rest and sleep. Some neurons generate action potentials at a fairly constant rate, which might or might not be modifiable by sensory stimulation. Other neurons show occasional bursts of activity, others again have a complicated "firing pattern". It was proposed that a low level of general neural activity might occasionally by chance reach a local threshold for a certain behavior pattern to be elicited (Roeder, 1968; Bullock, 1977).

Our proposition differs from this one not so much in the underlying mechanisms, which are unknown anyway, but in the functional significance for the animal. We will argue that such a stochastic process should not be regarded as an undesirable by-product of an imperfect apparatus but as part of a sophisticated algorithm providing better adaptation to an unpredictable future. If initiating activity is the origin of voluntary behavior, if it ultimately enables beings to produce new thoughts, the Darwinian paradigm seems to us the only rational way in which nature could provide this. Initiation and voluntary behavior will be discussed in the last chapter.

Even in insects vision is an active process. Although this book is largely concerned with homoeostatic functions of the brain, we will meet initiation at all levels. Most of the published work about visual behavior in *Drosophila* describes the relation between stimuli and behavioral responses. Little is known about endogenous behavior. Early attempts to measure ethograms as a correlate of central brain function in *Drosophila* (Manning, 1959) have been abandoned. It is much easier experimentally to study input–output relations than to wait until the animal does what it "wants" to do. Our own investigations began as an input–output analysis and for several years we were reluctant to recognize any initiating activity of the fly. As it stands now, we realize that even seemingly simple visual responses contain elements of activity. It is the interplay between actions and responses out of which behavior is made.

Input–Output Analysis

The prime sources of information about brain functions are introspection and behavior. Behavioral work starts in the field. Only the natural habitat provides the needs and goals for the behavioral repertoire. The ethologist describes behavioral units and subunits and their interrelations, but as the last 50 years have shown, mere observation hardly ever leads to a description detailed enough to match the underlying neural structure and its function. For this purpose a behavior sufficiently well known in its natural context must be taken to the laboratory where it can be scrutinized, tested under extreme conditions, and manipulated by surgical or pharmacological means. In this respect those behaviorists were right who stressed that with a freely moving animal in a natural or even experimentally simplified environment firm conclusions about behavioral input–output relations are very difficult.

We owe much to the school of E. von Holst, where more than 35 years ago the so-called open loop experiments with tethered insects were begun.

They have led in the meantime to most detailed descriptions of stimulus–response chains in insect behavior, for instance in the visual system of flies (e.g., Reichardt, 1970; 1973; Götz, 1964; 1968). To a large extent also this study is based upon open loop experiments.

The visual feedback loop is said to be opened when the behavior of the experimental animal (or human subject) is prevented from interfering with the stimulus situation. In insect studies on vision this generally implies that head and thorax are immobilized. The animal is firmly attached to a recording device where it can fly or walk without changing its position and orientation. The stimuli impinging on the surface of the animal can thus be quantitatively determined and the maneuvres the animal intends to execute can be continuously monitored. Not much of the behavioral repertoire of an animal lends itself to this treatment. One almost gains the impression that the animal refuses to display most of it in such a situation. However, behavioral functions analyzed under adverse laboratory conditions have in some cases been shown to match the properties of the underlying neural network. In fact, for some behavior to be discussed here, wiring diagrams have been proposed and neurons belonging to such pathways have been identified. Any behavior should eventually be understood at this level.

We intend to describe the visual system of *Drosophila* in such a way that all functions can in principle be imagined as neuronal circuits (and neurochemical processes). This is the reason why we had to introduce initiating activity right at the beginning. It provides the conceptual framework for terms such as voluntary behavior, intentions, and goals. We do not suggest that by introducing the concept of initiating activity we have explained what corresponds to it in terms of neuronal mechanisms. On the contrary, the understanding of any of these psychological terms at the physiological level will still have to wait for a major breakthrough; but with initiation as a basic brain property these terms seem less awkward. They can at least be discussed at the same operational level as stimulus–response chains, of which the physiological basis often is unknown as well.

Our presentation of vision in *Drosophila* is limited. The description is first of all necessarily biased by the large amount of detail known about laboratory behavior. Secondly, we are concerned about the fact that some of the *Drosophila* strains we are investigating have for more than 1500 generations been kept under extremely impoverished rearing conditions. Studies with wild-living flies are indispensable if one hopes to get an idea of the full potential of an insect brain. Still another limitation lies in the behavioral approach itself to which at present there is no alternative. Obviously, the visual system is not switched off while a fly refrains from behavioral activity. Thus behavior can tell us only part of what is going on in the fly brain.

The most stringent limitation of this account, we believe, is, however, still due to conceptual difficulties. Vision has a semantic side, i.e., is concerned with the significance of what is seen. Thus, ultimately, vision is tied in with the fly's "value system". Of this we have no operational understanding at the neuronal level as yet; and even if we knew all that, would we not have to admit our ignorance about the central question: "How does a fly see the world"? This is not a scientific problem.

Orientation and Orientedness

Behavior often involves orientation. We will be concerned with visual orientation of *Drosophila* in most of this book. We describe it as consisting of three levels. At the first level the animal tries to stabilize its gaze. This is a requirement for most other visual functions. Optokinetic nystagmus in vertebrates and optomotor responses in insects are aspects of this ability. In flying insects gaze stabilization is often tied to course control. It is this aspect of vision of which we know the most. In Chapters 3, 4 and 5 we summarize the present state of knowledge about optomotor responses and their structural correlates in *Drosophila.*

At the second level the animal chooses a certain orientation in respect to a point of reference. This problem is referred to by some people as orientation proper. It is conceptualized in the different types of taxes and recently in the theory of object fixation (Reichardt and Poggio, 1976). This level is introduced in Chapter 7 and, from there on, remains a central theme of the book.

The third level concerns the goal the animal pursues when choosing an orientation. Possibly, an animal may occasionally have a certain orientation without attaching a "meaning" to it, for instance when a housefly is idling under a lampshade. But in many cases the animal takes a certain orientation for some purpose, either to get to some place or to gather (or give) information, or to be ready for some later action. Although Chapter 13 addresses the problem of "meaning" in vision, we are aware that our knowledge is too scarce to do this third level justice.

Even with these three levels, however, orientation is not adequately described. The term *orientation* reflects only the observer's viewpoint. If one wishes to understand orientation on the basis of the neural circuitry in the brain of the animal which does the orienting, one needs to treat it as part of the procedure by which the animal gathers as much knowledge about its situation as possible. This knowledge we call *orientedness*. It represents the organism's relation to the world. It reaches into the past and into the future and is continually updated by the currently arriving sensory data. Orientedness is not to be structured primarily into sensory modalities but rather into matters of concern, such as cleanliness, provision of food, fighting, security, search for a mate, and care for the survival of the offspring. The same forest looks different for the forester and for the hunter. Orientation is for most of an animal's life the combination of its orientedness and of the goals it pursues.

We do not know off hand which aspects of our own mind may be similar to that of a fly. We assume that, if anything, orientedness may be such an aspect. The assumption that flies do "see" and the many parallels between human vision and fly vision to be reported in the following chapters are an expression of this view.

The second part of this book (with the exception of Chap. 7) contains first attempts to deal with orientedness in the context of visual orientation. In so far as orientedness is a mental state this is, in a way, a first approach to the study of the fly's mind.

Neurogenetics

This book is, after all, an exercise in neurogenetics. The possibility of applying methods from genetics is the main reason behind the choice of *Drosophila* for behavioral research. Throughout this century *Drosophila* has been the principal genetic research animal. The huge amount of genetic knowledge accumulated during that time, the thousands of mutant strains, the relative ease in obtaining new ones and the recent advances in recombinant DNA and transformation techniques make *Drosophila* a favorable choice even in brain research.

Neurogenetics links behavior to genes. Inborn behavior patterns are easily recognized in insects. We will use the term "phylogenetic information" for the properties of the fly brain reflecting the world in which the fly has to survive. This phylogenetic information has been collected by the millions of generations of ancestors of present-day *Drosophila* and is transmitted from generation to generation through the genome. How it is coded there is enigmatic at present. The answer to this question would include a causal description of brain development; but our book stays away from developmental problems, although we consider them most relevant for understanding brain function. All the currently accumulating evidence from neurogenetic studies at the network level in nematodes as well as in flies points to the intimate association between functional organization and genetically programmed development of the nervous system. Yet this aspect of neurogenetics is still in its infancy; only very few data are available.

For our present needs neurogenetics appears more like a set of new tools for the analysis of a complex biological system. It is now 15 years since, in his seminal paper on the isolation of nonphototactic mutants, Benzer (1967) heralded the genetic dissection of the *Drosophila* visual system. Many people were enthusiastic at first but soon became disillusioned and considered the progress of neurogenetics disappointingly slow. For a mutant approach to be successful, the basic structure of the system under study has to be known already to some extent. In this respect genetics cannot stand on its own.

We will discuss extensively throughout the book how brain structure and function can be genetically dissected in a meaningful way by mutations. The modification of a single gene may eliminate a certain behaviorial property and leave most other aspects of behavior unchanged. A mutation affecting vision mostly does not matter for olfaction. A block in the perception of visual motion may not abolish visual orientation. The impairment of plasticity in visuo-motor coordination may be irrelevant for color discrimination learning, etc. The genetic approach reveals the extensive parallel organization of the visual system in respect to behavioral subroutines. Course control, orientation toward landmarks, visually induced landing and color vision seem to be at least in part represented by distinct neural pathways in the optic lobes. A fly can lose 75% of the cell types in the optic lobes without losing several of its basic visual functions. Mutations may prove decisive for disentangling small brains. A list of all the mutants to be cited with short descriptions of their properties will be found as Appendix at the end of the book.

At present the visual system as a whole, however, is so little understood that still no use can be made of most visual mutants. Hundreds of nonphototactic mutants have been isolated and subsequently abandoned. It is one of the incentives for writing this book to try to picture, vaguely as it may be, the visual system in depth, even though admittedly the investigation of detail so far has only proceeded to the preliminaries. We hope that with a better general understanding of vision in *Drosophila* it will be possible to select more specific mutants with defects confined to one of the various brain activities involved. Analyzing a neurological mutant genetically and phenotypically is a major enterprise; but fortunately, progress in this field is cumulative. Mutants once characterized are available for all future studies.

There is a further aspect of neurogenetics in this book which deserves to be mentioned. It concerns the relation between brain and behavior. It leads back to the problem of initiation. We have mentioned above that the insect brain is considered a huge store of phylogenetic information which is the base of inborn behavioral patterns. Is behavior completely determined by the genes, the environment and the interaction of these two? The brain of a fly is largely the product of a genetically controlled developmental process. Furthermore the behavior of an animal is determined by its nervous system interacting with the environment. One might thus be inclined to believe that the behavior of a fly might, step by step and movement for movement, depend upon its genes and the environment.

However, as we will discuss in the last chapter, animals relying entirely on appropriate behavioral programs designed to match in detail all possible circumstances would require such a huge amount of genetical information that this evolutionary strategy is clearly impossible. In particular, such an animal would be at a loss if confronted with a new situation for which there was no blueprint in its genome. To cope with this dilemma, nature has equipped ambulatory organisms with initiating activity. Or to put it differently: active behavior is a type of strategy by which the genome with its most insufficient amount of information about the future can still grant the organism a reasonable chance of survival. Freedom, we learn in school, is to follow one's own directive in doing what has to be done. This definition holds all through evolution of behavior. In our context it applies to behavior which is not triggered by "releasing stimuli" and still has an adaptive value for the animal. As mentioned before, the Darwinian paradigm provides this combination of properties.

The relation between genes and behavior is marked by the evolution of freedom. Our genome does not only grant our prompt and reliable reactions to all kinds of dangerous or beneficial situations, it also enables us to decide, choose, think, invent and to try to change the world for the better. In a visual mutant of *Drosophila* the mutation does not only reduce the efficiency of a visual response it also reduces the options a fly has at its disposition. Most genetic defects lead to a decrease of behavioral freedom. We hope to convince the reader that for insects, as much as for humans, the interplay between behavioral freedom and behavioral constraints is one of the keys to central brain function and the basis for understanding the role of genes in behavior.

Part A
Eye, Brain, and Simple Behavior

1 The Compound Eye

How do we know that the faceted compound eyes of insects are eyes? No pupil, no iris, no eye ball, no lid, no brow. It is difficult to have eye contact with a hornet even if it is watching you. Whatever the answer may be, once we take for granted that insect eyes are indeed eyes, we realize that they must have many properties in common with our own. First of all, visual space for all organisms is a sphere with the viewing subject in the center. Humans can see only half of it at a time; insects may see all around. But for insects, as for ourselves, the distance between two points in the visual sphere is given as an angle with its apex in the viewing subject. Likewise the area of the projection of an object is given as a solid angle. The same projection may correspond to an object which is close and small or large and far away. Its distance from the viewing subject can be derived only from secondary cues (disparity, motion parallax, previous knowledge, etc.). Second, insects and humans exploit roughly the same spectral region of electromagnetic radiation. Both types of eye contain densely packed arrays of photosensitive cells which transform the local properties of the light in the visual sphere into an array of neuronal signals in the brain. A description of all the known similarities would fill many pages.

The eye determines what can possibly be seen. The parameters of the light which an organism perceives depend upon the properties of the photoreceptors in its eye. Also the shape, position, and motion of visual objects are given in relation to the optical axes of the photoreceptors. Thus, analysis of vision requires a detailed knowledge of the eye. This knowledge, then, allows some prediction of the neural events caused by a visual stimulus in the nervous system. We will briefly summarize those properties of the *Drosophila* eye relevant for the visual behavior to be discussed afterwards.

The Distribution of Optical Axes

The two bright red compound eyes are the most prominent features of the head of *Drosophila melanogaster* (Fig. 1). Consisting each of roughly 700 ommatidia together, they sample about 85% of visual space (E. Buchner, 1971). Only a vertical stripe of about 40° in width in the back of the fly is

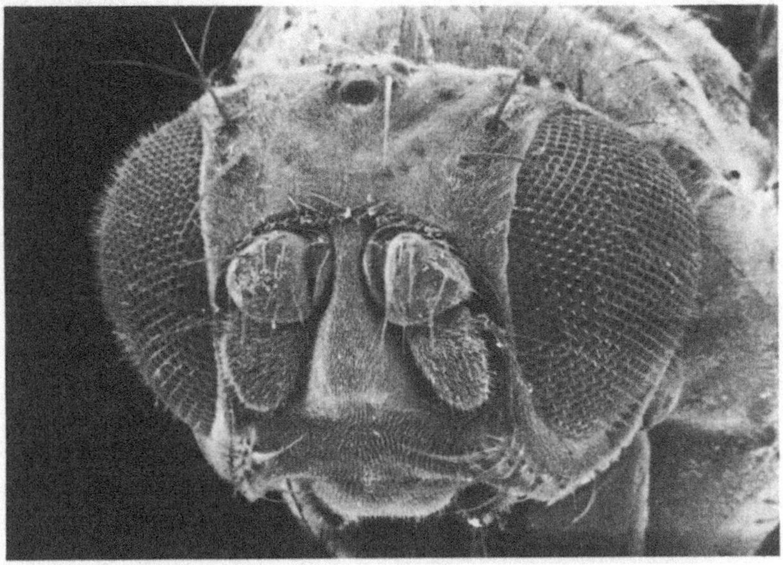

Fig. 1. *Drosophila melanogaster*; dorso-frontal view of head. (SEM photograph; courtesy L. Schneider)

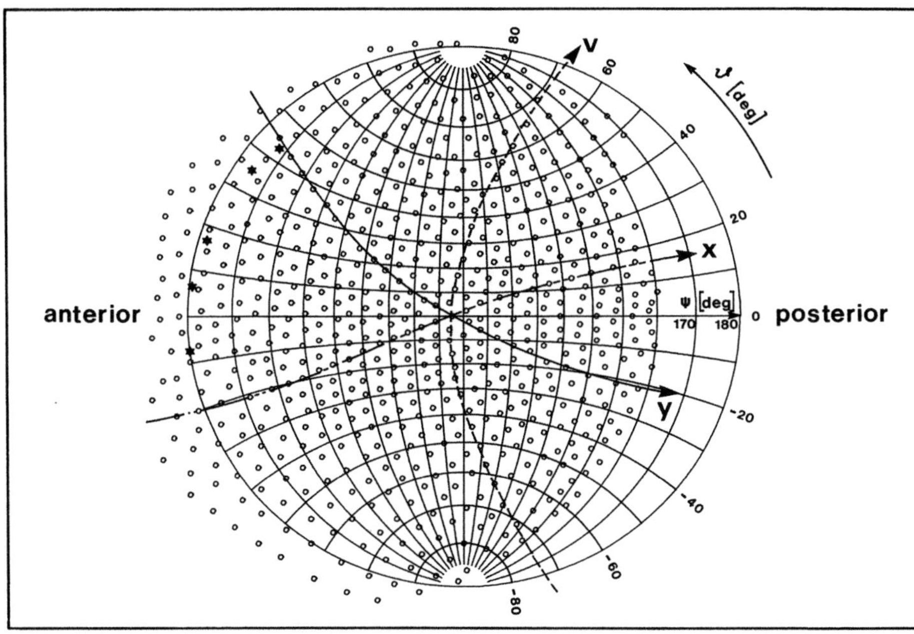

Fig. 2. Spatial distribution of optical axes (*small circles*) of visual elements[1] as shown in a stereographic projection. *Large circle* shows (approximately) left visual half-sphere of fly. (Optical axes marked with an *asterisk* are oriented in parallel to the fly's longitudinal body axis). The three main axes *v*, *x* and *y* of the hexagonal lattice of visual elements are marked as *arrows*. Optical axes outside the large circle are directed into the other half-sphere and mark (again approximately) the region of binocular overlap. (Data from E. Buchner, 1971)

---

[1] A visual element is the set of retinula cells of one eye having parallel optical axes (see below)

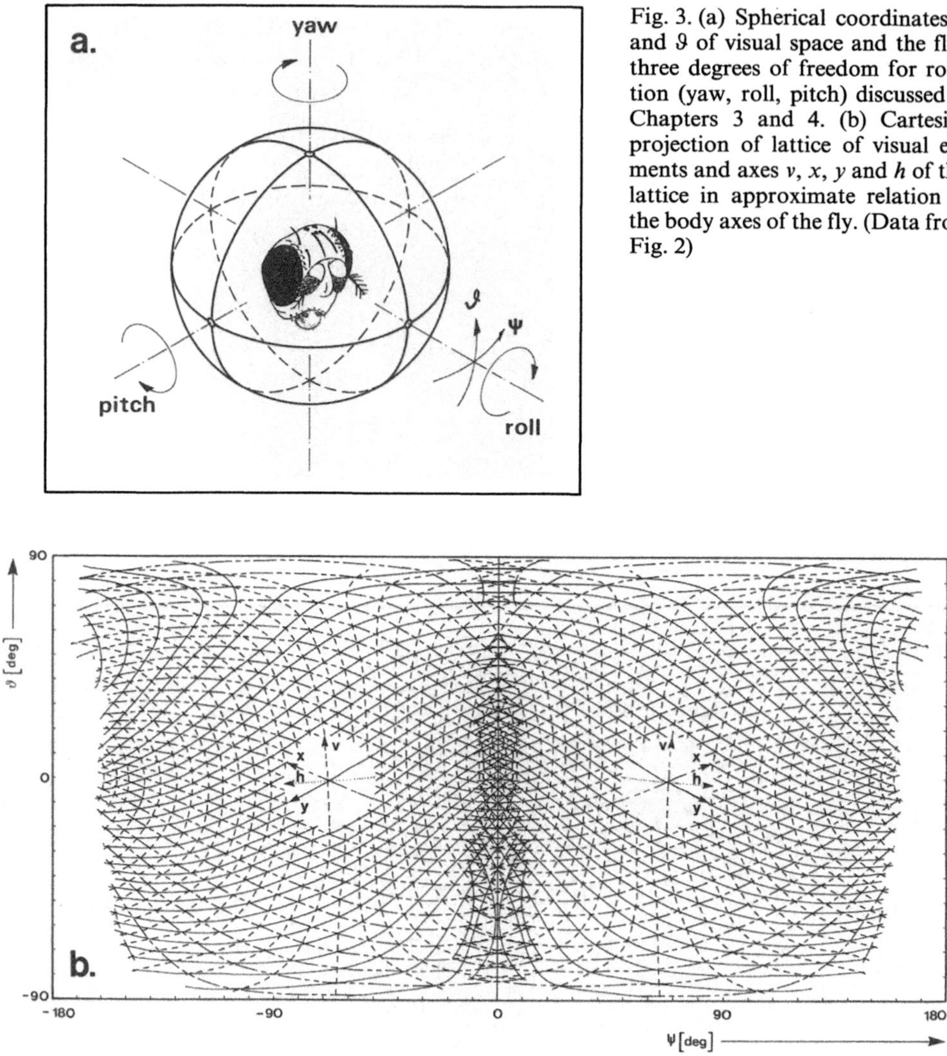

Fig. 3. (a) Spherical coordinates ψ and ϑ of visual space and the fly's three degrees of freedom for rotation (yaw, roll, pitch) discussed in Chapters 3 and 4. (b) Cartesian projection of lattice of visual elements and axes v, x, y and h of this lattice in approximate relation to the body axes of the fly. (Data from Fig. 2)

excluded (Fig. 2). In the one female fly which was thoroughly investigated, E. Buchner (1971) found an overlap between the visual fields of the two eyes of maximally three rows of visual elements in front, below and above the fly.

Figure 2 represents the intersections of the optical axes of the visual elements of one eye with a concentric half-sphere. The optical axis of a visual element marks the center of its receptive field (i.e., its angular sensitivity distribution, see below). The divergence angles ($\Delta\varphi$) between the optical axes of adjacent visual elements are quite uniform but decrease slightly toward the front. The number of visual elements per unit area is about 50% higher in this part than in the rest of the eye. Nevertheless, a real fovea as in other flies (e.g., Collett and Land, 1975a) cannot be distinguished in the *Drosophila* eye. The ommatidia are packed in a regular hexagonal array which, in the center of the eye, has its three main axes oriented vertically and at $\pm 30°$ with respect to the horizontal (Fig. 3b). In all other parts due to the curvature of the eye these axes are tilted by variable degrees, as can be deduced from Figs. 2 and 3b. This arrangement has far-reaching behavioral consequences.

Neural Superposition

Drosophila has a neural superposition eye (e.g., Kirschfeld, 1973; Franceschini, 1972), which is typical of cycloraphan flies (*Drosophila*, *Musca*, *Syritta*, *Eristalis*, *Calliphora*, *Phaenicia*). In this eye the units sampling the visual space are not the ommatidia, but the above-mentioned visual elements consisting of groups of eight photoreceptors which are directed toward the same point in space and are distributed in seven adjacent ommatidia (Fig. 4). Each retinula cell (photoreceptor) contains as photosensitive organelle an

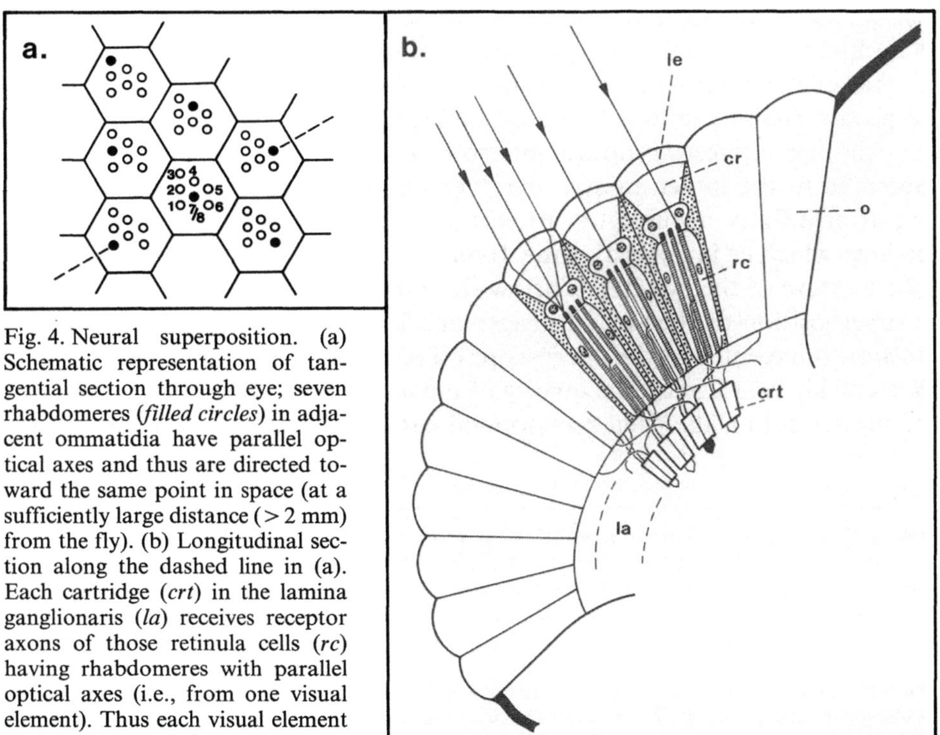

Fig. 4. Neural superposition. (a) Schematic representation of tangential section through eye; seven rhabdomeres (*filled circles*) in adjacent ommatidia have parallel optical axes and thus are directed toward the same point in space (at a sufficiently large distance (> 2 mm) from the fly). (b) Longitudinal section along the dashed line in (a). Each cartridge (*crt*) in the lamina ganglionaris (*la*) receives receptor axons of those retinula cells (*rc*) having rhabdomeres with parallel optical axes (i.e., from one visual element). Thus each visual element mediates visual information from one point in space. (*le*, lens; *cr*, crystal cone) *Numbers* in (a) refer to retinula cells R1–6 and R7/8

optically isolated rhabdomere which extends through the whole depth of the eye. As an exception cells No. 7 and 8 have their rhabdomeres arranged in tandem fashion with R7 distal to R8. Thus in each ommatidium one finds at any level seven rhabdomeres. They receive light through their distal endings which lie in the focal plane of the corneal lens. Using the same optical apparatus, the retinula cells of one ommatidium have different optical axes and, hence, belong to different visual elements (with the exception of cells R7 and R8 which have a common rhabdomere). This special principle of superposition requires that the divergence angles between the optical axes of adjacent ommatidia be the same as those of neighboring rhabdomeres in one ommatidium. This is in fact the case (Autrum and Wiedemann, 1962; Kirschfeld, 1967).

The lattice of visual elements can be observed in the living fly with antidromic illumination. By focusing light onto the head capsule one can get enough of it to be emitted by the distal rhabdomere endings for the far field radiation pattern of the eye to become observable through a telescope (Fig. 5c; Franceschini, 1975). This method has been elegantly used to monitor the precise application of visual stimuli (E. Buchner, 1976 see below p. 37–41).

Using antidromic illumination the optics of the neural superposition eye can be readily visualized in the living fly by focusing a microscope with an extremely small objective aperture on to the cornea (Franceschini and Kirschfeld, 1971). Only the one visual element having its optical axis parallel to that of the microscope is seen to illuminate the seven facets in which the respective rhabdomeres are located (reduced corneal pseudopupil).

Another interesting optical phenomenon of the *Drosophila* eye can be observed in the intact animal: the "deep pseudopupil". Each cornea lense forms at infinity an upright virtual image of the seven distal rhabdomere endings which lie in its focal plane. If one focuses the microscope at the center of curvature of the eye (180 μ below the cornea for *Drosophila*), one can see a superposition of the virtual images of all those ommatidia whose optical axes are intercepted by the microscope objective (Fig. 5a, b; Franceschini and Kirschfeld, 1971). Since the quality of the image depends upon the precision of superposition, the deep pseudopupil can be used as an indicator for the

→

Fig. 5. Ray diagrams (a) explaining formation of deep pseudopupil (b) and far field radiation pattern of visual elements (c). The dioptric apparatus (cornea lense, *c*) of the ommatidium generates a virtual image (*vi*) of the distal rhabdomere endings (*re*) at $-\infty$ and a real image (*ri*) at $+\infty$. The latter is transposed into the back focal plane (*bfp*) of the microscope objective (*mo*). In the center of curvature of the eye the (magnified), virtual images of the distal rhabdomere endings from each ommatidium are superimposed to form the deep pseudopupil (*dpsp*). In the back focal plane of the microscope objective the real images of rhabdomeres with parallel optical axes are superimposed as the far field radiation pattern of visual elements (*pve*) (E. Buchner, 1974). In (b) and (c) the eye is illuminated from within. Light enters rhabdomeres at the basement membrane and is emitted at their distal tips. *Numbers* in (a) refer to retinula cells R1–6 and R7/8

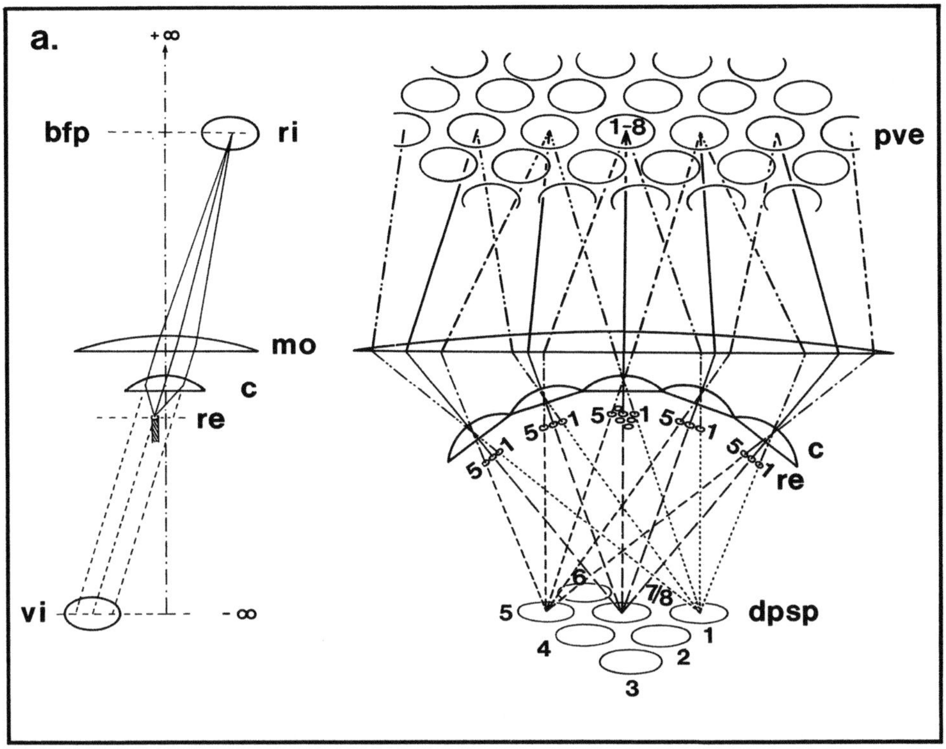
a.
+∞
bfp
ri
mo
c
re
vi
-∞
1-8
pve
5
1
c
re
6
7/8
dpsp
4
2
3

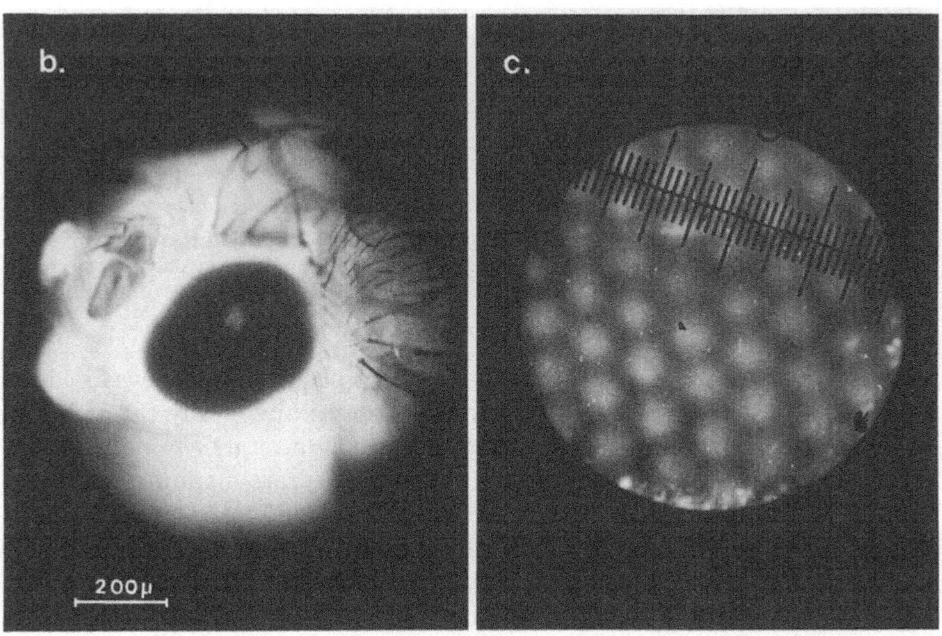
b.
200µ
c.

proper alignment of optical axes of the rhabdomeres. This is particularly useful in experiments in which the head capsule is opened or for the characterization of visual mutants.

The Equator

Typical for the cycloraphan eye is the asymmetrical arrangement of rhabdomeres (Fig. 6). Presumably, symmetry was sacrificed in order to provide

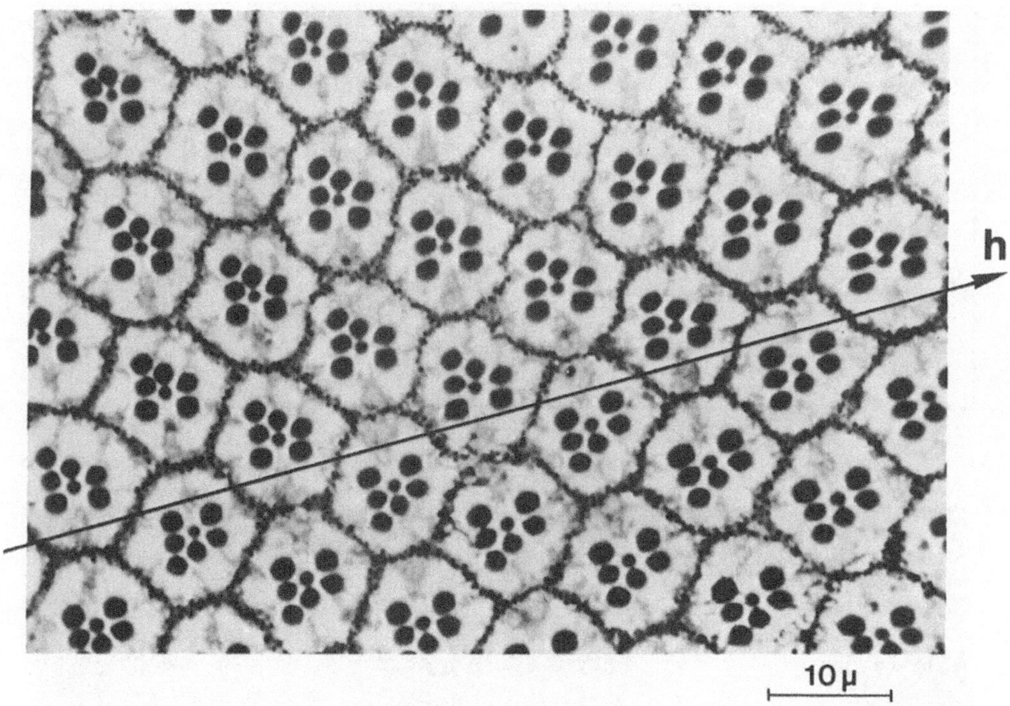

Fig. 6. Tangential section through compound eye. In this part of the eye the equator coincides with the h-axis of the lattice of visual elements (*arrow*). Note mirror symmetry of typical rhabdomere pattern. 1 μ plastic section stained with azur II blue. (Courtesy K. F. Fischbach)

space for the cell body of retinula cell No. 7. This peculiar shape reveals a line of mirror symmetry ("equator") extending about horizontally through the center of the eye. As shown in the tangential section of a *Drosophila* eye in Fig. 6, the trapezoidal configuration of rhabdomere cross-sections is pointing upward in the upper part of the eye and downward in the bottom part (Dietrich, 1909). The mirror symmetry extends through the first into the second optical neuropil, as was shown for large flies (Braitenberg, 1970; Campos-Ortega and Strausfeld, 1972).

Angular Sensitivity of Visual Elements

The resolving power of an eye depends not only upon the angular spacing of visual elements but also upon their angular sensitivity (acceptance angle). Most of the light reaching a rhabdomere in a wild-type *Drosophila* comes through the cornea. As mentioned above, the rhabdomere as an optical wave guide receives most of the light through its distal tip, which lies in the focal plane of the corneal lens. The acceptance angle of the rhabdomere is thus determined by the size of the rhabdomere tip and the optical properties of the cornea lens. Franceschini and Kirschfeld (1971) have shown that the optical quality of the cornea lenses in the *Drosophila* eye (diameter $D = 16\ \mu$; focal length $F = 25\ \mu$) is limited by diffraction and that for the angular sensitivity of a given retinula cell neither the size of its distal rhabdomere tip nor the properties of "its" cornea lense can be neglected. This is of functional importance since, as can be seen in Figs. 5b and 6, the distal endings of the central rhabdomeres R7/8 are significantly smaller than those of rhabdomeres R1–6. The angular sensitivity of rhabdomeres R7/8 should thus be slightly narrower and they should thus provide slightly better resolving power than the peripheral rhabdomeres R1–6. This should lead to two distinct contrast transfer functions for the lattice of visual elements, one adapted to high acuity mediated by the central retinula cells No. 7/8 and one adapted to high sensitivity mediated by the peripheral cells No. 1–6 (Kirschfeld and Franceschini, 1968). The different angular sensitivity of the central and peripheral retinula cells has been directly confirmed in large flies by intracellular recording (Hardie, 1979). In *Drosophila*, where electrophysiological experiments are still cumbersome, the acceptance angle of the peripheral retinula cells has been indirectly assessed from behavioral experiments. Its half width $\Delta\varrho$ lies between 5° and 6°, a value reasonably close to the angular separation of visual elements ($\Delta\varphi$) (see Chap. 4).

Screening Pigments

From all sides other than that facing the cornea an ommatidium is optically isolated by a dense sheath of pigment granules (diameter $0.4\ \mu$) apt to absorb scattered light in the wavelength range of the visual pigment. Since *Drosophila melanogaster* photoreceptors are insensitive to red light (see below) also the screening pigment does not need to absorb in the red. It is the nonabsorbed scattered light of the screening pigment granules which causes the red eye color, the "trade mark" of this species.

Most of the screening pigment is located in special pigment cells separating the ommatidia from one another and forming the basement membrane. The pigment is a mixture of pteridines and ommochromes. An abundance of mutants with reduced eye pigmentation have been isolated. These offer the opportunity to study the influence of scattered light on visual functions. Hengstenberg and Götz (1967) showed that the overall light sensitivity of the eye in the mutant *white* is about 20 times that in wild type, while concomitantly in the mutant pattern contrast for a grating with spatial wavelength $\lambda_{sp} = 360°$ is attenuated by a factor of 25. Nevertheless, in *white* the half-width of the angular sensitivity distribution ($\Delta \varrho$) is about the same as in wild type, since the sensitivity of the retinula cells to light from any part of the visual sphere far off the rhabdomere's optical axis is less than 5% of its on-axis sensitivity (Götz, 1964). However, the effect of scattered light which reduces subjective contrast makes itself felt more strongly in the high spatial frequency range.

Screening pigment in the retinula cells plays a special role. It controls the light flux in the rhabdomere (Franceschini, 1975; Franceschini and Kirschfeld, 1976). In the retinula cells the ommochromes are contained in small lipid granules (diameter 0.1 μ) in the cytoplasm. Upon illumination these pigment granules migrate toward the rhabdomere and, by interacting with the evanescent wave traveling just outside the rhabdomere reflect, absorb or scatter a considerable portion of the light incident upon the rhabdomere tip. The two states of adaptation can be easily observed in the deep pseudopupil. In orthodromic illumination the deep pseudopupil reflects green light of metallic appearance which results from the reflection by the pigment granules at the outside of the rhabdomere. After dark adaptation the rhabdomere endings appear black due to the lack of this reflectance. Each cell is autonomous for this process (Franceschini and Kirschfeld, 1975). The intensity threshold for pigment migration is about 100 times higher in retinula cell No. 7 than in cells No. 1–6.

Pigment migration is coupled to the intracellular concentration of Ca^{2+}-ions. The ommochrome granules can be kept dispersed in the cytoplasm as in the dark-adapted state by injecting the Ca^{2+}-complexing agent EGTA into the eye (Kirschfeld and Vogt, 1980). In *Drosophila* screening pigment mutants may serve to eliminate the pupil mechanism. In particular the mutants *vermilion*, *scarlet* and *cinnabar*, which lack most of the ommochromes but still have pteridines ensuring reasonable contrast transfer properties of the eye, are useful in this respect.

Electrical Responses to Light

So far we have only discussed the optical properties of the *Drosophila* eye. The main purpose of an eye is to transform the local quantum flux of the visual surround into electrical signals which can be processed by the nervous system. Fly retinula cells, like those of most invertebrates, have resting potentials of − 60 to − 70 mV in the dark mainly due to an active transport system in the membrane which accumulates Na^+-ions extracellularly. Light absorbed by the visual pigment (and helper pigments) causes a passive influx of Na^+. This can be recorded as the receptor potential (Pak et al., 1980). Under constant illumination an equilibrium must be reached between outward transport and influx. Until recently it had been assumed that, as in vertebrates, the chromophore of rhodopsin-like visual pigments in insects was retinal. Experiments of Vogt (1983) on large flies and on *Drosophila* (Vogt, in preparation), however, suggest it to be a similar but more polar aldehyde. Vogt proposes calling visual pigments of this type "xanthopsins".

The receptor potential is composed of quantal units which in the dark can be observed as small (0.2−1 mV), uniform depolarizations. These so-called quantum bumps can be triggered by single photons (Wu and Pak, 1975). The process leading from photon absorption to the receptor potential (phototransduction) is under intensive investigation. A variety of photoreceptor mutants have been found in which the process is altered or blocked at some stage (Pak, 1979).

One property of the *Drosophila* retinula cells is noteworthy in respect to visual behavior. Strong blue irradiation of receptors No.1−6 can cause the depolarization to last for many minutes after cessation of the stimulus. This "prolonged depolarizing afterpotential" (PDA) which is observed best in mutants with white eyes (lacking blue-absorbing screening pigments) is a consequence of the massive conversion of xanthopsin (X) to metaxanthopsin (M) and can be "switched off" by the reverse process, the photo-conversion of M to X by orange light (Cosens and Briscoe, 1972; Minke et al., 1975; Stark, 1975). As judged from behavior, a saturating PDA renders the retinula cells unresponsive to further visual stimulation if reversion from M to X is avoided (Heisenberg and E. Buchner, 1977). In retinula cells No. 7 a PDA can apparently be induced by UV-light and terminated by blue light (Harris et al., 1976), whereas for cells No. 8 no PDA has been found.

Both the receptor potential and PDA from many retinula cells at a time, can be extracellularly recorded in the electroretinogram (ERG) (Fig.7). Again the eye-color mutant *white* is most suitable for these experiments since it allows for the homogeneous illumination of all photoreceptors without much experimental sophistication (Hengstenberg and Götz, 1967). In the ERG measured with one electrode on the eye and the other at an arbitrary position in the haemolymph, the summed receptor potentials (retina potential) are superimposed on an "evoked potential" from the first optic neuropil region, the lamina ganglionaris (see review by Pak, 1979). Receptor- and lamina potentials can be partially separated by placing an electrode just beneath the retina proximal to the basement membrane. In this case only the retina potential is recorded in relation to the surface of the eye and the ERG minus the retina potential in relation to the reference electrode (Heisenberg, 1971 a).

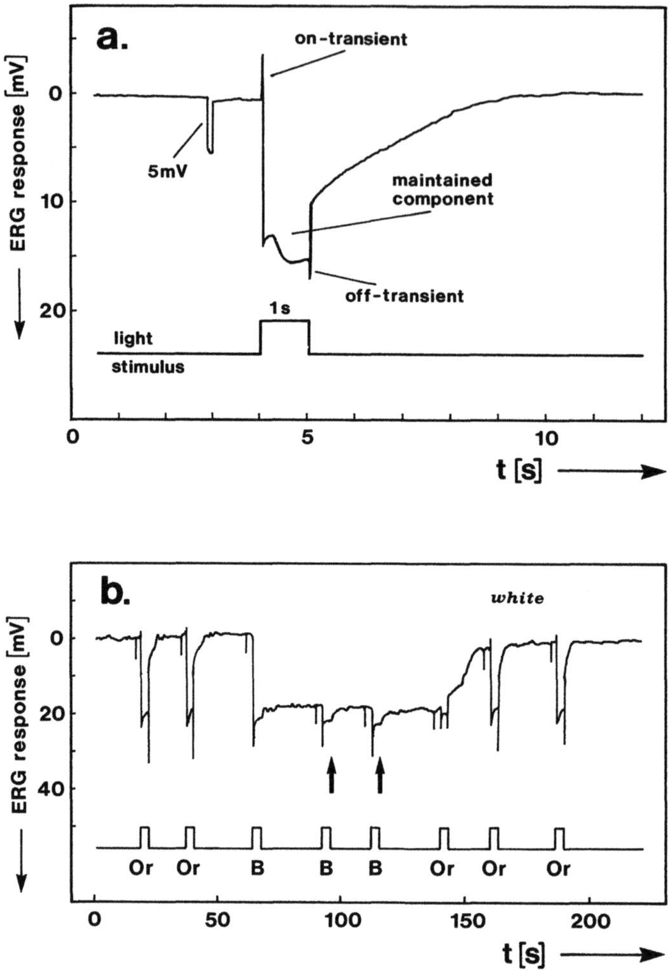

Fig. 7. (a) Electroretinogram (ERG) of eye color mutant *white* using white light for stimulation. On- and off-transients reflect electrical activity of lamina neurons, maintained depolarization is largely due to retinula cells. (b) Successive ERG's to orange (*Or*) and blue (*B*) light. Blue light eliminates contribution of retinula cells R1–6 and the lamina neurons to the ERG and "clamps" R1–6 cells in depolarized state (PDA, prolonged depolarizing afterpotential). Orange light restores sensitivity and normal dark potential of R1–6 cells. *Arrows* point to ERG of retinula cells R7/8. As they by-pass the lamina no lamina potential contributes to ERG of R7/8 cells. (After Pak, 1979)

Spectral Sensitivity

The ERG and the optical techniques described above are convenient diagnostic tools for the characterization of visual mutants obtained by behavioral screening procedures. Defects can tentatively be classified as peripheral or central and in the periphery optical abnormalities, retinula cell defects and failures in the lamina can be distinguished. A large number of mutants with altered ERG's have been isolated, most of them still waiting to be studied (for analysis of transducer process see Pak et al., 1980). However, three eye mutants have been successfully investigated in which whole classes of retinula cells are more or less completely missing. The most clear-cut one is the mutant *sevenless*^{LY3} (*sev*) lacking all retinula cells No. 7 with no gross impairment to the rest of the eye (Harris et al., 1976). The absence of cells No. 7, however, has secondary effects on the other photoreceptors. In particular, the visual input to retinula cells No. 8 is severely altered. Without rhabdomere R7 cell No. 8 is lacking a crucial part of its dioptric apparatus which normally channels light from the focal plane of the cornea lens to the rhabdomere. Thus, the "image" mediated by the lattice of retinula cells No. 8 must be blurred and must have a much lower sensitivity than that in wild type. In addition, the distal tips of rhabdomeres R1–6 come close together in the mutant, filling the space left by the central cell. Accordingly the optical axes in each visual element are less well aligned, leading to a slight blur in the "image" mediated by the large retinula cells as well. In the other two mutants rhabdomeres R1–6 are affected and rhabdomeres R7/8 left unimpaired. These are the strain *retinal degeneration B*^{KS222} (*rdg B*), a particular allele of the *rdg B* gene in which most retinula cells No. 1–6 degenerate in the light (Harris et al., 1976; Harris and Stark, 1977) and the mutant *outer rhabdomeres absent*^{JK84} (*ora*) which contains only rudimentary rhabdomeres in the large retinula cells No. 1–6 (Koenig and Merriam, 1977; Harris et al., 1976). It seems to be the gene for xanthopsin protein in retinula cells R1–6 which is affected in *ora*.

These mutants were used to determine the spectral sensitivity of the three different retinula cell types (No. 1–6, No. 7, and No. 8) by simple ERG measurements. The mutations just mentioned were combined with one another and with the eye color mutation *white* in order to compare the effects of the genetical elimination of retinula cells R1–6 and R7 with the functional suppression of their electrical activity by means of the PDA. Figure 8 shows the spectral sensitivities of the mutants *w*, *w sev*, *w ora*, *w sev ora* and *w rdg B sev* in the dark-adapted state and after intense blue or UV irradiation (Harris et al., 1976). The two methods give consistent results, showing that the retinula cells No. 1–6 have a two-peaked sensitivity curve with one maximum at $\lambda_{h\nu} = 480$ nm the other below $\lambda_{h\nu} = 350$ nm. In the ERG with white-eyed flies the spectral sensitivity is mainly due to stray light entering the rhabdomere from all sides. In the wild-type eye, most light reaching the rhabdomere R8 is filtered through rhabdomere R7. Hence the spectral sensitivity of retinula cells R8 will be shifted to longer wavelengths (for large flies, see Hardie, 1977). All receptors have steeply decreasing sensitivities at spectral wavelengths above $\lambda_{h\nu} = 600$ nm. Retinula cells R1–6 appear to be homoge-

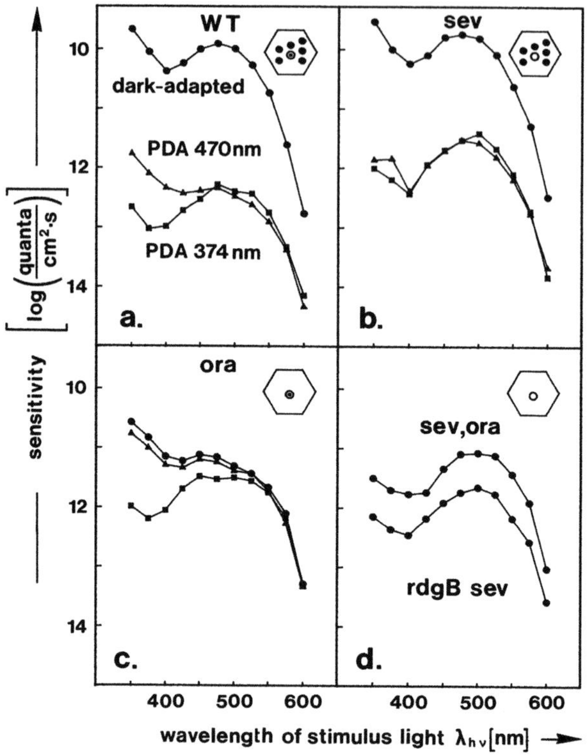

Fig. 8. Spectral sensitivity of electroretinogram (ERG) in wild type and mutants lacking certain retinula cell types (for details see text). All strains carry the additional eye color marker *white* (in order to facilitate homogeneous stimulation of eye). Contribution of retinula cells R1–6 to the ERG of the dark adapted eye can be eliminated by induction of a PDA (see Fig. 7) using intense 470 nm light for preadaptation (triangles). Preadaptation with 370 nm light induces PDA in R1–6 and R7 but not in R8 (*squares*). *Filled circles in hexagons* symbolizing ommatidia represent rhabdomeres No 1–6, *small dots* indicate rhabdomere No 7; *open circles* No 8. *Ordinate* gives number of quanta required to elicit an 0.5 mV ERG. See Appendix for description of mutants *sevenless* (*sev*), *outer rhabdomeres absent* (*ora*) and *retinal degeneration B* (*rdg B*). (After Harris et al., 1976)

neous as to their spectral sensitivity. In situ microspectrophotometry has revealed, however, that retinula cells No. 7 consist of two spectral types ("yellow" and "pale"), one transmitting yellow, the other white light. This is in agreement with intracellular recordings on R7 cells of large flies also showing two types, one with a single sensitivity peak in the UV, the other with an additional peak in the blue. In *Drosophila* the distribution of the two spectral types in the eye appears to be random (Franceschini, personal communication). Whether also in *Drosophila* retinula cells No. 8 consist of more than one type, as is the case in large flies, remains to be investigated.

Polarization Sensitivity

Rhabdomeres absorb dichroically when they are not twisted. In general, absorption is maximal for light with the E-vector parallel to the microvilli in the rhabdomeres. The orientation of the microvilli is given by the radial position of the retinula cells in the ommatidium. Thus each visual element (composed of seven rhabdomeres) can be regarded as consisting of differently oriented analyzers for linearly polarized light. This dichroic absorption is due to the primary photopigment (xanthopsin). Recently, however, two types of accessory pigments have been reported, a UV-absorbing pigment in retinula cells No. 1–6 which in large flies has been shown to absorb nondichroically, and a blue-absorbing pigment in the "yellow" rhabdomeres R7, which has a strong dichroism with its direction of maximal absorption perpendicular to that of the microvilli (Kirschfeld et al., 1977, 1978).

The situation is further complicated by the possibility that dichroic absorption in addition may be caused by the unilateral attachment of screening pigment to the rhabdomeres in the light adapted state (Wolf et al., 1980). On the other hand, at least some fly rhabdomeres seem to be twisted around their longitudinal axis. At present it is not possible to deduce their physiological responses from their histological appearance. Polarization sensitivity (PS) of retinula cells has been recorded electrophysiologically only in large flies. Not all cells seem to be polarization sensitive. If present, polarization sensitivity is small (PS $\approx$ 2) for retinula cells No. 1–6 (Jaervilehto and Moring, 1976). It can be considerable for retinula cells No. 7 and No. 8 (Hardie et al., 1979).

Ocelli

The three dorsal ocelli of *Drosophila* and the underlying neuropil have been investigated anatomically by Schmidt (1975). Labhart (1977) obtained intracellular recordings from ocellar photoreceptors and the spectral sensitivity of the ocellar receptors has been determined using the ERG recording technique (Hu et al., 1978). Finally, the role of the ocelli in visual behavior, particularly in phototaxis, was studied with mutants lacking ocelli or by occluding the ocelli with paint (Fischbach and Reichert 1978; Hu and Stark, 1980). For most of the experiments to be described in this book the influence of the ocelli has either not been investigated or more often has been found to be negligible.

Summary

The reader who is interested in more details about the *Drosophila* eye is referred to the respective chapters in recent review articles on invertebrate eyes (i.e., in *Handbook of Sensory Physiology*, Vol. VII/6A–C, 1979 and 1981). For our purposes it is sufficient to recall that the *Drosophila* eyes sample the visual space with about 1400 evenly spaced visual elements, each of which provides the brain with the electrical signals of eight retinula cells carrying, in parallel, time functions of the local quantum flux. By comparison of these primary functions in principle new time functions of the local properties of the incident light can be calculated, yielding some information about spectral composition, degree of polarization, and E-vector orientation. To which degree the brain uses or eliminates certain receptor properties can only be determined from behavioral experiments. Thus we will return to the eye from the point of view of visual behavior at various occasions, most extensively in Chapter 4. First we will take a brief glimpse at the neural architecture of the fly brain and in particular at the optic lobes.

2 Neuronal Architecture of the Visual System

Retinotopic Maps

The eye transforms the visual sphere into electrical signals in an array of neuronal channels. It is one of the appealing features of the arthropod brain that this property is immediately apparent in histological preparations. The orderliness of the compound eye as an array of identical subunits (ommatidia, visual elements) imposes itself upon the neuropil adjacent to it (Fig. 9). The optic lobe can be viewed as a two-dimensional neural crystal (Benzer, 1973). Neurocrystallinity is most apparent for the first and second neuropil regions (lamina and medulla) but a one-to-one relation between the visual elements in the eye and structural subunits in the neuropil has been established for all four regions (lamina, medulla, lobula, and lobula plate) (Power, 1943b; for *Musca* see Braitenberg, 1970; 1972; Preißler, 1974; Strausfeld, 1976).

As one proceeds from the periphery toward the central brain, the columnar organization becomes less and less obvious. Connections between columns, tangential large field neurons perpendicular to the columns, and columnar neurons of lower periodicity than that of the visual elements become more prominent (Braitenberg, 1972; Braitenberg and Strausfeld, 1973; Strausfeld, 1979). The structure suggests that close to the periphery the parallel processing in these columns is predominantly concerned with the local properties of the visual input (adaptation, temporal resolution, light intensity, spectral composition, E-vector orientation etc.), whereas the relations between points in space are abstracted at a later stage of the processing chain. The optic lobes can be understood as a stack of retinotopic maps, each concerned with a different aspect of the scenery, ordered in a hierarchical sequence.

Most neurons in the optic lobes belong to so-called "isomorphic sets" (Strausfeld, 1979), a term coined to indicate their structural and functional similarity. For instance, in the optic lobes many isomorphic sets are thought to consist of one neuron per column, all performing the same operation, each one for the column in which it resides. The neurons in some of these sets may send branches into other columns. These branches, then, are all oriented into the same direction – not in respect to the coordinates of the fly but in relation to the local axes of the array of visual elements. The property of neurocrystal-

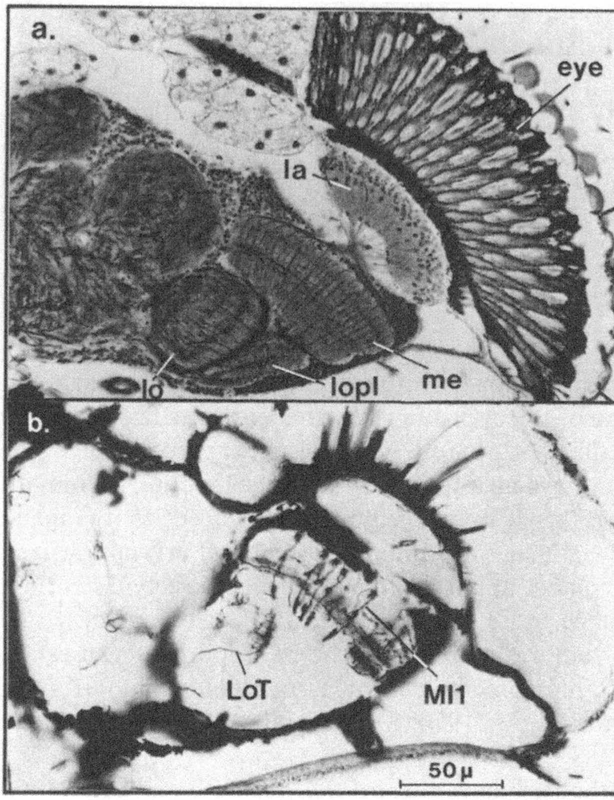

Fig. 9. Eye and optic lobes. (a) Horizontal 7 µ paraffin section stained according to procedure of Holmes-Blest (Blest, 1961) after Carnoy fixation. (b) Golgi stained (Colonnier, 1964) 35 µ Durcupan section (Courtesy K. F. Fischbach). *la* lamina; *me* medulla; *lo* lobula; *lopl* lobula plate; *MI1* medulla intrinsic neuron 1; *LoT* lobula tangential neuron

linity in the optic lobes reflects the developmental processes by which the neuropil is formed. It must be largely local properties rather than long distance effects (e.g., gradients) guiding fiber outgrowth of columnar elements in the optic lobes. This principle has important functional consequences to be discussed in Chapter 4.

Visual Foci

The fine grain of the retinotopic maps is lost in the transition from the optic lobes to the central brain (optic stalk). No set of isomorphic neurons at this level has the same periodicity as (or higher than) that of the visual elements (e.g., Fischbach and Lyly-Hünerberg, 1983). However, the total visual sphere is represented many times in parallel at this level, in part by single large field neurons, by small groups of them (e.g., 3 HS neurons, Fig. 13) or by sets of isomorphic small field neurons each receiving input through at least seven visual elements (Strausfeld and Nässel, 1980). The isomorphic neurons have their fibers joined in bundles or sub-bundles linking the optic lobes to certain "foci" in the central brain. They often preserve retinotopy on the way. Strausfeld and Hausen (1977; for *Drosophila*: Strausfeld and Singh, 1980) were able to stain such neuronal assemblies by filling a common postsynaptic neuron with Co^{2+}-ions. Co^{2+}, then, is taken up by all the members of such assemblies and can be traced there with a special reduced silver method (Timm, 1958).

The visual foci in the central brain seem to have two functions. From there large "descending" neurons relay signals directly to the ventral ganglion where they may participate in the initiation of motor programs. In addition, the foci are the "windows" through which the central brain gains access to the sensory information. Thus, if necessary, the sensory data flow takes the fast through-way to the effectors, avoiding the ponderous processing in the central brain, yet the brain shares the data and has the option to interfere with the motor output.

Lamina ganglionaris

Let us now take a closer look at the optic lobes. The lamina ganglionaris is a striking example for the "radical anatomist's" view maintaining that the functional properties of a nervous system will be reflected in its structural organization (Braitenberg, 1973). The visual elements, which in the retina are complicated, interlaced physical entities, are represented in the lamina as the basic structural building blocks (Fig. 10; neuro-ommatidia, cartridges), which, thus receive information about the luminance, each at just one point of the visual sphere. Each set of eight retinula cells with parallel optical axes from seven adjacent ommatidia send their axons to one lamina cartridge. The fibers form a most orderly plexus between eye and lamina which has attracted the interest of neuroanatomists and sensory physiologists for many decades (Vigier 1907; Braitenberg 1967; Kirschfeld, 1967). In fact, the properties of the neural superposition eye can only be fully appreciated if this special projection of the retinula cell axons to the lamina is taken into account. As mentioned above, the retinula cells No. 1–6 with their long and thick rhabdomeres seem to be specialized for high quantum efficiency. As we see now, in addition in each lamina cartridge the signals of the six cells with parallel optical axes are summed (Fig. 10), improving the signal-to-noise ratio by a factor of $\sqrt{6}$ without much loss of acuity (Scholes, 1969; Smola and Gemperlein, 1972). If, instead, in a single cell the area of the distal rhabdomere tip had been increased sixfold, the angular sensitivity distribution ($\Delta\varrho$) of the visual elements would broaden by a factor of $\sqrt{6}$ with a concomitant loss of spatial resolution.

The multisynaptic contacts of the six retinula cell axons with a set of columnar neurons linking the lamina to the medulla is the most obvious

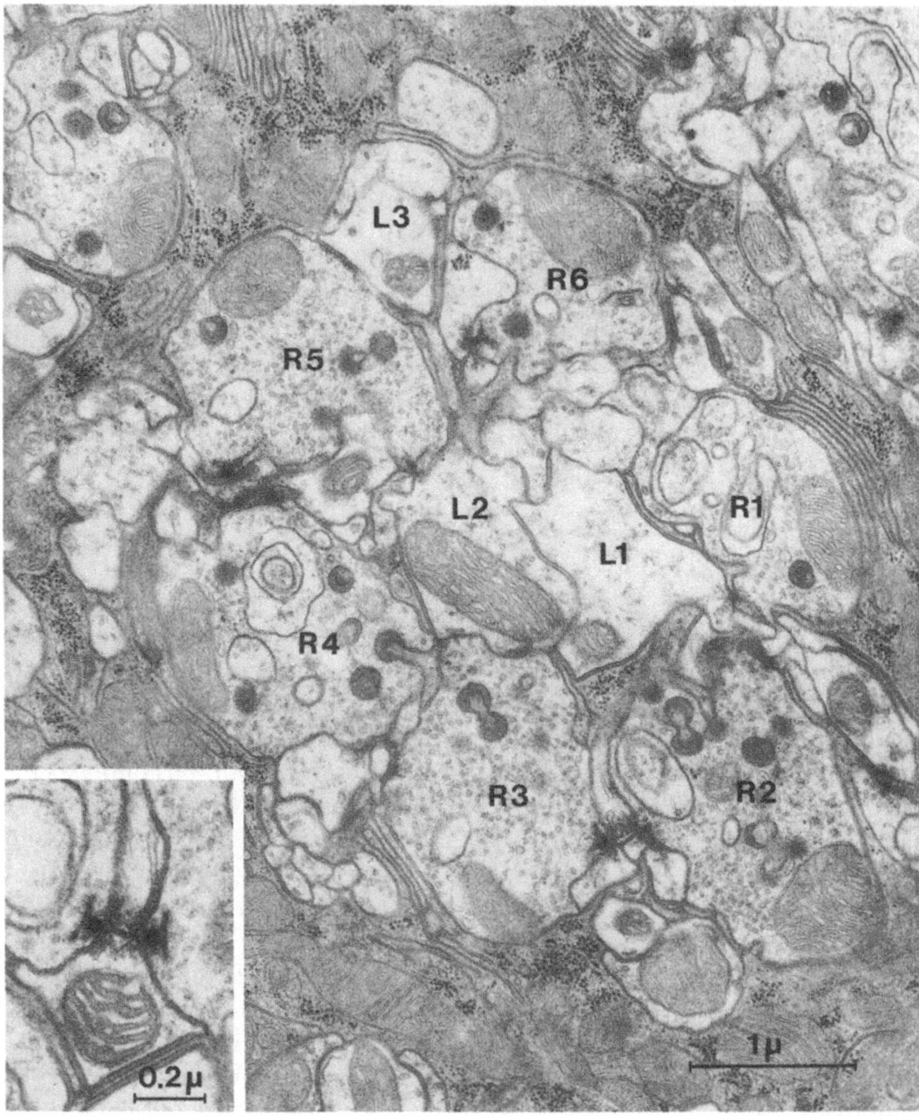

Fig. 10. Electron micrograph of tangential cross-section of lamina cartridge. Assignment of numbers to profiles of retinula cells R1–6 (*R*) and lamina monopolar neurons (*L*) is tentative. Between all six R fibers profiles of basket cell (*T1*) and amacrine cells (α) should occur (not identified). Note additional fibers adjacent to *L1* and *L2*. *Inset* shows two synapses with T-shaped ribbon. (Courtesy I. Meinertzhagen)

structural feature of the first optic neuropil region (Fig. 10). The thin axons of retinula cells No. 7 and 8 bypass the cartridge, apparently without forming any synaptic contacts. They project together with the columnar neurons of the lamina into the medulla, where they terminate in different layers (Fig. 11). The neural superposition of the signals from six retinula cell axons

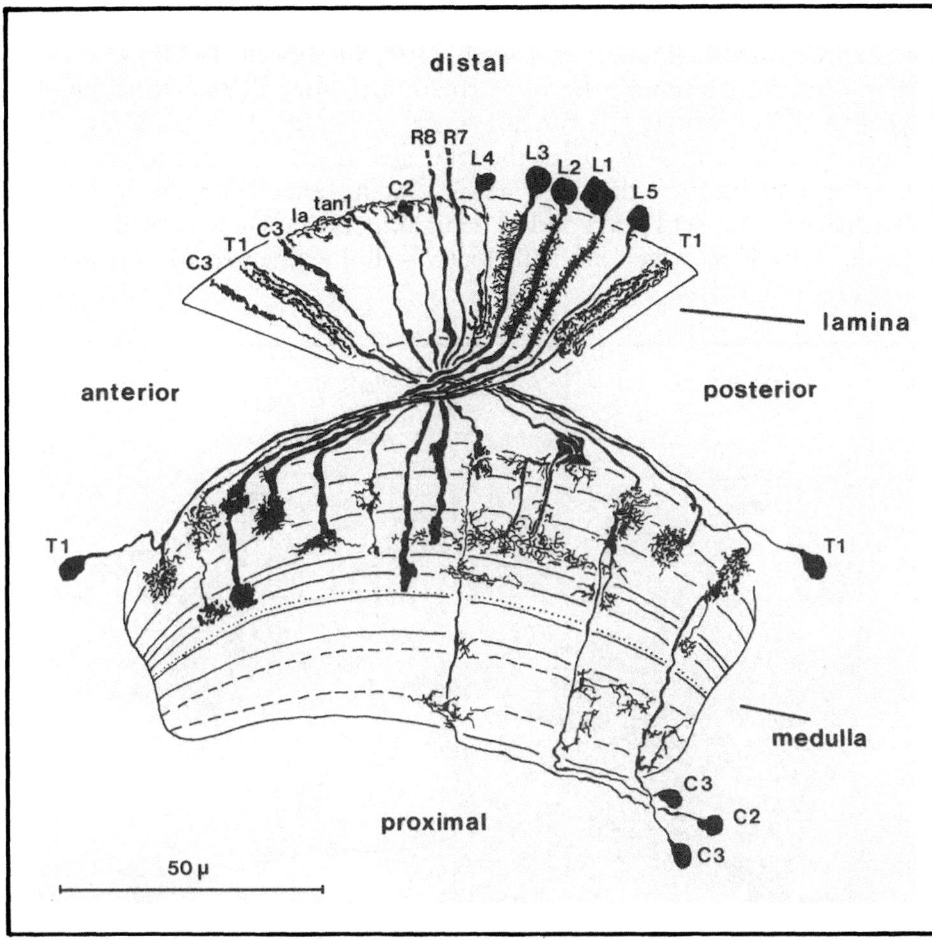

Fig. 11. Camera lucida drawings of neurons connecting lamina and medulla (Golgi preparations). Nomenclature according to Strausfeld (1976). (After Fischbach, 1983 b)

in the lamina leads to a reduction in the number of "input channels" to each column of the visual neuropil from 8 to 3 (which we will call R1–6 channel, R7 channel and R8 channel). Recent behavioral data require at least one additional channel carrying information about polarized light from retinula cells R1–6 (see Chap. 4). The structural correlate of this channel is not known. Also we have to keep in mind that the lattice of R7 cells (and probably that of R8 cells as well) is not homogeneous in respect to spectral sensitivity (see "yellow" and "pale" R7 cells, p. 22). It is not known, however, whether these spectral specializations have correlates in the connectivity of the neuropil.

The fine structure of the lamina has been elaborated in large flies (Trujillo-Cenoz, 1965; Braitenberg, 1967; Boschek, 1971; Braitenberg and

Hauser-Holschuh, 1972; Strausfeld, 1970) but only few data concerning *Drosophila* are available (Hauser-Holschuh, 1975; Fischbach, 1983 b; Meinertzhagen, personal communication). Furthermore, none of the electrophysiological work concerning the functional properties of the lamina has been performed with *Drosophila*. Thus the interested reader is asked to gain some general insight into the dipteran lamina, for instance from the reviews of Strausfeld and Nässel (1981) and of Laughlin (1981). Here we will concentrate on those facts which are sufficiently well documented for *Drosophila*.

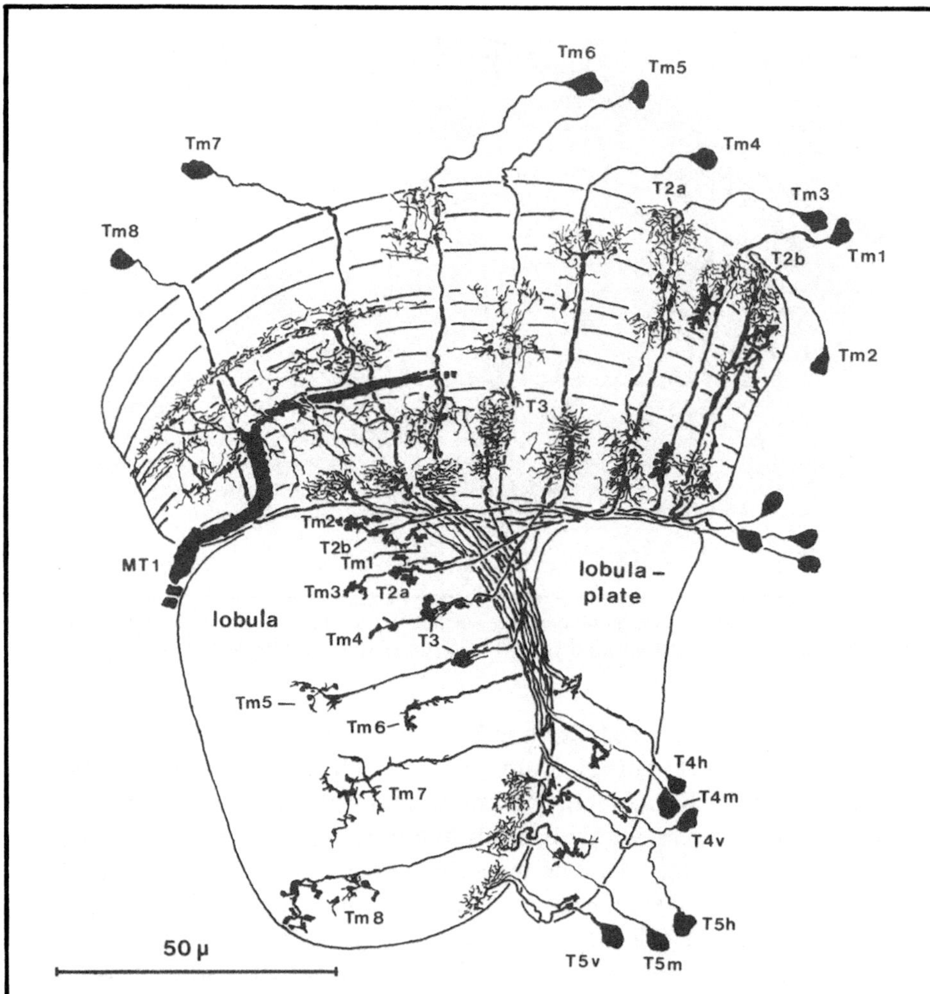

Fig. 12. Neuron types connecting: medulla to lobula (*Tm*) with cell bodies distal to medulla; lobula to medulla (*T2*, *T3*) with cell bodies between medulla and lobula; lobula plate to medulla (*T4*) and lobula plate to lobula (*T5*). *MT1* Medulla tangential neuron 1 (incompletely filled). Camera lucida drawing from Golgi preparations. (After Fischbach, 1983 b)

Most of the cell types constituting the lamina cartridge in *Calliphora* and *Musca* have been identified in Golgi preparations of *Drosophila* as well (Fig. 11). Their shapes are slightly different from those of the corresponding cells in large flies (Fischbach, 1983 b). The various parts of the neurons are scaled down by varying degrees, but also the branching pattern in the medulla and even the lamina suggest some species-specific structural adaptations already at this peripheral level. The synaptic relations between the fibers in the lamina are also similar to those in large flies (Hauser-Holschuh, 1975), but have not been analyzed in great detail. On the way from the lamina to the medulla the R1–6 channel is split up in at least five parallel channels represented by the columnar cells L1, L2, L3 and by the neurons T1 and L4 (via the amacrine cell α). The functional interpretation of this structural divergence will be given in the last chapter.

Medulla, Lobula, Lobula Plate

As in large flies, the lamina and the medulla in *Drosophila* are connected by a large chiasm, causing a back to front inversion of the retinotopic maps in the medulla. No such fiber crossing is observed along the vertical axis. A second horizontal chiasm links the medulla, lobula and lobula plate with each other. The lobula and lobula plate are oriented at right angles to the medulla. Thus the frontal part of the visual field is represented laterally in the lobula and lobula plate, the rear part medially (Fig. 12). This projection pattern is reflected in the thickness of the ganglia. Apparently, processing of infor-

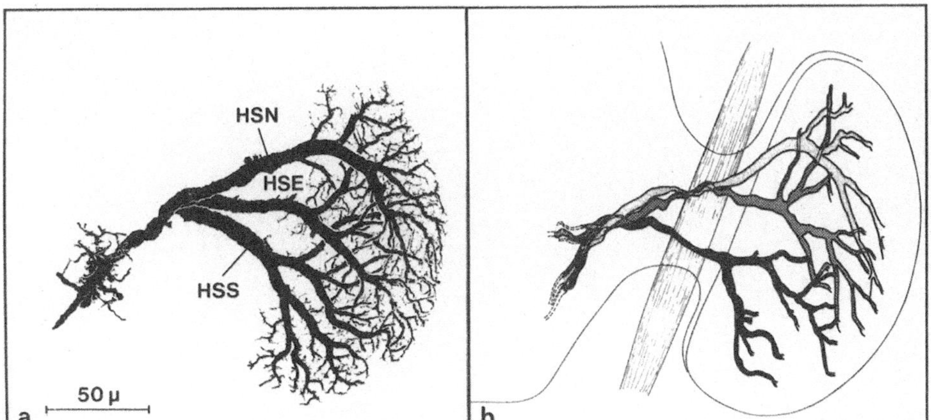

Fig. 13. Frontal view of horizontal giant neurons (*HSN*, *HSE*, *HSS*) connecting lobula plate to central brain from two only distantly related WT (Berlin) females. (a) Golgi preparation (Courtesy N. Strausfeld), camera lucida drawing from K. F. Fischbach (1983 b). (b) Reconstruction from 1 µ serial sections in Araldite. (Heisenberg et al., 1978)

mation in the frontal visual field is more thorough or faster than that for information in other parts (for gradients in the optic lobes of large flies see Braitenberg and Hauser-Holschuh, 1972). The optic lobes contain 60,000–70,000 neurons on each side of the brain (Fischbach and Technau, 1983). This is roughly the same number of cells per visual element as counted for *Musca domestica* (Strausfeld, 1976). Many of the cell types found in the medulla and lobula complex by Golgi staining can be homologized with neurons in *Musca* on the basis of their shape and position in the neuropil (e.g., Fig. 13 and Fischbach, 1983a, b). A detailed account of *Drosophila* neuroanatomy is being prepared by Strausfeld and Fischbach (eds.). We will mention some more of the neuroanatomical details in Chapter 4.

3 Motion Sensitivity Under Open Loop Conditions

Orientation

Adult *Drosophila* can fly, land, walk, rest, start flight again etc. They search for food, water, mates, optimal temperature, and shelter. They display elaborate courtship behavior and even fight with conspecifics; but often they just sit on the lower side of a leaf and appear to do nothing.

In most visual behavior orientation plays an important part. Thus we will be concerned with visual orientation in most of this book. In the tradition of entomological ethology, visual orientation is studied by correlating the visual stimuli reaching an animal with its changes in orientation or posture. The refinement of such experiments has been one of the main lines of progress in this field during the last 30 years. We would like to warn the reader, however, that the existing knowledge on such "input–output" relations constitutes a powerful bias, suggesting that visual orientation itself should be an input–output process. We hope to show that orientation in most instances is not induced by visual patterns. The animal directs itself toward such patterns. Orientation is based on orientedness.

At this point, however, we are not yet ready to discuss orientedness. We need to take advantage of input–output relations under so-called open loop conditions (see Introduction), which prevent the animal from acquiring its normal state of orientedness. In our case the fly is firmly attached by its thorax to a rod, wire, or piece of cardboard, and is kept stationary while flying or walking. In addition, the head is glued to the thorax. This technique has the advantage that well-defined stimuli can be applied. It has led to the systematic investigation of the basic constituents of orientation. What are these constituents? One is the ability to counteract forced changes of orientation; the second is the ability to choose and maintain a certain orientation, and the third, the least understood one, is the ability to link such behavior to meaning, i.e., to pursue a goal by choosing and maintaining a certain orientation. Obviously, with freely moving animals it is difficult to separate these three levels. The main advantage of the experiments with tethered, stationarily walking and flying insects is that these levels can be studied separately. In this chapter we review the open loop experiments concerning the first level: visual course control.

Optomotor Yaw Torque and Turning Behavior

For a moving subject in a patterned environment the visual sphere is a flow field. If the subject rotates, it experiences coherent rotatory motion of all pattern elements in the visual sphere except at the poles of the rotatary axis. During translatory motion it is confronted with expansion and contraction of the pattern elements in the visual sphere at the poles of the translatory axis. Secondly, during locomotion incoherent movement of foreground and background provides an important depth cue. Thirdly, visible events in the world are often associated with additional motion. Thus the evaluation of motion must be among the basic functions of any visual system.

In most of the work with tethered flies, motion perception has been studied by measuring the turning tendency, or the forces the fly exerts in the attempt to control its course. The first measuring device for stationarily flying *Drosophila* was the yaw torque compensator (Götz, 1964) shown in Fig. 14. It is used in most of the experiments to be described here. Various other devices have been developed since. Figure 15 shows a very simple torque meter which can be used to measure roll and pitch torque as well (Blondeau and Heisenberg, 1982). Similar instruments have been developed to measure translatory flight forces (e.g., Götz, 1968). Devices for recording the turning tendency of walking *Drosophila* flies have been designed by Götz and Wenking (1973; "tread compensator") and by E. Buchner (1976; "styrofoam ball"). The locomotion recorder shown in Fig. 16 is an improved version of Buchner's apparatus (Bülthoff, 1980). We do not intend to give a complete list of recording procedures here. (For technical details the reader is referred to the literature cited above. Several further devices will be mentioned below where this information seems necessary for understanding the respective experiments).

Flies respond to the movement of visual patterns with turning behavior apt to reduce the relative motion in the visual field. If the perceived flow field is due to movement of the animal itself, this will produce stabilization of its track. *Drosophila* is sensitive to motion in any part of its visual field. In flight the yaw torque response saturates already at low stimulus strength, implying a large open loop gain. With sufficiently weak stimuli the steady-state response is proportional to the fraction of the visual field stimulated. For instance, two motion stimuli with opposite direction presented to different parts of the visual field may cancel each other (Götz, 1964). We will come back to the relation between the extent of the stimulus pattern and the yaw torque response in Chapter 9.

Direction-sensitive movement perception requires at least two input channels derived from different visual elements and with different properties. These channels must interact in a nonlinear way (Hassenstein, Reichardt, Varju, 1951–1967; Kirschfeld, 1972). In fact, a multiplication of the signals

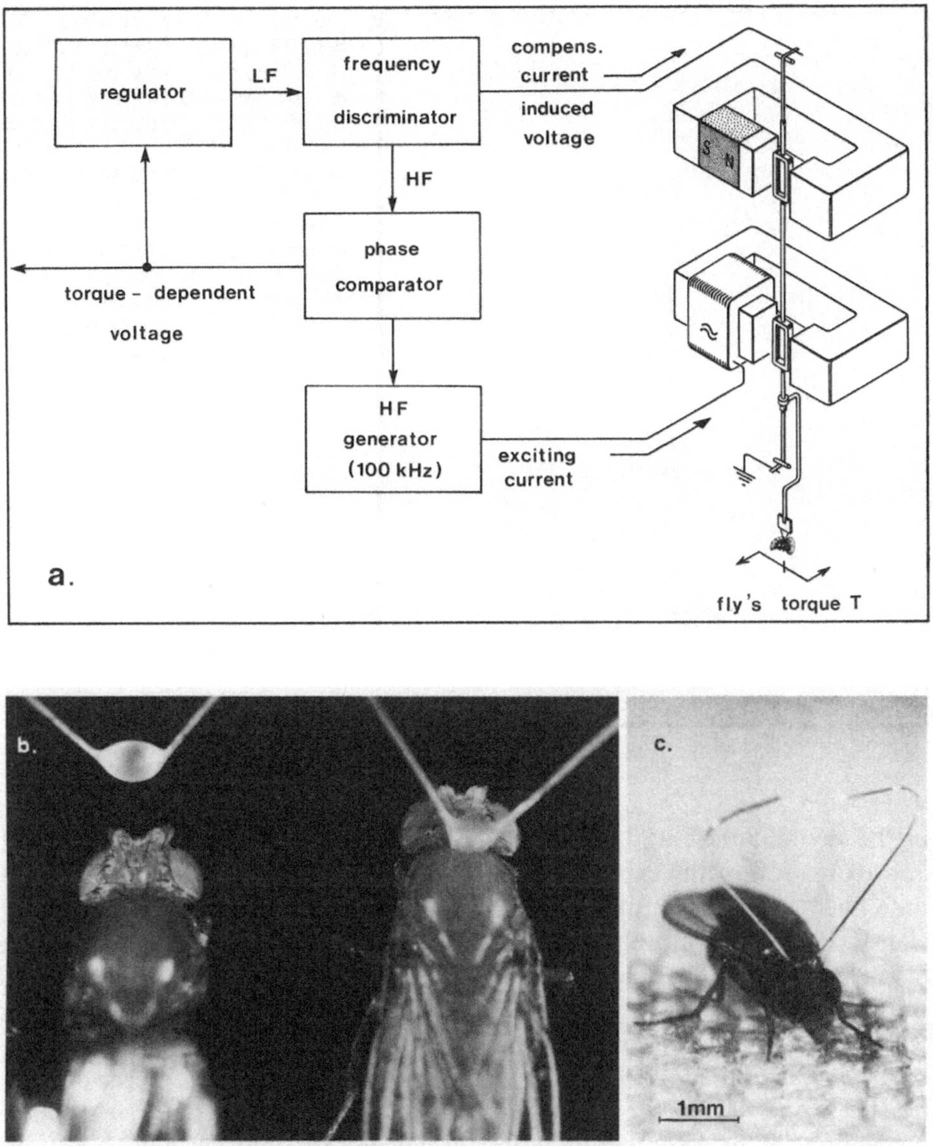

Fig. 14. Torque compensator for recording yaw torque of *Drosophila* in stationary flight. (a) Block diagram of mode of operation. *HF* high frequency; *LF* low frequency; redrawn after Götz, 1964, where mode of operation is fully explained). (b) Fly is attached to silver wire (50 μ) by a droplet of a mixture of beeswax and double bass rasin (2:1) applied between head and thorax by heating the wire with a pulse of current. (c) Flies are little impeded by the wire; they can stand, walk, and fly with it

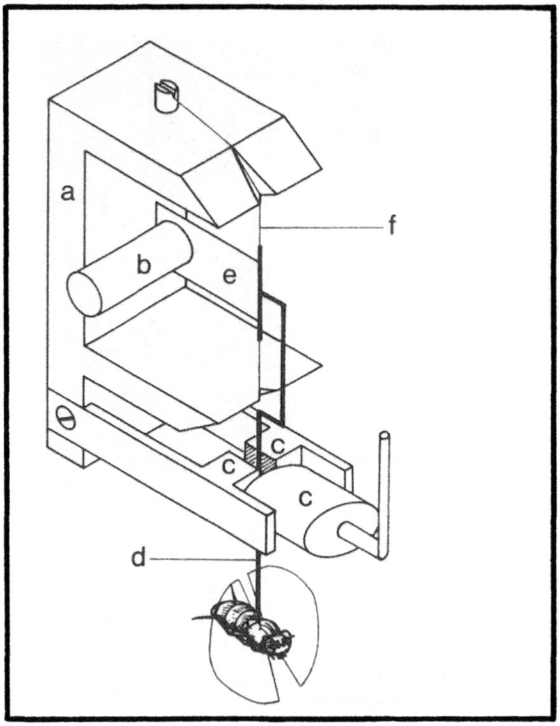

Fig. 15. Simple torque meter. Depending upon the attachment of the fly to the axis (*d*) yaw, roll or pitch can be recorded (see Fig. 24). (*a*) frame; (*b*) coil of inductive distance transducer (Philips PR 6431/PR 7451); (*c*) clamp for protecting the axis while flies are attached; (*d*) stiff axis; (*e*) aluminum plate; (*f*) torsion spring (tungsten wire). *Drosophila* flight torque on the axis causes minimal torsion of the wire and displaces the aluminium plate in front of the distance transducer. (Blondeau and Heisenberg, 1982)

in the two channels and subsequent integration (Fig. 17) leads to a good approximation of the experimental data if only uniform movement and the time average of the optomotor response are considered (Poggio and Reichardt, 1973 b; a comprehensive description of the correlation model is given in Appendix A of E. Buchner, 1976, and in E. Buchner, 1983). In much of the early work large field patterns of regularly spaced black and white stripes were used as stimuli, for instance vertical cylinders rotating around the rotatory axis of the torque meter. With such gratings as stimuli three characteristic properties of the correlation model for direction-sensitive motion perception can be demonstrated: first, at low pattern contrast $\Delta I/\bar{I}$ (I: pattern luminance) movement detection depends upon ΔI^2 (McCann and MacGinitie, 1965; Hengstenberg and Götz, 1967; E. Buchner 1976; Heisenberg and E. Buchner, 1977). Second, the response is not based on the measurement of angular velocity (w) but rather temporal frequency w/λ_{sp} (contrast frequency), i.e., the frequency at which the pattern elements pass at a position of the visual field (Fig. 18). Third, the strength of the (time-averaged) steady-state optomotor response is the same for all patterns composed of the same pattern frequencies and amplitudes irrespective of their phase relations (Fig. 19). This had been predicted on the basis of the correlation model and was

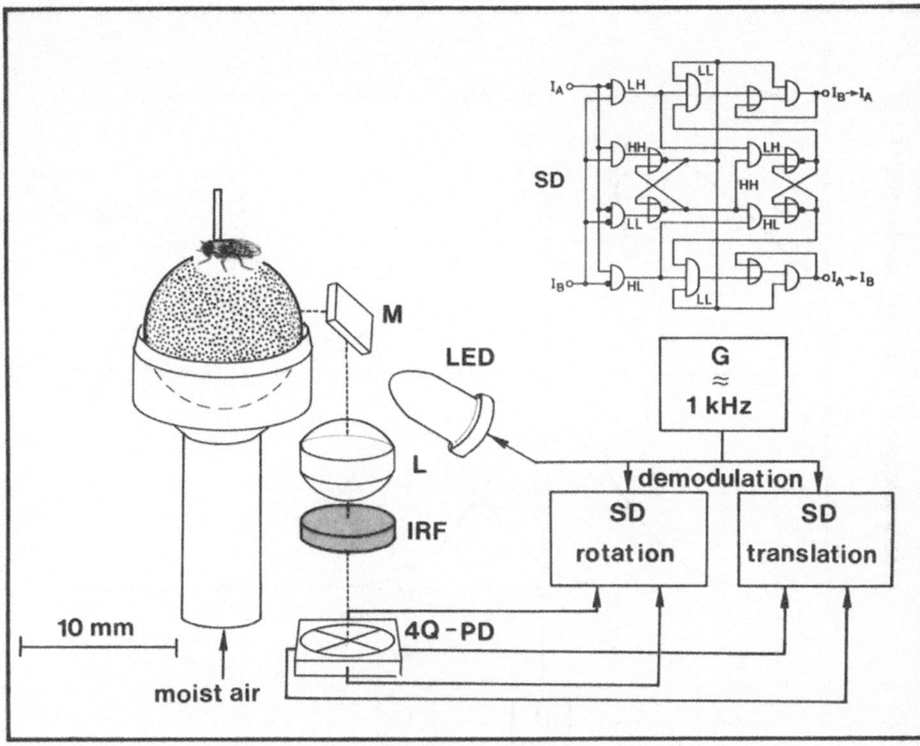

Fig. 16. Locomotion recorder. *Drosophila* glued to a rod or hook walks on top of a styrofoam ball floating on a stream of air. Rotations of the ball are recorded optoelectronically. Black dots (D = 0.2 mm) on red background of ball are projected by a mirror (*M*) and a lense (*L*) onto a four-quadrant photodiode (*4Q-PD*). The ball is illuminated by an infrared light emitting diode (*LED*) and only infrared light is allowed to reach the 4Q-PD (infrared filter, *IRF*) to make recording independent of experimental situation. The magnified images of the black dots cover about 75% of the sensitive area of the 4Q-PD. Opposing quadrants of 4Q-PD are evaluated by sequence detecting logic (*SD*, note *inset*) for rotatory and translatory locomotor activity of fly. The LED is modulated with f = 1 kHz to increase maximum output; synchronous demodulation of optoelectrical signals diminishes disturbances from other light sources. (Bülthoff, 1982 b; inset (SD) Heisenberg and Götz, 1975)

subsequently confirmed for the beetle *Chlorophanus* (Varju, 1959; Varju and Reichardt, 1967) and for *Drosophila* (Zimmermann, 1973).

The "movement detector" shown in Fig. 17 is considered a "minimal" model, i.e., an interaction between input channels which is necessary and sufficient to account for the basic observations. At the neuronal level, however, a large variety of realizations for this function seem possible. Based on earlier work of Hassenstein (1951) and Götz (1964, 1965), E. Buchner (1976) showed that visual course control in *Drosophila* is, indeed, mediated by "elementary movement detectors" (EMD's) which consist of two-channel interactions as the "minimal" model suggests. By selectively stimulating

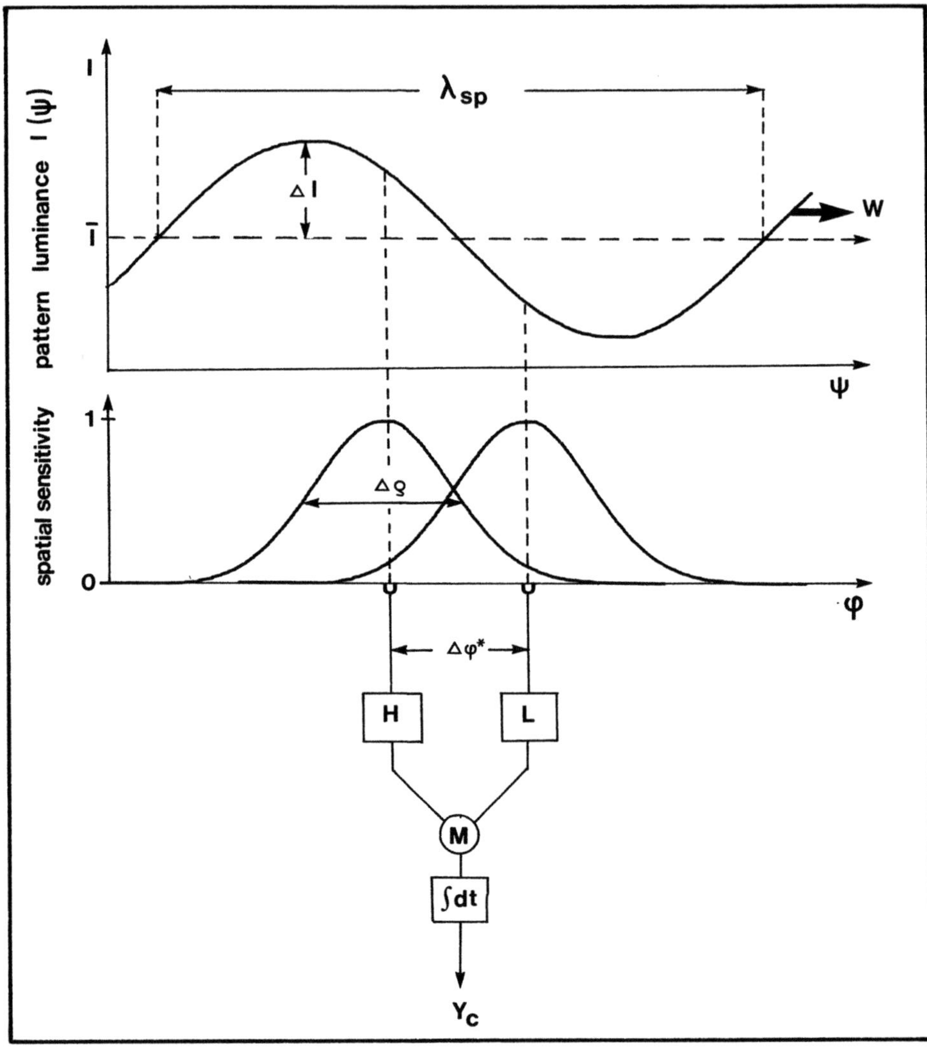

Fig. 17. Correlation model of direction-sensitive motion perception. *M* multiplication of the time functions arriving through the two input channels; *H* and *L* high pass and low pass filters in the input channels which have to be tuned to a 90° phase difference for the system to be insensitive to flickering light; $\int dt$: the product of the two time functions has to be integrated over at least one complete light/dark cycle to make the output y_c of the model proportional to the mean steady state optomotor response; λ_{sp} spatial wavelength of stimulus pattern consisting of a periodic grating with sinusoidally changing luminance I; $\bar{I}$ average luminance; ΔI spatial modulation of luminance; $\Delta \varrho$ half-width of angular sensitivity distribution of visual elements; w angular velocity of pattern; $\Delta\varphi^*$ projection of angular distance between visual elements $\Delta\varphi$ onto the direction of motion; φ angular coordinate of eye; ψ angular coordinate of visual sphere (see Fig. 3a, where ψ designates horizontal space coordinate). (After E. Buchner, 1983; Kirschfeld 1972.) Recent experiments with *Musca* reveal that this model does not account for responses to accelerating and decelerating patterns whereas the original model of Hassenstein and Reichardt (1956) does (Reichardt, personal communication)

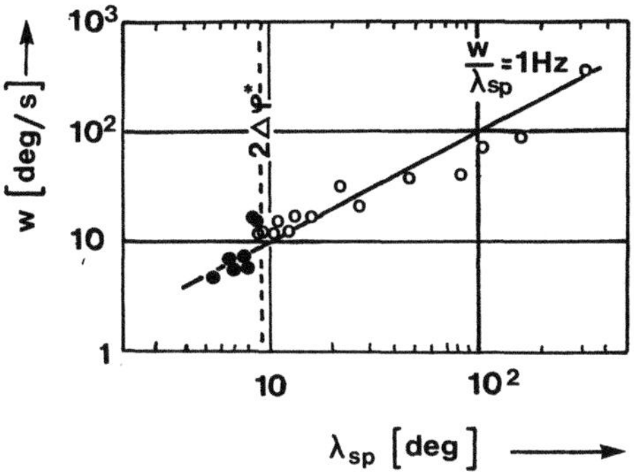

Fig. 18. Optomotor yaw torque responses in stationary flight were recorded as functions of angular velocity (w) using periodic gratings of various spatial wavelengths (λ_{sp})as stimuli. For explanation of symbols see Fig. 17. *Data points* are the maxima of the spatial wavelength versus velocity functions. *Filled circles* show torque responses against the direction of stimulus motion. This inversion of the optomotor response is due to geometrical interference between the lattice of visual elements (sampling base $\Delta\varphi^*$) and the grating (spatial wavelength λ_{sp}). The spatial wavelength range in which optomotor responses *against* the direction of motion can be measured is $2\Delta\varphi^* > \lambda_{sp} > \Delta\varphi^*$ (Götz, 1964). (After Götz, 1972 b)

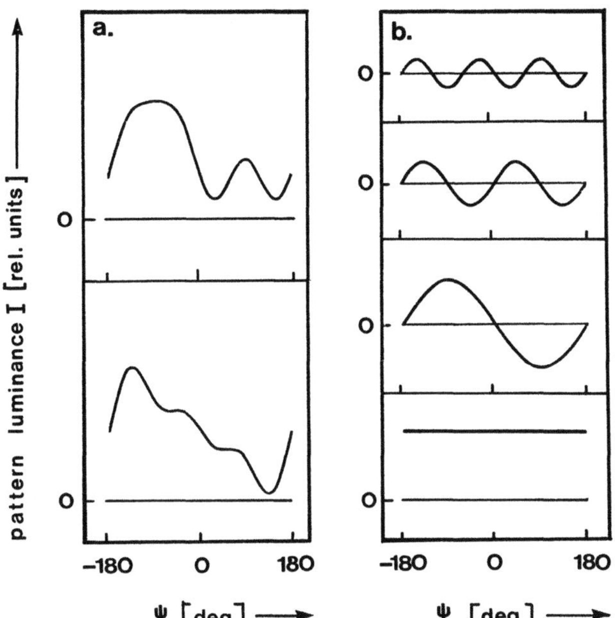

Fig. 19. (a) Angular distribution of luminance of two patterns composed of the same Fourier components shown in (b). The two patterns in (a) elicit the same average yaw torque response in stationary flight. (Zimmermann, 1973; cited after Götz, 1972 b)

restricted arrays of visual elements by sinusoidal gratings of variable, precisely defined spatial wavelengths and orientations (Fig. 20), he determined the contributions to the optomotor response of 18 potential interactions (EMD's) between different positions in the lattice of visual elements (first to fourth neighbors). Most of the response can be attributed to two interactions (EMD's) receiving their input from adjacent visual elements along the x and y axes of the lattice oriented at + and $-30°$ to the horizontal (Fig. 3b; Fig. 21). Buchner's analysis was restricted to the upper frontal part of the visual field. However, pilot experiments by Götz (1964) and E. Buchner (personal communication) suggest that most, if not all, of the visual field is

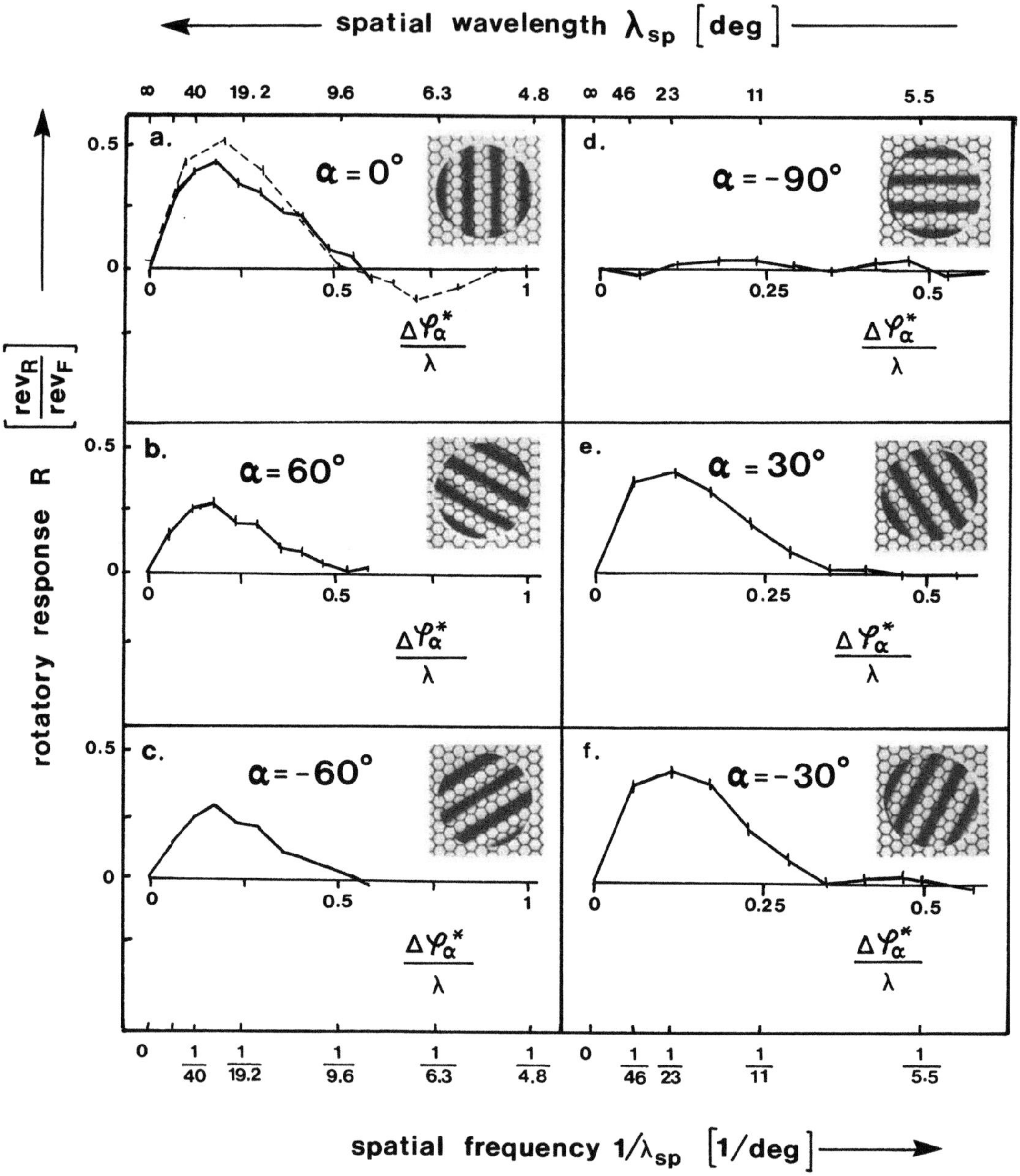

Fig. 20. Optomotor turning responses on the locomotion recorder as functions of pattern wavelength λ_{sp} (plotted as spatial frequency $1/\lambda_{sp}$ in units of $\Delta\varphi^*$ of a moving sine-wave grating presented in a circular area of the upper frontal visual field covering about 50 visual elements. As in Fig. 17, effective sampling base $\Delta\varphi^*$ is the divergence angle between adjacent rows of synchronously stimulated visual elements. For each of the six directions of pattern motion α (a–f) Fourier coefficients can be calculated from the response functions. Contributions of the various combinations of pairs of visual elements (potential EMD's) to the overall response can then be calculated from the Fourier coefficients. (E. Buchner, 1976)

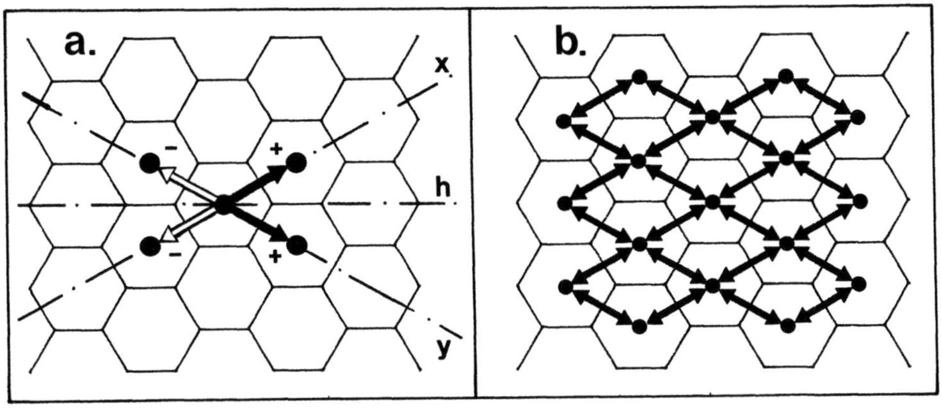

Fig. 21. Elementary movement detectors (EMD's) mediating most of the optomotor yaw torque reponse are oriented along the x- and y-axes of the array of visual elements. Each EMD receives its input from two adjacent visual elements. Presumably EMD's are unidirectional (see discussion on pp. 46 f and 124 f) as shown in (a). They form a regular lattice (b) since each column in the visual neuropil is connected to its neighbors by EMD's (cf. Fig. 25). (After E. Buchner, 1976)

sampled for local movement along these two directions by a homogeneous lattice of such EMD's and that the signals of these EMD's are summed to generate yaw torque or the turning response.

In addition to the EMD's just mentioned, Buchner finds that a third interaction between nearest neighbors along the h-axis contributes significantly to the optomotor response but only with low "weight". Second-neighbor positions along the x- and y-directions in the lattice of visual elements also appear to interact weakly. However, as Buchner points out, the interpretation of this finding is not clear. The analysis assumes pure second-order interactions (e.g., simple multiplications) for all EMD's. Nonlinearities of higher than second order in the dominant EMD's would show up as additional interactions between more distant positions in the lattice.

The possibility of a third type of EMD contributing to the optomotor response raises an important matter of discussion. The term "optomotor response" suggests a simple, uniform item of the fly's behavioral repertoire; but actually it is a term defined by the measuring procedure. Everything in the fly contributing to yaw torque or turning as a function of the movement stimulus is summarized as optomotor response. Thus an important result of Buchner's analysis is the small number of classes of EMD's involved. It is at least compatible with the notion (but, of course, by no means a proof) that, after all, a single uniform behavior, namely the need to counteract the fly's own involuntary rotations, might account for most of the optomotor response.

A second problem must be addressed here. The equal weights of the two principal EMD's cause movement along the horizontal direction in the lattice

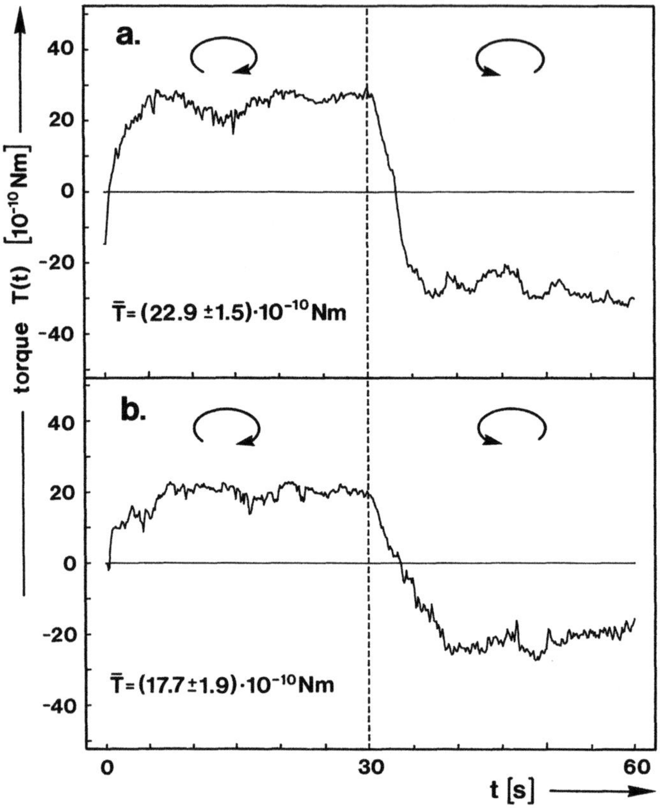

Fig. 22. Time course of yaw torque in stationary flight in response to a striped drum (a) ($\lambda_{sp} = 18°$ in the equatorial plane; pattern contrast $m \approx 1$; $w/\lambda_{sp} = 1$ Hz) and a zebra stripe pattern (b) (LT 134; see Fig. 45 b) at the same angular velocity. Each measurement consists of 30 s of clockwise and 30 s of counterclockwise rotation of the panorama. Traces are averages of ten sweeps from the same ten female flies (WT Berlin). *Inset numbers* ($\bar{T}$) represent mean and standard error of the mean of the time-averaged responses

to be the most effective stimulus eliciting yaw torque or the turning response. However, due to the curvature of the eye, the lattice is tilted in the upper and lower front and back of the eye in respect to the horizontal of the fly, the tilt reaching as much as 60° at some positions (see Figs. 2 and 3 b). Thus, if the lattice of EMD's is in fact homogeneous over the whole projection of the visual field, these areas should contribute less to the stabilization of the fly's orientation in the horizontal plane (see Fig. 35). In addition, vertical movement in the respective parts of the visual field would cause erroneous course correction responses around the flies' vertical body axis. This problem has not been sufficiently investigated (but see Hausen, 1981). Its interest, from the neurogeneticist's point of view, lies in the compromise which will be found to be achieved between the perfection of the neural apparatus providing visual course stabilization on the one hand and the simplicity of the developmental program generating this apparatus on the other (see Chap. 4).

One of the outstanding properties of the optomotor responses is their persistence. After 48 h of stimulation with a saturating stimulus, the turning response of a walking fly is not much smaller than at the beginning. This may be one of the reasons why so much work has been devoted to this behavior.

The optomotor yaw torque response in stationary flight at the torque compensator is shown in Fig. 22. It takes about 5 s to reach a steady state level under the conditions given in the figure legend. It is sensitive to a wide range of contrast frequencies. Significant responses are obtained between 0.002 and 50 Hz (Götz, 1964).

Optomotor Equilibrium

The study of optomotor responses with tethered animals under open loop conditions is meant to elucidate, as the next step, the visual course control behavior of these animals when they are free to move around. However, this step is no slight one. In much of the second part of this book we will be concerned with the relation between open loop and closed loop behavior and we will show that it is the fly which treats these situations differently. But

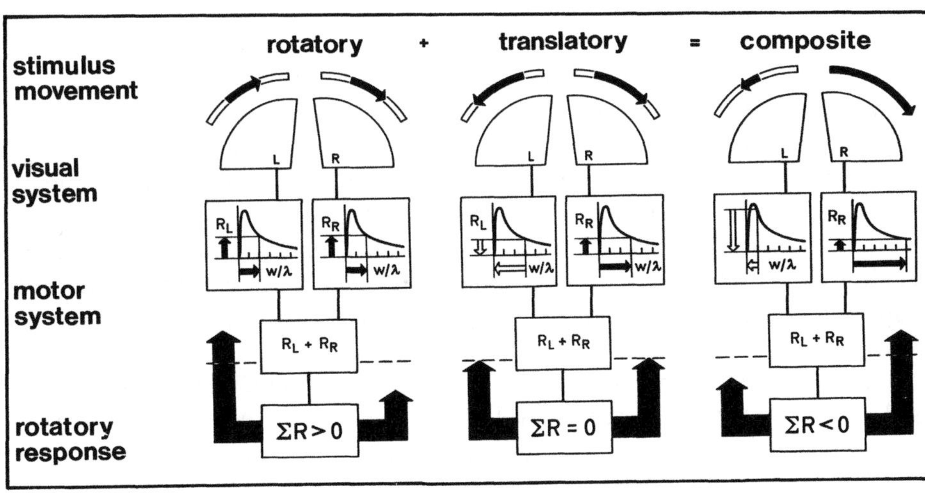

Fig. 23. Optomotor equilibrium. During forward flight and walk the perceived motion field in a stationary surround is composed of two components generated by the fly's rotatory and translatory self-motion. The two components have opposite sign in the visual half-field to which the fly turns and the same sign on the opposite side. Since at high contrast frequency the rotatory response decreases with increasing contrast frequency of the stimulus pattern, the overall response to the two motion stimuli in the visual half fields may be negative, i.e., against the direction of the stimulus, and may thus destabilize the course. In this case optomotor equilibrium is a curved trajectory. (Götz, 1972 a; here λ is λ_{sp} and R indicates rotatory response)

already, if one applies what has been learned so far from open loop studies to describe the track of a moving fly in a stationary environment, one encounters unexpected effects (Götz, 1975 a). The forward velocity and the simultaneous angular course fluctuations during locomotion (Reichardt, 1973; Horn and Mittag, 1980; Wehner, 1979) cause, respectively, front-to-back and rotatory components of the pattern motion in the visual field. The fly responds to the sum of these components on either side of the visual field. Since the correction response of the fly declines toward slow as well as toward fast angular velocities of the stimulus, the net effect of pattern motion may amount to a rotation in the direction opposite to that of the rotatory component alone. Thus with a large translatory component which occurs in the vicinity of objects, positive feedback will destabilize the course and drive the fly to a stable optomotor equilibrium at a certain angular velocity. Under open loop conditions the fly is unable to decompose the motion field into its rotatory and translatory components (Fig. 23; Götz, 1972 a; 1975 a). Also with freely walking flies, some experimental results have tentatively been explained on this basis (Götz, 1972 a, 1975 a; Wehner and Horn, 1975). To what degree this property of optomotor behavior determines the trajectory of an unrestrained fly depends upon the means a fly has to cope with self-induced pattern motion (see Chap. 6).

Optomotor Roll and Pitch Responses

As compared to the optomotor yaw response, little is known about other visual course control mechanisms. Recently a pilot study of roll and pitch optomotor responses in *Drosophila* (Fig. 24) has been published (Blondeau and Heisenberg, 1982), indicating that many characteristics of the yaw torque response also apply to the other two rotatory responses. They can be elicited by local movement in large parts of the visual field. They are mediated by homogeneous arrays of EMD's consisting predominantly of nearest-neighbor interactions presumably along the vertical axis of the lattice. The responses are based on contrast frequency (w/λ_{sp}) and not on angular velocity.

It is, however, worth noting also the differences. The high open loop gain and subsequent low saturation level of the yaw torque response (Götz, 1964) has no parallel in the roll and pitch responses. We have argued (Heisenberg and Wolf, 1979) that this property of the yaw response may be part of a mechanism providing a zero torque reference (see Chap. 5). Basic orientation of the body in respect to the roll and pitch coordinates may be maintained already in response to gravity (Blondeau and Stumpf, unpublished; see, however, the contradictory evidence of Hengstenberg (1984) for large flies).

The importance of the gravitational vector for pitch optomotor responses is evident from the observation that the average torque measured without

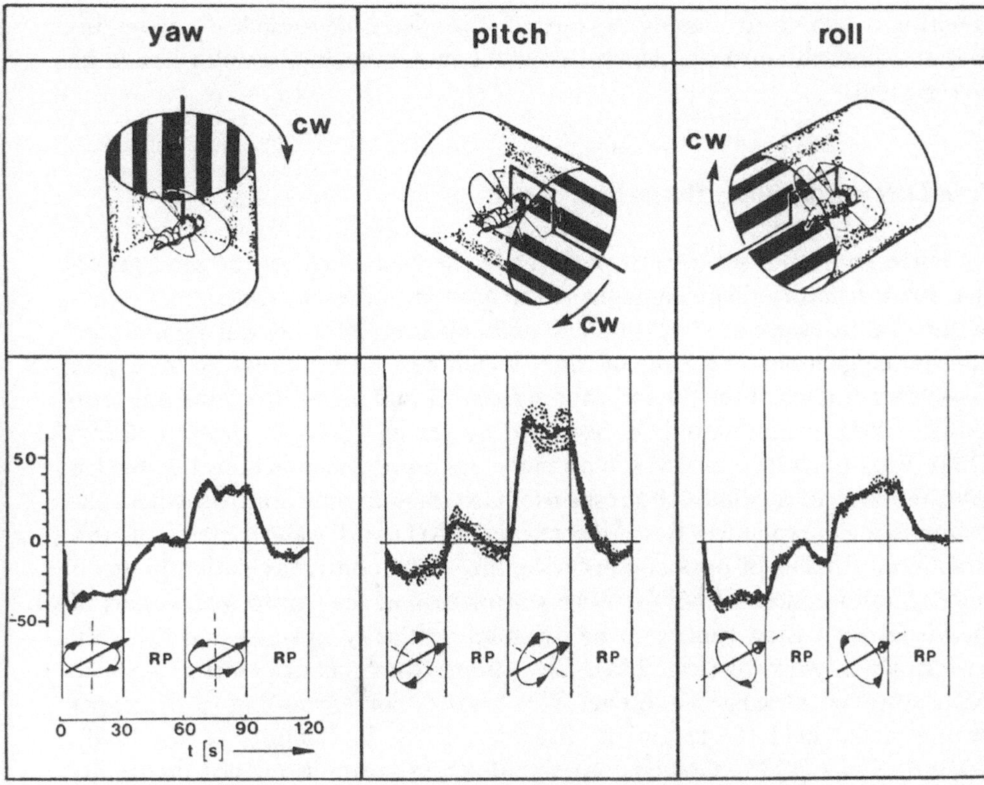

Fig. 24. Optomotor torque responses for yaw, roll, and pitch in stationary flight as recorded with the torque meter shown in Fig. 15 and a striped drum as stimulus, $\lambda_{sp} = 36°$. The traces are averages of 25–30 experiments with 5–6 flies (WT Berlin). *Dotted envelopes* indicate standard errors of the means. Note asymmetry of pitch response to clockwise (*cw*) and counterclockwise rotation. This depends upon the fly's orientation in the gravitational field. *RP* (rest period) indicates 30 s intervals without motion. (Blondeau and Heisenberg, 1982)

stimulation can be shifted between the torque levels reached in response to clockwise and counterclockwise stimulation by shifting the fly into different orientations with respect to gravity. The pitch response is about equally strong to both directions of rotation when the fly is tilted head down by 45°–90° (Blondeau and Stumpf, unpublished). Oriented horizontally the fly responds much more to counterclockwise than to clockwise rotation, as can be seen in Fig. 24.

A remarkable additional feature of the pitch response is its negative sign in the bottom part of the visual field (data not shown). Front-to-back motion underneath the fly – a situation which occurs when the fly approaches the ground in forward flight – elicits upward pitch (data not shown). Since in real flight this would be a positive feedback loop, the stimulus would cause an all-or-none reaction and would turn the fly rear down. The fly would start hovering or would fly steeply upward (J. Blondeau, unpublished). How this special aspect fits into the general course control mechanism still has to be investigated.

Translatory Optomotor Responses

These are responses to flow patterns in the visual field of the kind generated by translatory displacements of the animal. Different translatory optomotor responses are at work in *Drosophila* during flight and during locomotion on the ground. The force of flight which has a fixed direction in respect to the coordinates of the fly and thus affects lift and thrust at a constant ratio (Götz, 1969) is modulated in response to vertical pattern motion (Götz, 1968). Again, a lattice of EMD's possibly less homogeneous than the one for rotatory control mediates the response. Contrary to yaw control, where the results can be explained by bi-directional EMD's as well as pairs of unidirectional EMD's of opposite polarity, lift/thrust control is evidently mediated by unidirectional EMD's. With stimulation in the fronto-lateral part of the visual field they appear to be oriented vertically upward and at −30° posteriorly downward (Fig. 25a). Deviations from this orientation seem to exist, however, at least in the part of the lattice corresponding to the upper frontal visual field (Götz and E. Buchner, 1978; E. Buchner et al., 1978). Recently Götz (1983) showed that the lift/thrust response is not invariably "wired" to this lattice. When *Drosophila* is flying in an air current the response is mediated by two sets of EMD's oriented vertically and antiparallel (Fig. 25b). Since blowing at the fixed flying fly from the front also causes a

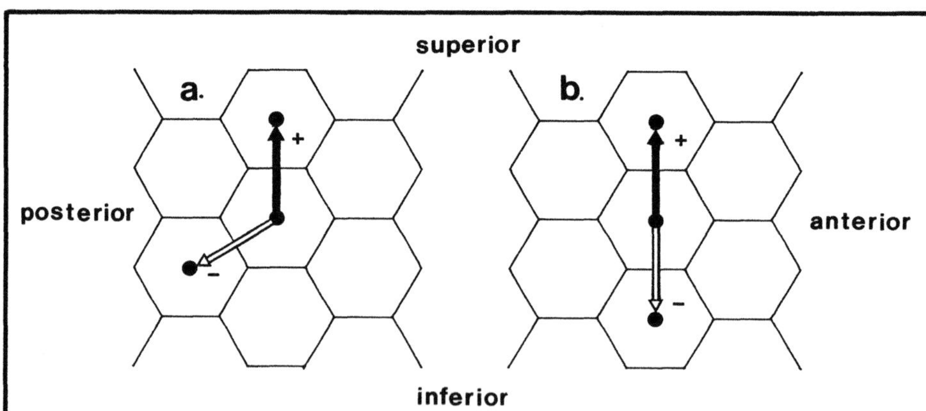

Fig. 25. Unidirectional elementary movement detectors mediate lift/thrust control in response to vertical pattern motion in the fronto-lateral part of the visual field. During stationary flight in still air EMD's are oriented vertically upward and along the x-axis downward (a). Flying in an air current the fly uses the two vertical EMD's shown in (b). (Götz, 1983)

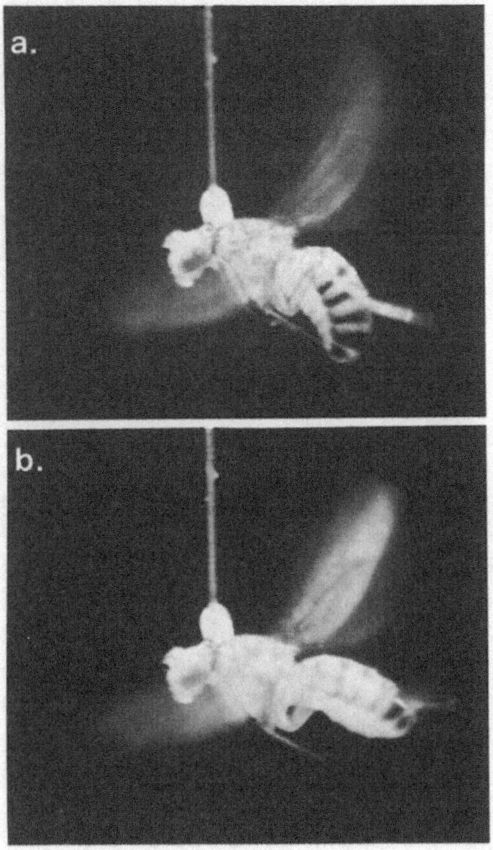

Fig. 26. *Drosophila* attached by head and thorax to a stiff rod is flying (a) in still air, (b) in an air current of $V_{air} = 30\ cm\ s^{-1}$. (Courtesy K. Vogtmann)

change in body posture (Fig. 26; Vogtmann, unpublished) the changing weights of the EMD's in the lattice may reflect a different flight mode. Possibly the air current favors the illusion of fast cruising flight, whereas the lack of it causes hovering flight even if the orientation of the body is forced to remain horizontal. This conjecture is supported by the striking asymmetry of the optomotor pitch response to downward and upward motion in the frontal part of the visual field (see Fig. 24) and by the fine structure of yaw torque in closed loop experiments (see Chap. 6).

Altitude control is useless for the walking fly. Indeed, walking speed is insensitive to vertical pattern motion. However, horizontal motion of vertical stripes from back to front increases, and motion from front to back decreases the propulsive force of the walking fly. Curiously, the strain *In(1)sc*^{8}, *sc*^{8}*w*^{a} happens to show the inverse relation: front-to-back movement accelerates the fly, back-to-front movement slows it down (Fig. 27, Götz and Wenking, 1973). The role of this response in course control of freely walking flies is not clear.

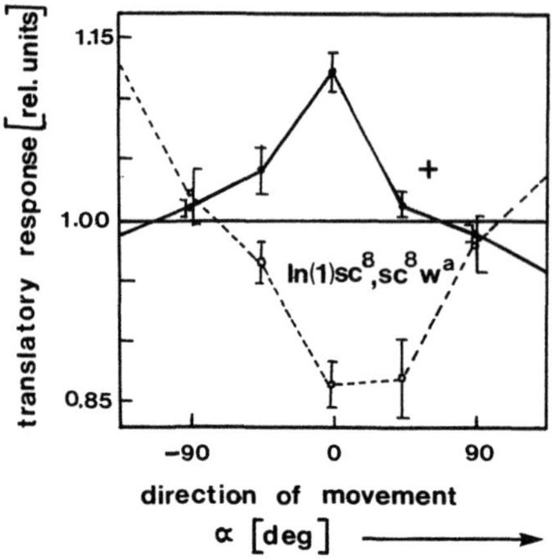

Fig. 27. *Drosophila* modulates its walking speed in response to horizontal pattern motion. (α indicates direction of pattern movement relative to horizontal.) While WT (Berlin) responds equidirectionally, i.e., by a maneuvre which during free locomotion would reduce relative motion, the strain *In(1)sc*^{8}, *sc*^{8}*w*^{a} responds with opposite polarity, a peculiarity of this special stock not related to the reduced screening pigment effect of *w*^{a}. (Götz and Wenking, 1973)

Visual control of velocity over ground seems important in flight but no such mechanism has been found with tethered flies. Experiments in the wind tunnel, however, provide good evidence that *Drosophila* modulates its forward velocity mainly by changing the inclination of its longitudinal body axis (David, 1978). In this way it can maintain a certain over-ground speed irrespective of its air speed by keeping the angular velocity of image motion constant. In response to a stepwise change in the distance of the pattern, it changes to the angular velocity which would give about the same over-ground speed as before (David, 1982). The hoverfly *Syritta pipiens* is able to control translational course deviations in any direction (Collett, 1980 a) without any apparent change in body posture. The tethered moth, *Lymantria dispar*, flying in a kind of merrygoround, shows elaborate over-ground control of flight speed (Preiss and Kramer, 1984).

Optomotor Posture Control

Dependence on visual motion is evident also in some reactions regarding movements and posture of body appendages. The head follows visual motion independently of the thorax. The prosternal organ, a field of small sensory hairs at the neck joint, monitors mechanically the relative orientation of the two. (For control of head posture see also Chap. 6, p. 98 f). Götz (1968) observed that in stationary flight the optomotor yaw torque and lift/thrust responses can well be accounted for by the fly's wing beat amplitude. Wing beat frequency is independent of pattern movement and neither wing pitch nor stroke plane are consistently modulated with the stimulus. Measurements of yaw torque and thrust on the one hand and wing beat amplitude on the other as functions of the direction of movement give qualitatively similar results. The difference between the wing beat amplitudes on either side is roughly proportional to yaw torque, the sum of the amplitudes to lift/thrust (Götz, 1968; Götz et al., 1979; Götz, 1983). [A discrepancy originally observed for back-to-front movement was due to a difference in the experimental procedure. In one case, the head had been glued to the thorax, in the other it had been free (Götz, 1983 and personal communication)]. Body posture, in particular lateral deflection of hind legs and abdomen, is also influenced by visual motion and may contribute to the optomotor turning reaction in free flight (Götz et al., 1979). Motor control of optomotor behavior is just beginning to be investigated (Götz, 1983; Heide, 1983). Obviously, torque in yaw, roll and pitch as well as lift/thrust could not be controlled independently through the modulation of wing beat amplitude alone.

Visually Induced Start of Flight and Landing

All the motion-sensitive behavior described so far may be assumed to serve the walking or flying fly as a continuous guidance system in visual course control (but see Chaps. 5 and 8). Take-off and landing to be described in the last part of this chapter may be considered as fixed action patterns in which motion sensitivity is part of the releasing mechanism.

a) Start of Flight

Although *Drosophila* is able to initiate flight voluntarily, take-off, as for instance as an escape maneuvre, often occurs also in response to visual stimuli, in particular visual motion. So far no satisfying paradigm for measuring it in tethered wild-type flies has been found. Very specific stimuli may be required. The response may be fast habituating or experimental situations may have a suppressive effect in general. In mutant flies lacking eye-screening pigments a strong, sudden change in overall light intensity or the shadow of an object moving overhead are moderately effective stimuli. Moreover, in white-eyed flies carrying certain other mutant alleles (Kaplan and Trout, 1974; Nakashima-Tanaka and Matsubara, 1979; Thomas, 1980) these stimuli elicit responses very reliably. As we will see in the next chapter, parts of the neuronal pathway mediating the start reflex are being intensely studied in *Drosophila*. Thus, the general conditions and appropriate visual stimuli for eliciting it in wild-type flies would be of considerable interest and experimental value.

b) Landing

Visual stimuli involved in landing behavior on the other hand have been studied extensively with large flies (i.e., Goodman, 1960; Braitenberg and

Taddei-Ferretti, 1966; Eckert et al., 1979; Wehrhahn et al., 1981) and recently also with *Drosophila* (Heisenberg and E. Buchner, 1977; Heisenberg et al., 1978; Fischbach, 1981; 1983 b). A fly may be ready to land and may even search for a landing site, but the final approach and the leg extension just prior to the touchdown must be under strict sensory control. It is conceivable that in flight landing is initiated as a disposition to use the next landing site. We may assume that even dispositions are subject to initiation in the sense expounded in the introduction.

Most experiments on landing behavior have been performed with tethered flies. In steady stationary flight flies hold their legs close to the body with the front legs folded underneath the thorax, the middle and hind legs stretched backward. When an object is approaching from the front, the fly unfolds its forelegs, lowers the middle and hind legs, and finally lifts the forelegs in front and above the head. Leg extension has been scored to define the visual requirements controlling it. In addition flight maneuvres just prior to landing have been studied by motion picture analysis (for hoverflies: Collett and Land, 1975 b; for *Musca*: Wagner, 1982).

Only the leg extension behavior in response to visual motion has been studied in *Drosophila*. Although each leg responds differentially to the various patterns of motion, only foreleg extension has been evaluated as a criterion. The directions of motion in the various parts of the visual field which are effective in eliciting landing may be related to the flow field during forward flight (for *Musca*: Wehrhahn et al., 1981). Expansion of a pattern, horizontal front-to-back motion, upward motion above and downward motion below the horizontal midline of the visual field are effectively stimulating foreleg extension. Back-to-front motion suppresses the response to simultaneous or successive front-to-back motion in the contralateral visual half-field. Also, downward motion of a stripe in the upper frontal part of the visual field suppresses the response to subsequent upward motion. Interestingly, downward motion preceded by upward motion is an effective stimulus (Fischbach, personal communication). It may well be that foreleg extension serves as a safety measure in ambiguous situations. This is corroborated by spontaneous landings occurring in dim light or displayed by certain partially blind mutants.

The motion perception properties of the landing response cannot be readily compared to those of the course control responses since the latter normally refer to steady state conditions of stimulation, while landing behavior is recorded as an all-or-none release following the onset of motion. Even with this note of caution in mind, however, one is inclined to attribute the motion sensitivity of the landing reponse to a different perceptual subsystem. The reasons are the following:

First, Hengstenberg (cited in Kirschfeld, 1979; see also Eckert and Hamdorf, 1981; Wehrhahn et al., 1981) discovered that the optimal contrast frequency eliciting foreleg extension is shifted to higher values as compared to that for the optomotor yaw torque response. In the model of motion perception (Fig. 17) the contrast frequency dependence is determined by the filters H and L in the input channels. Thus separate sets of EMD's would be required for the responses. Second, EMD's mediating the landing response to front-to-back motion in the fronto-lateral part of the visual field appear to have different sampling bases from those mediating optomotor responses. Gratings of spatial wavelengths $6° < \lambda_{sp} < 20°$ readily stimulate yaw torque in flight but are quite ineffective triggering leg extension (Heisenberg and E. Buchner, 1977).

Finally, the sensitivity of the landing response to front-to-back motion of a single stripe depends heavily upon the area of the visual field stimulated. The response probability laterally (at $\psi = 100°$) is less than 10% of what it is in the front (at $\psi = 20°$) (Fischbach, 1983 b). The decline of the optomotor yaw torque response toward lateral parts of the visual field is much less pronounced.

The notion of different perceptual subsystems providing the motion dependence of course control and landing responses is also supported by differential impairments of these behaviors in visual mutants. The landing response will be further discussed at the end of the next chapter.

It should be unnecessary to point out that the motion sensitive responses summarized in this chapter are only a small fraction of the fly's behavioral repertoire. Visual guidance of freely flying and walking flies is an interactive affair in which the pieces of the repertoire are normally concatenated by the motion of the fly. This is abolished when the fly is attached to the torque meter. The investigator should be satisfied if some of the basic constituents can be studied at all with tethered flies. This problem will be pursued in the second part of this book.

4 Toward Correlating Structure and Function

One aspect of understanding behavior is to spell out its underlying functional structure. The descriptions we gave in the last chapter are first attempts at "isolating" certain functional subunits and at defining them more quantitatively. As a next step in the analysis one might try to explain such functions as the properties of neuronal circuits, neurons, receptors, muscle fibers, neurosecretory cells and glia. To be sure, the physiological machinery only makes sense within the constraints of the body and of the ecological niche occupied by the animal. In this chapter we will summarize the efforts to find and investigate the neural constituents of visual behavior, in particular optomotor behavior. An important topic of this chapter is the visuomotor interface of visual course control. This is the step in the pathway of the optomotor torque responses where the coordinates of visual space are transformed into those of the motor system. But the story starts at the periphery and will proceed toward the center. We begin with the relation of the optomotor responses to the eye.

Retinal Subsystems

Many of the properties of the eye as described in Chapter 1 had first been deduced from optomotor experiments. Only afterwards were they compared with optical and electrophysiological data. The contrast transfer characteristics of the photoreceptors ($\Delta\varrho$) and the angular spacing of the optical axes of the visual elements ($\Delta\varphi$) were determined by recording optomotor yaw torque (Götz, 1964; 1965) and turning responses (E. Buchner, 1976; Heisenberg and E. Buchner, 1977) as functions of pattern contrast and spatial frequency. The values for the divergence angle ($\Delta\varphi$) in different eye regions are in good agreement with optical measurements (E. Buchner, 1971; 1976). In a restricted area of the visual field $\Delta\varphi$ can be determined most precisely by optical means through the visualization of the lattice of visual elements. As mentioned already in the legend of Fig. 18, the optomotor response is negative, i.e., against the movement of a pattern, when the spatial wavelength λ_{sp} is $2\Delta\varphi^* > \lambda_{sp} > \Delta\varphi^*$ due to geometrical interference between the lattice of visual elements and the stimulating pattern. E. Buchner (1976) found that the first zero crossing of the λ_{sp}-function (Fig. 20) coincided very well with $2\Delta\varphi^*$

($\Delta\varphi^* = 4.8° +/- 0.06°$ in the upper fronto-lateral part of the eye). For the angular sensitivity of the visual elements ($\Delta\varrho$), no optical or electrophysiological data are available for comparison in *Drosophila*.

As described in detail in Chapter 1, the lattice of the visual elements consists of at least three distinct transfer functions for each point in space represented by the three receptor types R1–6, R7 an R8. The retinal subsystem of R1–6 is specialized for high sensitivity due to the large size of the rhabdomeres and the optical and neural superposition of the six cells. It has a broad spectral sensitivity with maxima in the UV and blue-green. The first neural processing for this subsystem takes place in the lamina. The retinal subsystems of R7 and R8 having smaller-sized rhabdomeres and each consisting of only one cell per visual element have a lower overall sensitivity. However, the smaller diameter of the distal tips as compared to rhabdomeres No. 1–6 (Fig. 6) should give these two subsystems a better spatial resolution (see p. 17). The subsystem of R7 has a sensitivity peak in the UV, the subsystem R8 is predominantly green-sensitive (Fig. 8; Harris et al., 1976). Their first neural relay stations are in different layers of the distal medulla.

The relation between the optomotor responses and these three subsystems has been a matter of controversy for several years. Meanwhile, for *Drosophila* the mutant analysis has provided conclusive evidence that the subsystem of retinula cells R1–6 is necessary and sufficient as input for various movement-dependent responses. This has been shown in great detail for the optomotor yaw torque and turning responses. Suppression of the activity of the retinula cells R1–6 either by the induction of a PDA (prolonged depolarizing after-potential) in the mutant *white* (Fig. 28 a) or by the use of the retina mutants *ora* and *rdg B* deficient in this subsystem (Fig. 28 b) invariably suppresses the optomotor response as well. The mutant *sev* on the other hand, in which retinula cells R7 are missing and input through subsystem R8 should be drastically reduced [Willmund (1979) has provided circumstantial evidence for this conjecture], shows normal optomotor responses (Fig. 28 c). The angular sensitivity of the visual elements ($\Delta\varrho$), calculated from recordings of the spatial resolution of the optomotor response, is not significantly different from that of wild type ($\Delta\varrho_{WT} = 5°$, $\Delta\varrho_{sev} = 5.5°$). This is a critical test since, if subsystems with different $\Delta\varrho$ contribute to the input, the method gives a value for the smallest $\Delta\varrho$ involved. Thus, if the subsystems of R7 and R8 participated additively in the wild type, their $\Delta\varrho$ would be measured. Finally, the spectral sensitivity of the optomotor response is similar to that of the ERG and hence of retinula cells R1–6 (Heisenberg and E. Buchner, 1977).

Color Vision

Optomotor responses do not appear to be stimulated by moving color contrast. This was first demonstrated by Kaiser (reviewed 1975) for large flies

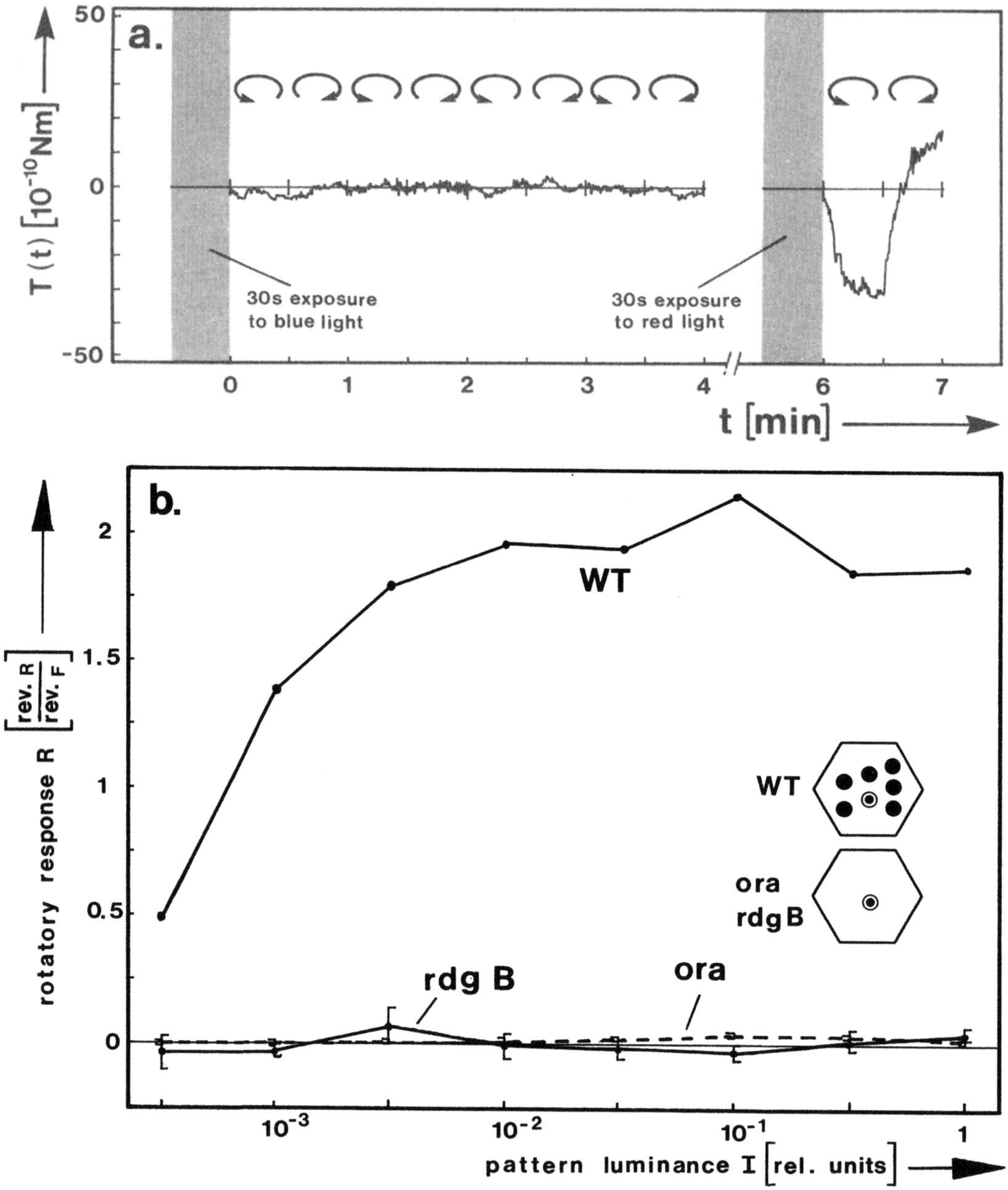

Fig. 28. (a) Inactivation and reactivation of retinula cells R1–6 in the eye color mutant *white* by preadaptation with intense blue (470 nm) and red (600 nm) light (PDA). The optomotor yaw torque response is switched off and on respectively (Heisenberg and E. Buchner, 1977). (b) Optomotor turning responses as functions of light intensity of WT (Berlin) and single mutant flies *ora* and *rdg B* lacking most of rhabdomeres R1–6. Flies were stationarily walking in the center of a striped drum; pattern wavelength $\lambda_{sp} = 45°$; contrast frequency $w/\lambda_{sp} = 1$ Hz (After Heisenberg and E. Buchner, 1977). (c) Optomotor turning responses of walking WT (Kapelle) and *sev* flies as functions of light intensity. Flies were stimulated in a small part of the upper fronto-lateral visual field covering about 50 visual elements; $\lambda_{sp} = 25°$; $w/\lambda_{sp} = 1$ Hz (from Wolf et al., 1980). Close similarity of responses between WT and *sev* is obtained after extensive back-crossing of the mutant to WT

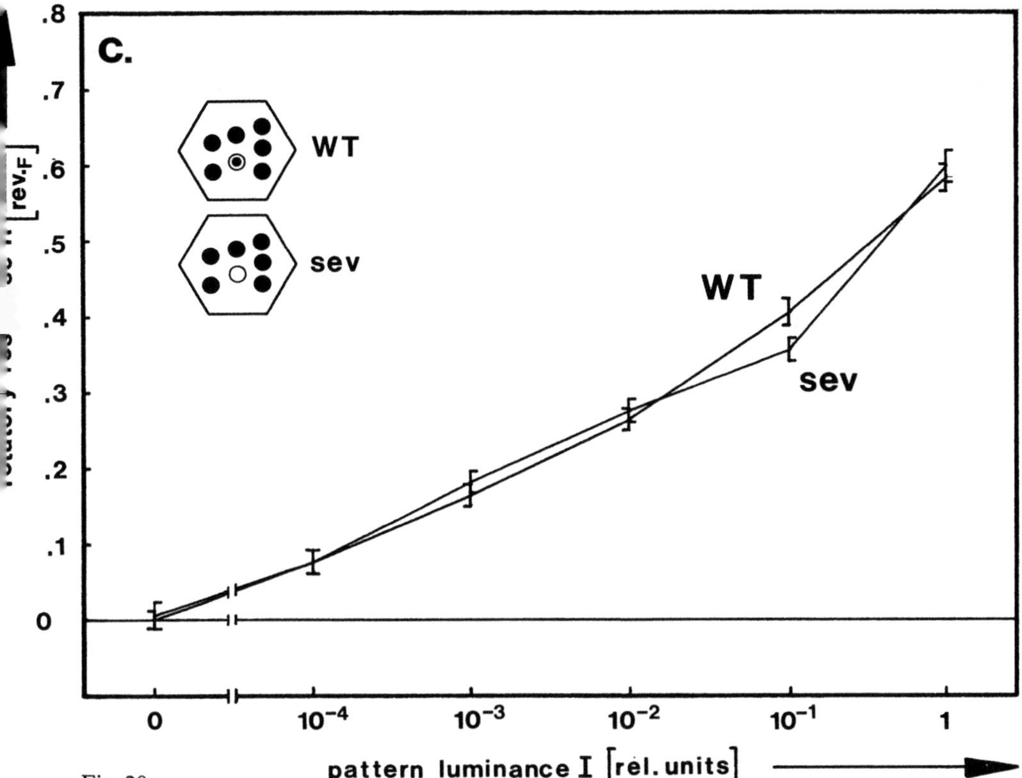

Fig. 28c

as well as bees. A similar conclusion, based on different experiments however, has also been reached for *Drosophila* (Heisenberg and E. Buchner, 1977).

The influence of the input channels R7 and R8 is only seen in orientation tasks, so far mainly in phototaxis of freely walking flies. UV light in the presence or shortly after the presentation of visible light is specially attractive for wild-type *Drosophila* but not for the mutants *sev* or *ora* (Schümperli, 1973; Heisenberg and E. Buchner, 1977; Fischbach, 1979 a). Even the double mutant *sev ora* in which only retinula cells R8 are functionally retained is not totally blind phototactically. The difficulty with the interpretation of the mutant results is that neither in *ora* nor in *rdg B* must suppression of the R1–6 input channel necessarily be complete since remnants of rhabdomeres are found in all flies. Quantitative arguments are required to prove that a certain behavior is independent of this input channel.

In a different phototaxis paradigm ("fast" phototaxis), the response is clearly dominated by the subsystems R7 and R8 at sufficiently high intensity of the stimulus light (Hu and Stark, 1980). In still another paradigm, modification of phototactic behavior by long-lasting exposure to a dim light after induction of a PDA appears to be mediated by subsystem R8 (Willmund and

Fischbach, 1977; Willmund, 1979). In all cases studied so far, the behavioral effects attributed to retinula cells R7 and R8 are observed at intensities about a 1000 times above the intensity threshold of phototactic behavior (Fischbach, 1979a; Bicker and Reichert, 1978; Jacob et al., 1977). This comparatively low sensitivity of subsystems R7 and R8 observed in behavior is in agreement with careful ERG recordings of their sensitivity after induction of a PDA in cells R1–6 (Minke et al., 1975).

Spectral wavelength-specific behavior in wild-type and retinula cell mutants may prove the involvement of certain retinal subsystems and may even reveal neural interactions between different input channels. For instance, an inhibitory effect of the R8 channel on the R1–R6 channel has been postulated by Jacob et al. (1977) on this basis. Color vision defined as hue discrimination independent of intensities and of spectral composition of the over-all illumination is a different matter. For fast phototaxis Menne and Spatz (1977) were able to demonstrate intensity-invariant hue discrimination using conditioning experiments. Whether this color vision is indeed trichromatic as suggested by the recent study of Hernandez de Salomon and Spatz (1983), whether it serves for object discrimination, and which neural interactions are involved remains to be seen.

Polarization Sensitivity in Behavior

In *Drosophila*, polarization sensitivity of the retinula cells has not been directly measured. Drawing the analogy to large flies, one would assume the R7 and R8 input channels to be polarization-sensitive (Kirschfeld et al., 1978; Hardie et al., 1979) whereas the R1–6 channel, according to the measurements of Jaervilehto and Moring (1976) should be polarization-insensitive. Summation at the level of the lamina cartridge of the electrical signals from the six large retinula cells with differently oriented microvilli in their rhabdomeres should give the same signal amplitude for all E-vector orientations.

In *Drosophila* the lamina does mediate polarization sensitivity of the large retinula cells since two distinct optomotor responses both mediated by the R1–6 input channel are sensitive to E-vector orientation (for one of them see Fig. 29; Wolf et al., 1980). The mutant *sev* lacking retinula cells No. 7 and

Fig. 29. Yaw torque responses of WT and mutants to the rotating E-vector plane of linearly polarized light and to a rotating four-quadrant pattern with low pattern contrast. The fly sees the pattern (or the luminous surface emitting polarized light) above its head. p_e degree of polarization; *m* pattern contrast. Impairment of retinula cells R1–6 [mutant *ora*, preadaptation with blue light (PDA) in the mutant *white*] strongly reduces the response; elimination of retinula cells R7 (*sev*) has no effect. Traces in e–h are averages from experiments with 4–7 flies and of 5–10 sweeps. (Wolf et al., 1980 and unpublished). *Inset values* (T̄) represent mean and standard error of the mean of the time-averaged responses

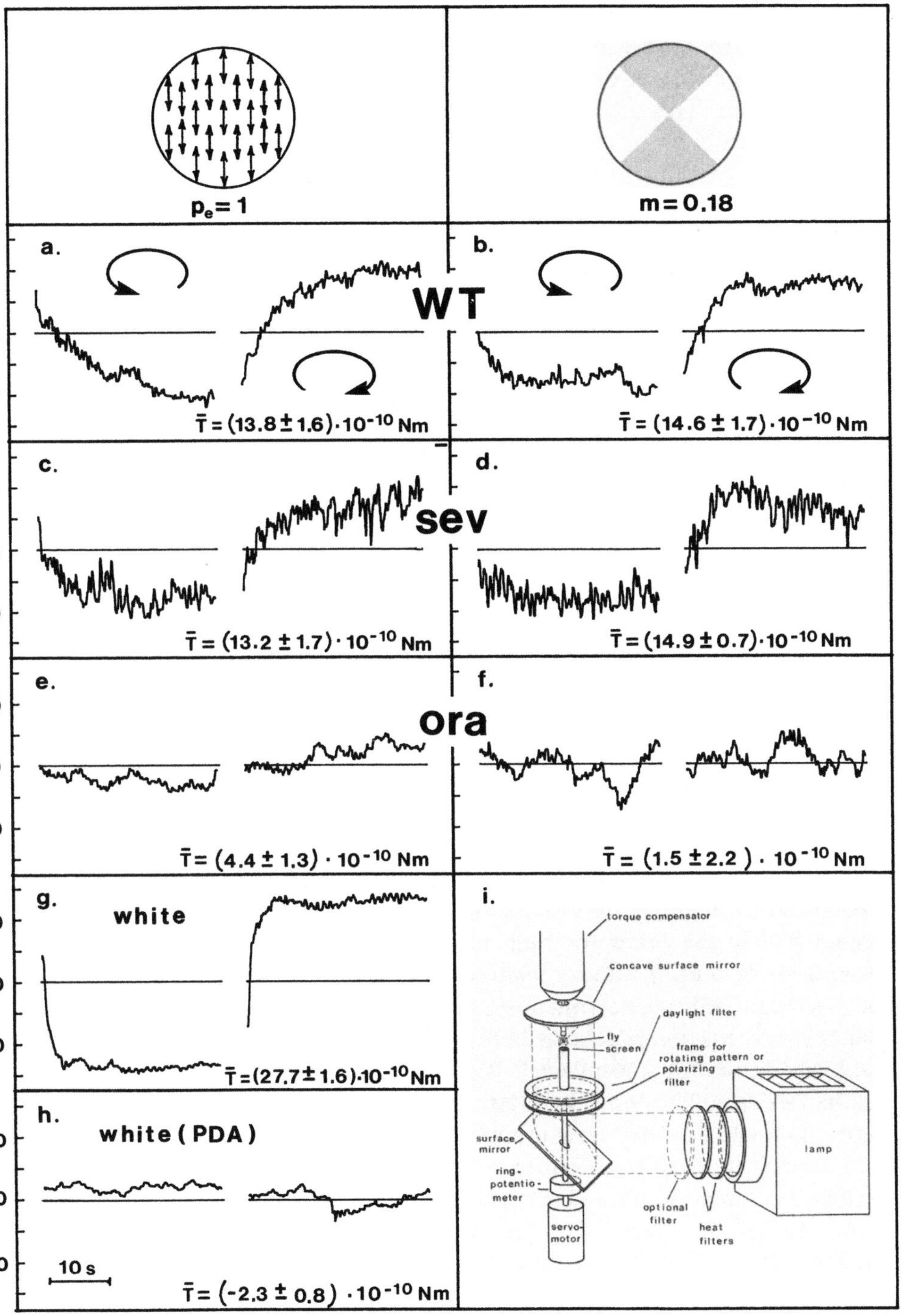

$p_e = 1$
m = 0.18
a.
b.
WT
T̄ = (13.8 ± 1.6) · 10⁻¹⁰ Nm
T̄ = (14.6 ± 1.7) · 10⁻¹⁰ Nm
c.
d.
sev
T̄ = (13.2 ± 1.7) · 10⁻¹⁰ Nm
T̄ = (14.9 ± 0.7) · 10⁻¹⁰ Nm
e.
f.
ora
T̄ = (4.4 ± 1.3) · 10⁻¹⁰ Nm
T̄ = (1.5 ± 2.2) · 10⁻¹⁰ Nm
g.
white
T̄ = (27.7 ± 1.6) · 10⁻¹⁰ Nm
h.
white (PDA)
10 s
T̄ = (-2.3 ± 0.8) · 10⁻¹⁰ Nm
i.
torque compensator
concave surface mirror
daylight filter
fly
screen
frame for rotating pattern or polarizing filter
surface mirror
ring-potentiometer
servo-motor
optional filter
heat filters
lamp
20
0
-20

proper visual input to the rhabdomeres of No. 8 is unimpaired. Thus, either the six large retinula cells of a visual element differ significantly in their polarization sensitivity or interneurons in the lamina are differentially connected to them. This alternative should be a challenge for neuroanatomists since structurally no piece of neuropil is known as well as the lamina of the fly.

Dark Adaptation on Account of Acuity

A further property of the R1–6 input channel revealed by visual mutants presumably has its structural basis in the lamina. Some of the body color mutants of *Drosophila* have neurological and behavioral defects. Many alleles of the mutants *tan* and *ebony* are lacking the lamina component of the electroretinogram and show abnormal phototaxis and optomotor responses. Similar neurological mutants but without body color abnormalities, were isolated in screens for ERG-mutants (Pak et al., 1969), nonphototactic mutants (Hotta and Benzer, 1969) and mutants with abnormal optomotor responses (Heisenberg, 1971 b). Many of the strains still show a substantial retina potential but an abnormal or reduced lamina potential can be attributed to one of two groups: the *ebony* group or the *tan* group. Mutants in these groups are antagonistic in certain aspects of their optomotor responses, visual orientation and spectral wavelength specific phototaxis. In the *ebony* group mutant flies show normal visual acuity in optomotor turning responses at high light intensity but they are optomotor blind at low light levels at which wild-type flies still respond. In the *tan* group mutant flies have the normal low intensity threshold in optomotor responses to coarse patterns but their visual acuity is reduced. On this basis two systems were postulated, one specialized for high resolution at sufficiently high light intensity (high acuity system, HAS) the other for high sensitivity in dim light, operating at the expense of resolution (high sensitivity system, HSS) (Fig. 30 a). Originally, the HAS and HSS were erroneously attributed to the retinal subsystems R7/8 and R1–6 respectively (Heisenberg and Götz, 1975). But since it is firmly established that only retinula cells R1–6 mediate optomotor turning responses the HAS and HSS must be interpreted as adaptational states of the R1–6 input channels. Careful measurements of optomotor turning in *Drosophila* wild type (Fig. 30 b) and *Musca* at high and low light levels indicate the existence of such an adaptation process which at low light levels sacrifices acuity on behalf of sensitivity (Heisenberg and E. Buchner, 1977; Pick and Buchner, 1979). The two groups of mutants appear to be "frozen" states of this mechanism. Since most of these mutants have electrophysiological defects in the lamina it is indicated to look there for the neural basis of this adaptation.

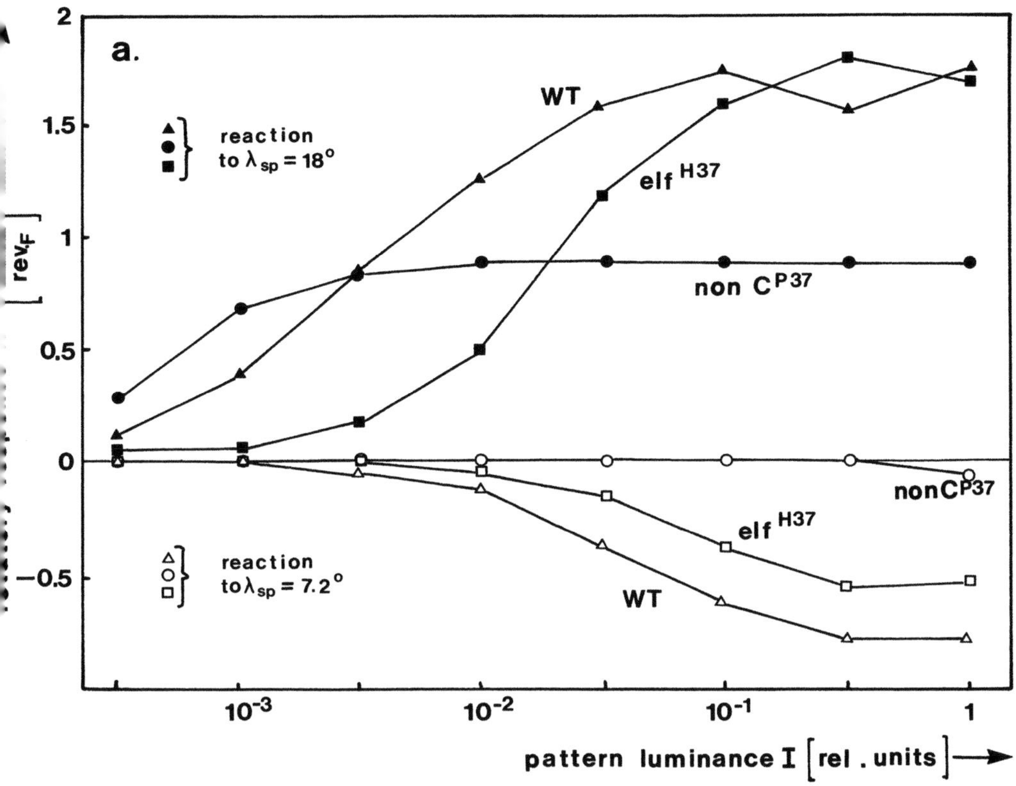

Fig. 30. (a) Optomotor turning responses of WT (Berlin) and dark adaptation mutants *extra lamina fiber*^{H37} (*elf*) (HSS-deficient) and *no on-transient* C^{P37} (*nonC*) (HAS-deficient) as functions of light intensity. Striped drum (w/λ_{sp} = 1 Hz) is used as stimulus. (b) Optomotor turning response of WT (Berlin) as function of spatial frequency $1/\lambda_{sp}$ (cf. Fig. 20). Average pattern luminance $\bar{I} = 3$ cd m^{-2} (*filled circles*) and $\bar{I} = 0{,}01$ cd m^{-2} (*open circles*). *Dotted curve* shows predicted spatial frequency function calculated to fit the response values at $\lambda_{sp} = 18°$ and $\lambda_{sp} = 15°$ (using theoretical description from E. Buchner (1976) and Heisenberg and E. Buchner (1977)). The larger than expected responses at $\lambda_{sp} > 18°$ may be due to special dark adaptation mechanism (HSS). (After Heisenberg and E. Buchner, 1977)

Elementary Movement Detectors

One of the obvious goals in the attempts to correlate structure and function is the search for the elementary movement detectors (EMD's) involved in large field course control. As will be discussed below, the pathway for optomotor turning and torque responses leads to the lobula plate. Thus, somewhere between the lamina and the lobula plate those nearest-neighbor interactions in the columnar lattice of the neuropil have to be identified which correspond to the multiplication-like step in the formalism outlined above (Fig. 17).

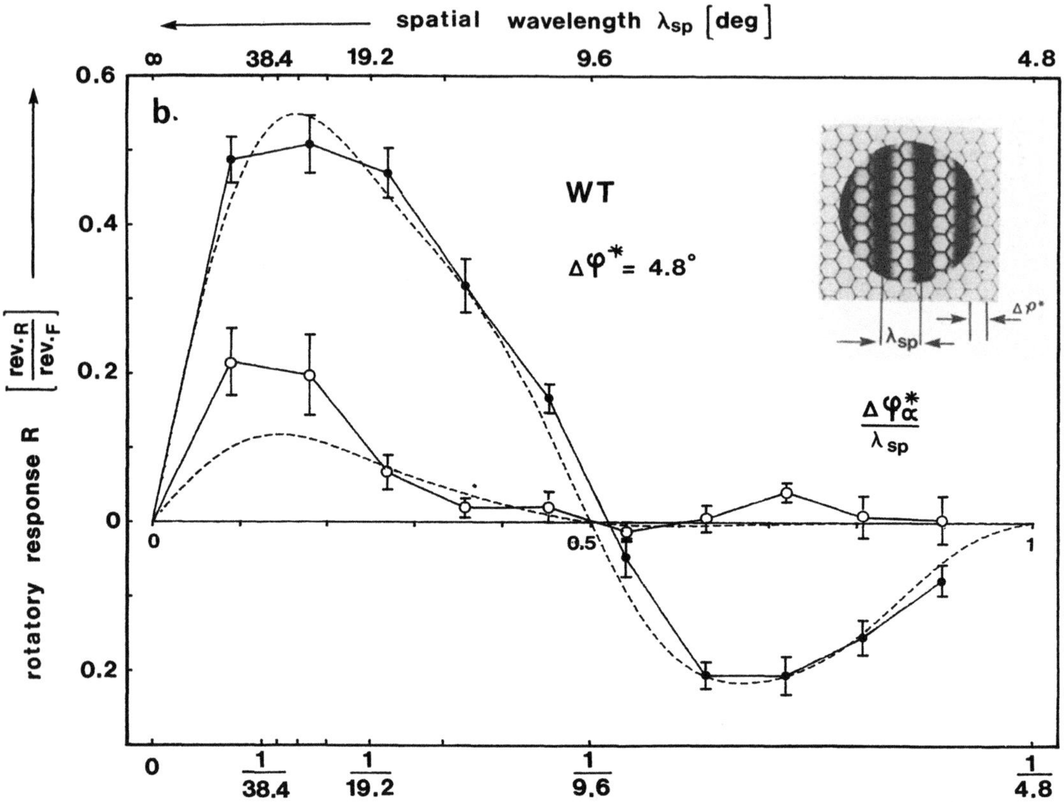

Fig. 30 b

In the lamina of large flies, the monopolar cells L4 send two collaterals backward to adjacent cartridges along the two main axes (x, y) in the lattice of visual elements. These connections would be suited to be part of unidirectional movement detectors for front-to-back motion. Arnett (1972) recorded extracellularly from fibers in the large chiasm between lamina and medulla of *Phaenicia*. Of the two cell types he distinguished, the so-called on-sustained unit may well have been the L4 monopolar neuron since it had a receptive field with an excitatory center and two lateral inhibitory areas adjacent to it. Recording from a giant fiber in the lobula plate, Srinivasan and Dvorak (1980) showed that the input to the movement detector for front-to-back motion undergoes lateral inhibition similar to that found by Arnett (1972). They suggest that in the lamina optomotor responses may be mediated by L4. Whether this lateral inhibition is due to the same mechanism as the dark adaptation of the optomotor turning response mentioned above needs to be investigated.

The regular net of L4 collaterals in the proximal lamina of large flies prompted (Braitenberg and Debbage, 1974) to investigate their synaptic relations. Distally in each cartridge the L4 fiber is postsynaptic to the amacrine cell (α). Proximally it is connected to the two collaterals from the L4 cells of adjacent cartridges by reciprocal synapses. Also, the collaterals have reciprocal synapses between them. This structure suggests a higher degree of symmetry than that observed for the inhibitory "surround" of the on-sustained cells, but since the properties of the synapses are not known the above hypothesis is still compatible with the ultrastructural results. (The only other candidate for the on-sustained cell would be T1, however, its synaptic contacts in the lamina suggest even less that it should be inhibited laterally along the x- and y-axes. Note that neither L4 nor T1 have been shown to be polarization insensitive in large flies). From Golgi studies the shape of the L4-neuron in *Drosophila* appears distinctly different from that in *Calliphora* and *Musca* (Fig. 11; Fischbach, 1983 b). Comparative studies may be of help in this question.

In the medulla of Lepidoptera and large flies, neurons directionally sensitive to moving patterns have been recorded electrophysiologically (Collett, 1970; DeVoe, 1980; Mimura, 1972). This suggests looking in the medulla for the EMD's.[2] The main cells connecting the medulla to the lobula plate are sets of different TmY- and Y-cells which have branches in the medulla, lobula and lobula plate (Strausfeld, 1976; Fischbach, 1983 b) and the T4-neurons (Fig. 12) of which three cells per column have now been identified (Fischbach, 1983 b) all having one branch in the most proximal layer of the medulla and the other terminal in one of the three layers of the lobula plate. As we will see below, these layers are occupied by different sets of large field tangential neurons, at least some of which are mediating optomotor torque and turning responses. Thus, on purely anatomical grounds the layer of the T4-neurons in the medulla is likely to be an important relais station in the optomotor pathway (see Fischbach, 1983 b).

^{3}H-Deoxy-Glucose

A direct approach toward identifying the EMD's has been taken by E. Buchner et al., 1979; S. Buchner and E. Buchner, 1980, using ^{3}H-deoxy-glucose as a histological marker for neuronal activity (Sokoloff, 1978). By adapting the method to the dimension of the insect brain they were able to identify in the most proximal layer of the medulla neural activity

[2] While this book went to press Riehle and Franceschini (1984) demonstrated an important property of elementary movement detectors in *Musca*. They combined recording from the so-called H1-neuron in the lobula plate (see below) with single-photoreceptor stimulation. Illumination of a photoreceptor led to an on-off response in the H1-neuron only if a particular neighboring photoreceptor had been illuminated before

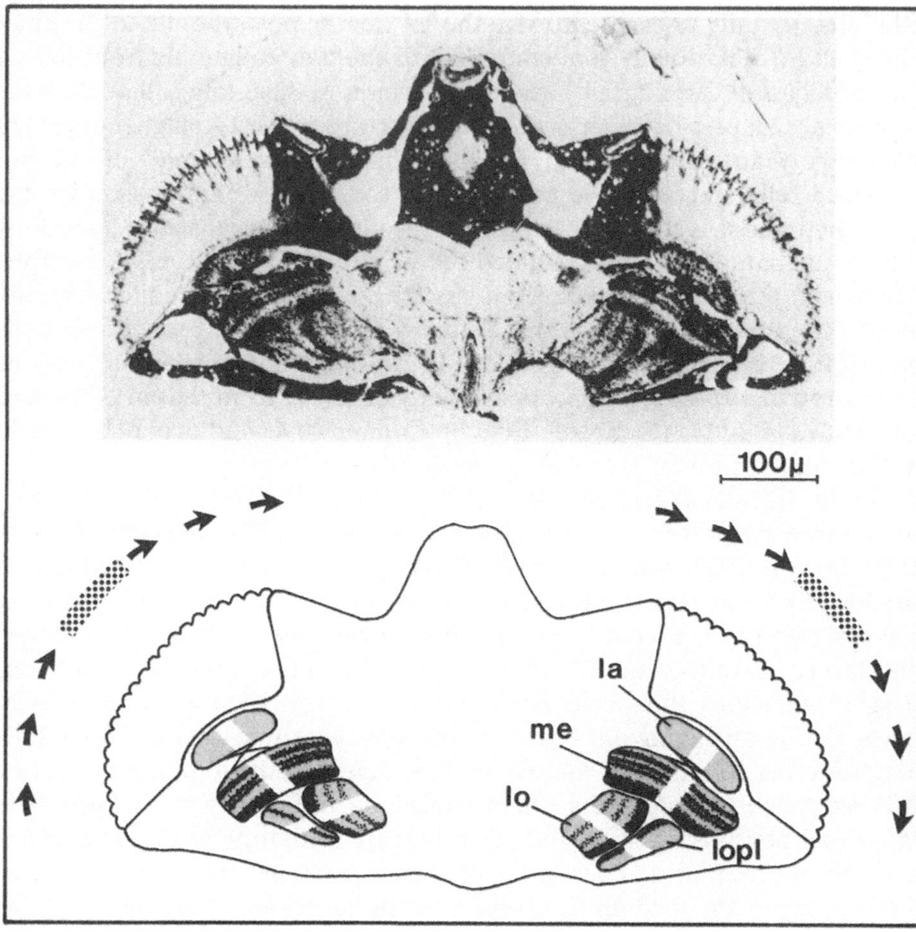

Fig. 31. Autoradiograph of horizontal section (Epon, 3 µ) through *Drosophila* head. Fly had been fed ^{3}H-deoxy-D-glucose just prior to an experiment in which it was stimulated by a grating of vertical stripes moving horizontally in clockwise direction around the fly. Label specific for neural activity has accumulated in certain tangential layers of the medulla, lobula and lobula plate. Note that anterior-most layer of lobula plate is labeled if ipsilateral eye is stimulated by front-to-back motion (*right side*). With back-to-front motion a layer slightly further back in the lobula plate is marked (*left side*). During stimulation two stationary screens had been inserted into the drum (*stippled areas*). Corresponding neuropil regions in the optic lobes are less heavily labeled. *la* lamina; *me* medulla; *lo* lobula; *lopl* lobula plate. (Courtesy E. Buchner)

specific for motion in the corresponding part of the visual field (Fig. 31). This is the input layer to the T4-neurons. The method is approaching electron microscopical resolution (E. Buchner and S. Buchner, 1984) and may thus eventually reveal the cell types carrying the activity label. It is not yet excluded that the multiplication-like interaction for movement detection takes place at the dendritic synapses of the lobula plate giant fibers. For the remaining discussion, however, we will assume that the EMD's are realized in the medulla and send their information about local movement to the lobula plate.

Lobula Plate Giant Neurons

In addition to the label in the medulla, the ^{3}H-deoxy-glucose autoradiographs show a conspicuous layer of motion specific radioactivity in the lobula at the level of the T5-terminals and in anterior and posterior layers of the lobula plate where two of the three T4-neurons arborize (cf. Fig. 12). E. Buchner and S. Buchner (1984) even succeeded in differentiating these layers further. Close to the anterior surface of the plate two layers specific for front-to-back and back-to-front motion respectively can be distinguished and close to its caudal margin one specific for upward and one for downward motion. These layers are the site of large field tangential neurons which are being intensively studied in big flies. By recording from them extra- and intracellularly with microelectrodes and by injecting dyes for their subsequent anatomical characterization, 21 classes have been identified. All of the cells possess directionally selective motion sensitivity and several cell classes are functionally interrelated (for review see Hausen 1981). In *Drosophila* none of the cells has yet been identified electrophysiologically. However, neurons obviously homologous to the two main classes, the HS- and VS-cells (Braitenberg, 1970; Pierantoni, 1973), are present in *Drosophila* as well (Heisenberg et al., 1978). We will briefly summarize some of the properties of these cells in large flies before we come to discuss their functional role in behavior.

The VS- and HS-cells are the main output pathways from the lobula plate to the so-called posterior slope of the central brain. They have large dendritic arborizations in certain layers of the lobula plate where they receive input from small field columnar neurons (possibly T4, TmY, Y, T5) and they in turn form the input to descending neurons leading from the posterior slope through the cervical connective to the ventral ganglion (e.g., Strausfeld, 1976). The two sets of cells were originally named horizontal system (HS) and vertical system (VS) according to the shape of their dendrites (Braitenberg, 1970; Pierantoni, 1973) and it was proposed that HS-neurons were mediating optomotor responses to horizontal motion, VS-neurons to vertical motion. The horizontal system in all Diptera studied so far consists of three cells [north (HSN), equatorial (HSE), south (HSS), see Fig. 13] with their main dendrites oriented roughly horizontally, together covering the entire anterior surface of the ganglion (e.g., Dvorak et al., 1975; Eckert, 1981; Hausen, 1982). The vertical system in large flies (*Calliphora*, *Phaenicia*) consists of 11 cells (i.e., Eckert and Bishop, 1978; Hengstenberg et al., 1982), whereas in *Drosophila* so far only six cells have been identified. Their main dendrites are oriented vertically and are staggered along the posterior surface of the lobula plate, overlapping considerably with their smaller side branches.

The HS- and VS-cells respond to movement in the visual field of the ipsilateral eye and some of them are also sensitive to motion in the other visual half-field. In general, HS-cells generate a graded depolarization consisting of summed EPSP's [3] as well as action potentials in the response to front-to-back motion in the visual field of the ipsilateral eye and a graded hyperpolarization (IPSP's [3]) to back-to-front motion. The HSE-neuron in addition responds (via a further large field tangential neuron, H2) to movement on the contralateral side in the opposite manner, i.e., with depolarization and action potentials to back-to-front motion, with hyperpolarization to front-to-back motion.

The preferred direction for VS-cells is downward motion in the visual field of the ipsilateral eye. To this direction the cells respond with depolarization and action potentials, whereas movement in the opposite direction again causes hyperpolarization. Recently, Hengstenberg (1981) has shown that some VS-cells also show weak responses to motion in the visual field of the contralateral eye, upward motion causing depolarization, downward motion the opposite.

The fact that the HS- and VS-cells cover large areas of the retinotopic projection of the visual field favors the idea that they mediate optomotor torque and turning responses. Only self-rotation of the fly invariably causes coherent motion of the same angular velocity in large parts of the visual field. Thus, the local signals of the appropriately oriented EMD's can be summed over large areas without much loss of information for rotatory course control.

Meanwhile a fair amount of evidence has accumulated supporting this hypothesis. A strong argument is provided by two neurological mutants of *Drosophila*. The first one, *optomotor-blind*^{H31} (*omb*) is lacking the HS- and VS-cells (as far as can be judged from histological preparations stained with reduced silver or from 1 μ plastic sections; Heisenberg et al., 1978 Fig. 34b). Blondeau (unpublished) has found no cell degeneration in the area of the HS- and VS-cell bodies during development. These cell bodies, which are quite conspicuous in the wild type appear to be missing right from the beginning in *omb*. The columnar organization of the visual ganglia including the lobula plate seems to be unimpaired and, while in some flies of the stock the architecture of the inner chiasm is distorted, in many flies the loss of lobula plate giant fibers is the only structural defect observed.

On the behavioral side the mutation primarily affects the optomotor responses to rotatory stimuli (Heisenberg et al., 1978; Blondeau and Heisenberg, 1982).

---

[3] EPSP: excitatory postsynaptic potential; IPSP: inhibitory postsynaptic potential

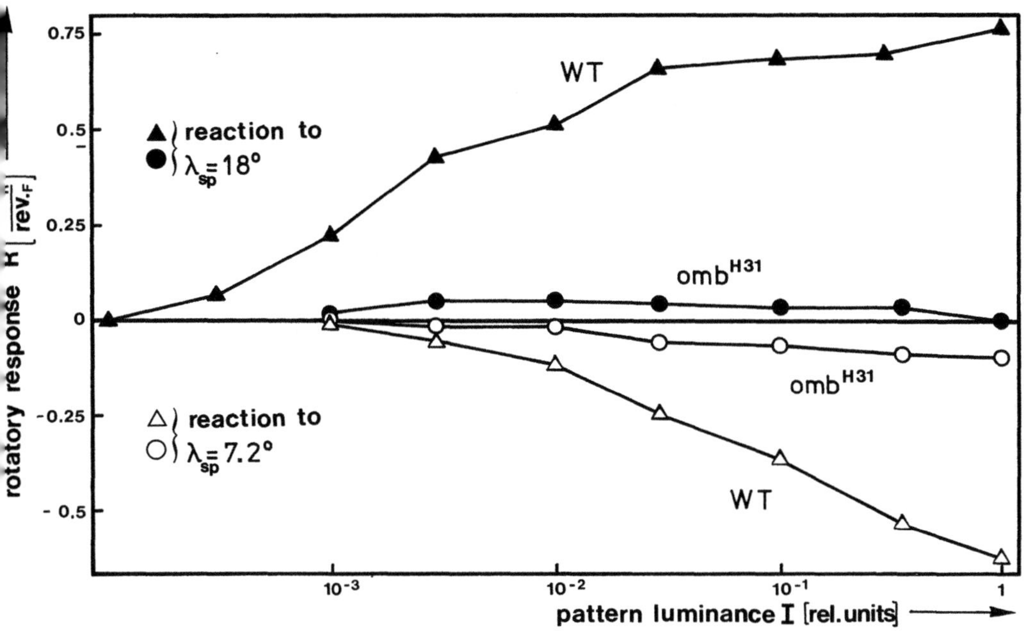

Fig. 32. Optomotor turning responses as functions of pattern luminance of WT (Berlin) and *optomotor-blind*^{H31} (*omb*) flies walking inside a striped drum ($w/\lambda_{sp} = 1$ Hz); WT measurements from 1 female fly at each pattern wavelength, *omb* curves are averages from 4 ($\lambda_{sp} = 18°$) and 3 ($\lambda_{sp} = 7.2°$) flies. (Heisenberg et al., 1978)

Visual control of yaw, roll and pitch in flight are all reduced to 30% of the normal response amplitude or less (Fig. 33). Most of the remaining yaw torque response has recently been shown to have different properties from the dominating response in wild type. The "large field course control response" to horizontal movement in flight (see p. 125 ff) seems to be suppressed by more than 90% in agreement with the optomotor turning response in walking *omb* flies, of which also less than 10% seems to be left (Fig. 32 and Chap. 7).

The coincidence between structural and behavioral deficits in a single mutant is not a strong argument that in wild type the structure in question mediates the behavior impaired in the mutant. Other structural and behavioral defects in the mutant might have been overlooked. It is a much stronger argument that a certain structure missing in a mutant is not crucial for all the behavior for which the mutant proves to be normal. Along this line it can be stated from experiments with *omb* that the HS- and VS-cells are neither crucial for the lift/thrust response to vertical motion in the lateral part of the visual field nor for the landing response induced by horizontal front-to-back motion of vertical stripes, nor for the object response which will be discussed in Chapter 7.

The second mutant of relevance in this context is *lobulaplateless*^{N684}(*lop*). In this strain the columnar organization of the lobula plate is missing and its volume is reduced to about 20% (Fig. 34c). The remaining neuropil is tightly apposed to the lobula and of the inner chiasm only the fibers linking the medulla to the lobula can be seen. The cellular cortex normally coating the lobula plate neuropil is largely missing, due to massive cell degeneration during the early pupal period (Fischbach, personal communication). Recently in the cellular cortex of the medulla of *lop* slightly larger than normal cell bodies have been observed which might constitute ganglion mother cells being left over from the developmental period.

A striking structural abnormality of this mutant are the lobula plate giant fibers. While the HS-cells still arborize in a slightly disordered manner in the rudiment of the lobula plate, a bundle of about 15 fibers, some of which presumably are homologous to VS-cells in the wild type grow past the lobula plate into the dorso-frontal medulla where they grow downward in an orderly spaced sequence as VS-cells would in the lobula plate (unpublished; Paschma, 1982; Fischbach, personal communication).

Optomotor yaw torque and turning responses in this mutant are not, or are only slightly reduced, more strongly in the frontal than in the lateral part of the visual field. The remaining response still has many properties of the normal optomotor response of the wild type. Optomotor roll and pitch responses, however, are much more severely impaired (Fig. 33). One would like to conjecture that the VS-cells in *lop* do not carry the relevant movement information for compensatory roll and pitch maneuvres and that functional VS-cells are not crucial for the optomotor yaw torque and turning responses.

	yaw	pitch	roll
	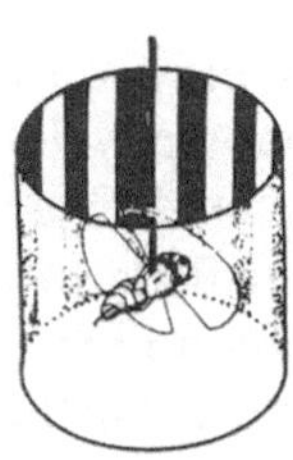	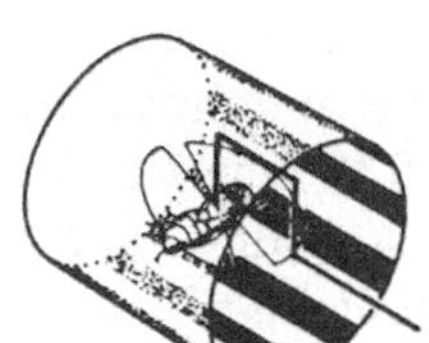	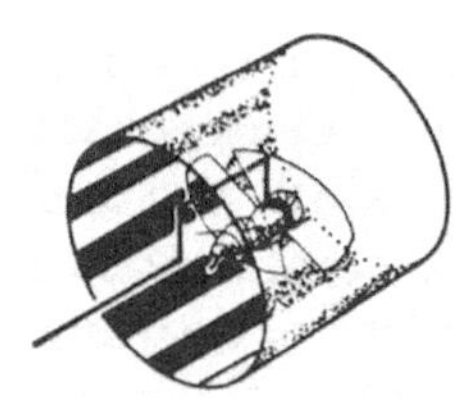
	$\bar{T}$ [% of $\bar{T}$ (WT)]	$\bar{T}$ [% of $\bar{T}$ (WT)]	$\bar{T}$ [% of $\bar{T}$ (WT)]
omb^{H31}	32.1 ± 4.2	31.0 ± 6.3	14.9 ± 16.3
lop^{N684}	100 ± 13.5	18.8 ± 12.5	0 ± 23.8

Fig. 33. Average optomotor yaw, pitch and roll responses of the mutants *optomotor-blind*^{H31} (*omb*) and *lobulaplate-less*^{N684} (*lop*) as % of wild-type responses. Experimental conditions as indicated in the legend to Fig. 24. (Data from Blondeau and Heisenberg 1982 and Paschma, 1982)

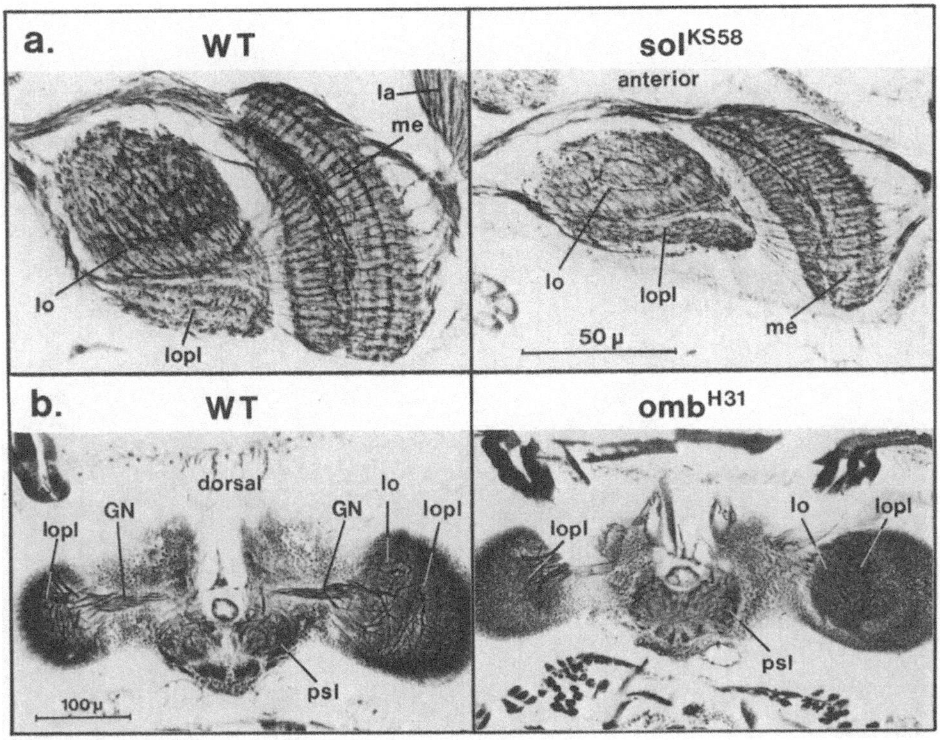

Fig. 34. (a) Horizontal section of optic lobes of WT and *small optic lobes*^{KS58} (*sol*) mutant fly. (Fischbach, 1983 b). (b) Frontal view of posterior slope (*psl*) and lobula plates (*lopl*) of WT and *optomotor-blind*^{H31} (*omb*) (WT preparation from Heisenberg et al., 1978). (c) Horizontal sections of optic lobes of WT and *lobula plate-less*^{N684} (*lop*) mutant fly at four roughly comparable levels. (Fischbach, 1983 b). *la* lamina; *me* medulla; *lo* lobula; *lopl-rud* rudiment of lobula plate; *GN* giant neurons of lobula plate. 7 µ paraffin sections, Carnoy fixation, reduced silver staining. (Blest, 1961)

One of several other mutants of interest is *small optic lobes* (*sol*) which is lacking about 50% of the neurons in each column of its medulla, lobula, and lobula plate (Fig. 34 a). The HS- and VS-cells, as well as the rotatory optomotor responses, nevertheless seem to be intact (Fischbach and Heisenberg, 1981; Fischbach and Technau, 1984). Of course, this mutant alone is only a weak argument for the involvement of these cells in these behavioral responses; but it may eventually be most valuable for identifying the neural pathway for "large field course control" in the optic lobes. In different mutants with reduced optic lobes different sets of columnar neurons are missing. By the combination of the appropriate genes, strains with only a small number of neurons per column can be obtained.

In large flies independent evidence has accumulated showing that the lobula plate giant fibers are involved in rotatory optomotor responses. Blondeau (1977, 1981) was able to elicit yaw, roll and pitch in *Calliphora* by applying weak electrical current pulses to one lobula plate. Geiger and Nässel (1981,

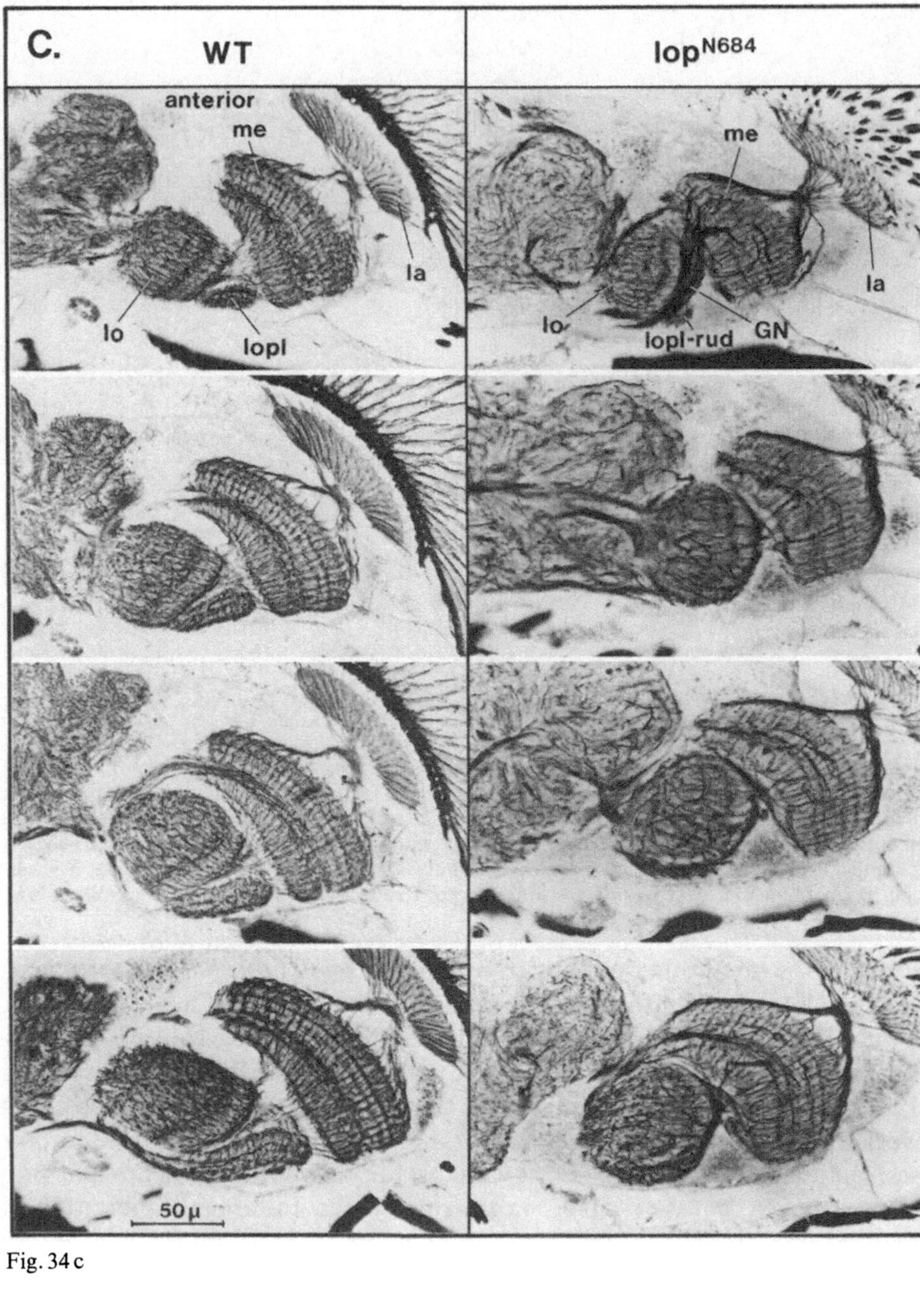

Fig. 34 c

1982) using laser-microbeam surgery in the late 3rd instar larva of *Calliphora* ablated a certain neuroblast of the inner optic anlage. This led to a "phenocopy" of the *omb* defect. Flies were lacking the lobula plate giant fibers. The optomotor yaw torque response (to periodic gratings) of such flies was reduced to about 50%, the remaining torque being due to the so called position function (see Chap. 7), which appeared to be still normal. Thus, the similarity between *Calliphora* and *Drosophila* in the behavioral defects correlated with the absence of VS- and HS-cells is striking. Surgical elimination of these cells in adult *Calliphora* is now in progress (Hausen and Wehrhahn, 1983).

The role of the VS-cells in behavior is less well documented in large flies. As mentioned above for some of the VS-cells, the preferred direction of movement sensitivity is downward in the ipsilateral and upward in the contralateral visual half-field. Hence, they respond maximally to rotations of the scenery around the fly's longitudinal axis, as during roll. Furthermore, some of these cells (VS1, VS7–9) have different preferred directions of movement sensitivity in different parts of their receptive fields. These properties make them well suited to mediate visual pitch control (Hausen, 1981; Hengstenberg, 1984).

In order to present this argument in detail we have to come back to the notion that the different tangential layers in the visual neuropil represent certain properties of the visual surround. The electrophysiological recordings from the lobula plate of large flies as well as the ^{3}H-deoxy-glucose experiments of E. Buchner and S. Buchner (e.g., 1984), show that the most anterior layer of the lobula plate represents a retinotopic map of front-to-back movement immediately followed by a layer representing back-to-front movement. The most posterior layer of the lobula plate represents downward motion and a layer just anterior to it upward motion. This, however, holds only for the lateral part of the visual field in which the vertical and horizontal axes (v, h) of the lattice of visual elements coincide with the vertical and longitudinal axes of the fly.

It is worth noticing at this point that the organization of these layers poses a developmental problem. Neurocrystallinity of the visual neuropil and with it of its functional properties is brought about by a set of repeating developmental processes. Thus, the movement sensitivities in the four layers of the lobula plate are generated developmentally by the establishment of neural interactions between certain neighboring positions in the lattice of the visual elements. Thus, the preferred directions of movement sensitivity being defined by neighborhood relations follow the tilt of the axes of the lattice of visual elements in the different parts of the visual field (see Fig. 2). The two anterior layers of the lobula plate receive direct or indirect peripheral input from the EMD's oriented along the x- and y-axes of the lattice of visual elements and, hence, represent motion in the direction of the h-axis of the lattice, but this is not always oriented in the direction of the horizontal of the fly. For instance, in the most frontal part of the visual field the h-axis may be inclined by up to 60° (in respect to the horizontal of the fly). This systematic variation in the directionality of the movement detectors is transferred directly to the four layers of the lobula plate, i.e., the anterior layers reflect movement sensitivity along the h-axis of the lattice and the posterior layers movement sensitivity along the v-axis.

It is in these layers that the HS- and VS-cells pick up the relevant information about local motion which enables them to contribute to a certain course control maneuvre. As mentioned before, the HS-neurons are confined entirely to the anterior layer and Hausen (1981) has shown recently that when stimulated in small areas of the visual field, their preferred direction of motion sensitivity coincides closely with the h-axis of the lattice of visual elements.

Let us for the sake of simplicity assume that each of the HS- and VS-cells is part of only one of the three rotatory course control responses and let us now, as an example, consider the pitch cells. Ideally, these cells should, in the area of the retinotopic projection they are covering, make synaptic contacts with those EMD's which are as closely as possible oriented perpendicularly to the axis of rotation for pitch. For those parts of the lobula plate corresponding to the frontal and rear part of the visual field at the height of the equator, the neurons should grow into the posterior layers of the ganglion in order to collect movement information from EMD's oriented along the v-axis of the lattice, whereas in those parts corresponding to the dorso-lateral visual field they should grow into the anterior layers and pick up information about movement along the h-axis of the lattice.

Some of the VS-cells in *Drosophila* (Heisenberg et al., 1978) as well as in large flies (Hengstenberg and B. Hengstenberg, 1980), do in fact have side branches in the anterior layers of the neuropil and it has recently been shown electrophysiologically in *Calliphora* that the neurons VS1 and VS7–9 according to the criteria discussed above do, in fact, qualify as pitch cells. For instance the cell VS1, in agreement with its dendritic arborizations, is depolarized by downward motion in the medio-frontal and by back-to-front motion in the dorso-lateral part of the visual field, a situation arising during upward pitch as outlined above (Hengstenberg, 1981).

Generalizing these results, we may interpret the giant fiber system of the lobula plate as the visuo-motor link for rotatory course control. The interface between the homogeneous array of EMD's and a specific motor command compensating for rotation around one of the fly's three body axes seems to be represented by the dendritic synapses of a set of HS- or VS-cells.

If this interpretation is correct we may, for instance, ask how well the interface for yaw control which, as we conjectured, is represented by the dendritic arborization of the three HS-neurons and thus collects the movement signals along the x- and y-axes of the lattice, fits the directions of motion experienced by the fly during yaw. A comparison of the left and the right map in the upper row of Fig. 35 shows that only at the rim of the (unilateral) visual field EMD's along the vertical axis in the lattice would be suited to support yaw control effectively. An improvement of the real interface (by the growth of dendrites of the HS-cells into the appropriate areas of the posterior layers of the neuropil) apparently has not been realized. In the case of roll (Fig. 35, middle row) an interface summating the EMD's oriented only along the v-axis would be close to optimal except for the region of binocular overlap. Here the EMD's along the x- and y-axes would contribute significantly. Thus only the most frontal VS-neurons for roll should be expected to have side branches in the anterior layers of the lobula plate. Figure 35, bottom row, suggests that this is different for the pitch interface. For pitch the visual half-sphere can be roughly divided into four sectors: frontal, dorsal, caudal, ventral. In the frontal and caudal sectors an efficient pitch interface should use the EMD's oriented along the v-axis. In the other two sectors the EMD's oriented along the x- and y-axes should be used, always with opposite polarity in opposing sectors. Unfortunately the shape of the VS-neurons in *Drosophila* is not known well enough to attribute them to either the roll or pitch interface.

Calculations like the ones leading to Fig. 35 also illustrate that most EMD's oriented along the y-axis are of little use in mediating yaw control for hovering flight (see p. 95), where the fly's longitudinal axis is turned by 60° rear down. It would be interesting to know whether the yaw interface can be functionally modified according to body posture as seems to be the case for lift/thrust control (Götz, 1983; see p. 46). It would be most interesting to compare the interfaces in different species of flies. One would like to predict that under large pressure for optimal visual course control the shapes of the dendritic trees of the HS- and VS-cells should be more elaborate regarding the different layers of the lobula plate, and the interfaces should match the optimal ones more closely.

At this point we would like to emphasize, however, that the discussion of these interfaces has ignored several constraints which in reality may modify their properties. The fly has to distinguish rotatory from translatory movement, it can acquire different body postures during flight and walking and its forward motion causes a "translatory bias" for the rotatory control system (see Chap. 3). Thus, the real interfaces may be closer to the optimal ones than it seems at first sight.

In *Drosophila* the shape of the dendritic trees of the HS- and VS-cells is the result of a "predetermined" developmental process. It does not rely on the experience of the individual fly. We have compared the optomotor torque responses of visually naive flies with those of normal ones and have found no difference. But the establishment and refinement of this developmental process in the natural history of our flies did rely on the successful orientation

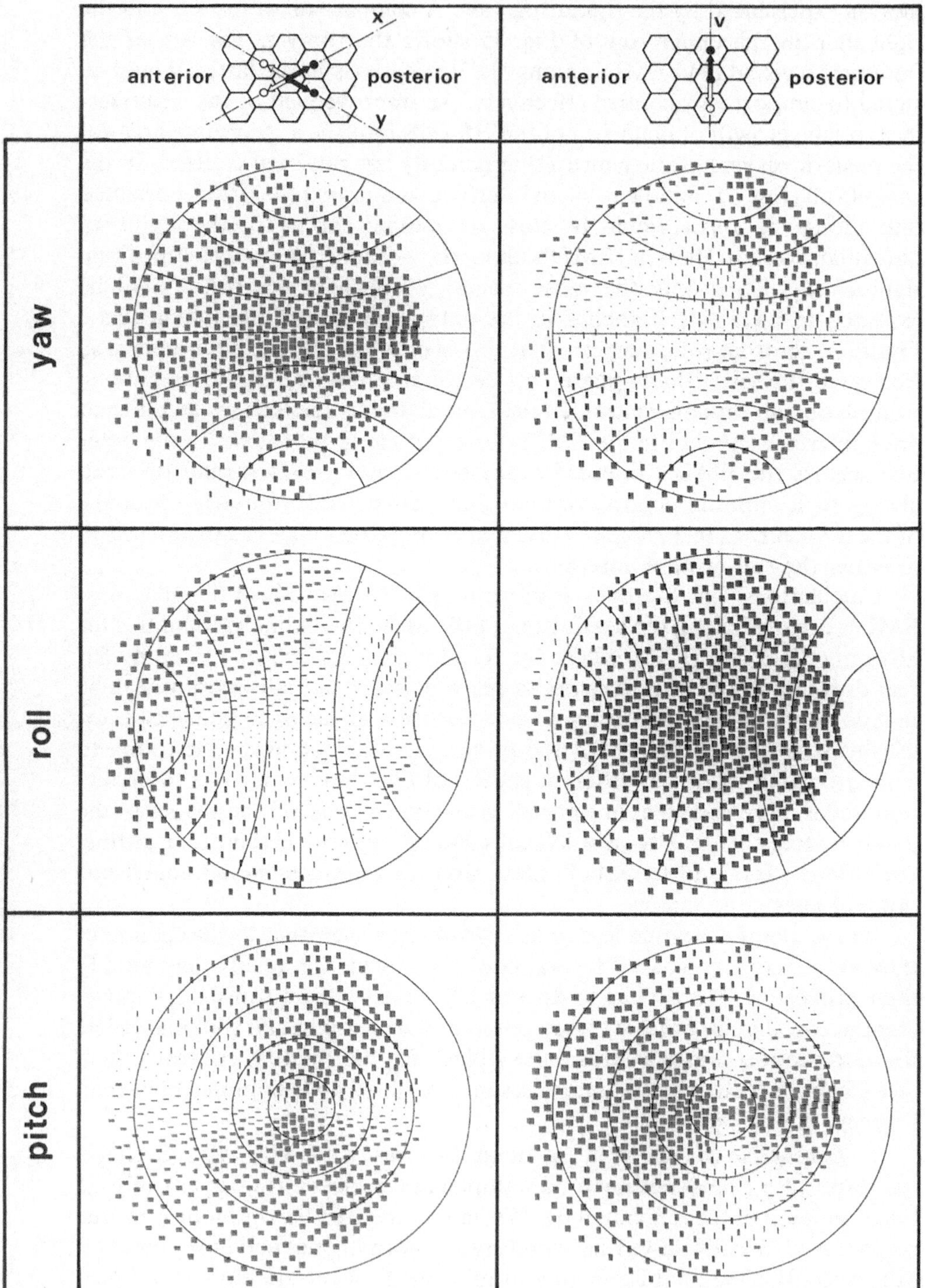
x
y
anterior
posterior
v
anterior
posterior
yaw
roll
pitch

of their ancestors during yaw, pitch, and roll. It is in this sense that the shapes of these cells reflect "phylogenetic information" as proposed in the introduction.

It can easily be seen that it may become very costly to build a perfect visuo-motor interface using the principle discussed above, i.e., relying entirely on phylogenetic information for a predetermined developmental process. In this case it may be simpler to connect the large field tangential neurons designated to serve, for example, pitch control to all EMD's, but with labile synapses which are only maintained permanently if they are appropriately activated while the motor program for pitch is executed. This scheme of development seems not to be used in the lobula plate. (Actual sensory information is required for the proper development of the mammalian visual system, e.g. for the establishment of binocular depth perception.)

We have dwelt on rotatory course control as the function of the lobula plate of which we know the most; but lift/thrust responses and landing maneuvres can be elicited by electrical stimulation in the lobula plate as well (Blondeau, 1977, 1981). In addition to the large field tangential neurons, columnar output elements terminating in the posterior optic foci (Strausfeld, 1976) are described anatomically. This should remind us how much of visual course control still remains to be understood.

If we follow the optomotor pathway from the lobula plate "downstream" we get to the telodendritic arborizations of the HS- and VS-cells which contact intrinsic neurons of the central brain and different sets of descending neurons which relay the information about rotatory course control to the ventral ganglion. The cervical connective as the main communication line between brain and ventral ganglion can be tapped using a hook electrode (Hengstenberg, 1973). Action potentials from many neurons are thus recorded simultaneously. This mass activity is greatly increased in response to front-to-back motion and much less by motion in the opposite direction and by upward motion. Interestingly, the dynamics of the response to front-to-back motion is different from that to back-to-front motion. Whereas at high

◄

Fig. 35. Local efficiency of EMD's in the control of yaw, roll and pitch. Due to the curvature of the eye the axes of the lattice of visual elements (see Fig. 2) change orientation with respect to the coordinates of the fly and – even more important – with respect to the flow field of visual motion the fly experiences during self-motion. The six representations of the lattice roughly indicate for each point in the lattice the contributions of the EMD's shown above, to the control of each of the three kinds of self-rotation. In the left column the contribution of the EMD's oriented along the x- and y-axes are added with equal weights. *Large circles* indicate visual half-spheres. *Drawn lines* (yaw and roll) and *smaller circles* (pitch) give the directions of motion in the flow fields. *Small vertical bars* represent positive contributions, *horizontal bars* negative ones, *dots* indicate zero contributions. *No. of bars* represent strength of contributions covering the whole range from −100% to +100%. Contributions are estimated as the cosine of the angle between the local flow field vector and the orientation of the EMD. (Data for the calculations from E. Buchner (1971) (see Fig. 2)

pattern luminance the latter is largest at a contrast frequency of about $w/\lambda_{sp} = 1$ Hz, front-to-back motion is most effective at contrast frequencies above $w/\lambda_{sp} = 9$ Hz. We have discussed already on p. 50f that the optimal contrast frequency of front-to-back motion eliciting the landing response is higher than that stimulating the optomotor yaw torque response. In Chapter 7 we will discuss a further behavior relying upon front-to-back motion. The cervical connective response indicates the special significance of this type of stimulus.

At least at the level of descending neurons, the data on visual motion are combined with data from the antennae dealing with wind and gravity. The HS-neurons and neurons from the mechanosensory centers of the antennae converge onto a group of descending neurons ventrally in the posterior slope. More dorsally, VS-neurons and ocellar neurons converge. A separation of pathways for roll and pitch at the level of the descending neurons is not apparent (Strausfeld and Bacon, 1983).

Recently, Heide (1983) and Götz (1983) have started to unravel the *Drosophila* optomotor system from the rear end. On the basis of earlier studies with large flies (Heide, 1971a, b; 1978; Spüler and Heide, 1978), they recorded action potentials from indirect flight muscles during flight, while stimulating visually with moving stripes. Of the muscles recorded so far, three seem to be involved in yaw and lift/thrust control, a fourth one is unresponsive. The few data available at present clearly demonstrate that the approach from the motor side will provide most important insight into the organization of the visual system.

Other Visual Pathways

It has been mentioned above that under certain conditions a jump or start reflex can be elicited visually. In white-eyed flies (*brown*; *scarlet*) a sudden

Fig. 36. (a) The left giant descending neuron (*GDN*) of *Drosophila melanogaster* (shown *black*) cobalt-filled from a glass microelectrode implanted in the neck connective. It is shown coupled to the contralateral GDN via heterolateral grand commissure intrinsic neurons (*GCI*) whose cell bodies are shown toward the top of each protocerebral lobe. (They appear disconnected because the neurites leading from the cell bodies to the main cell components are very thin and contain too little cobalt to be resolved clearly. In addition, the cell bodies of two unidentified neurons are stained). The dendritic and axonal components of Col A cells in the left lobula can be made out, although too little cobalt has reached their cell bodies for these to be also visible (see Strausfeld and Singh, 1980 for antennal connections). (b) In this preparation the right GDN has been filled transsynaptically from the right antenna. The illustration shows the axon entering the thoracic-abdominal ganglion (*th1*, *th2* 1st and 2nd neuromer) from the neck connectivce (anterior). The primary filled GDN is coupled to the contralateral GDN and to the ipsilateral tergotrochanteral muscle motorneuron (*TTMm*). It is also coupled to the ipsi- and contralateral peripherally synapsing interneurons (*PSI*). These are shown *black* for clarity, but in fact contain less cobalt than the GDN. Interestingly the PSI's have extensive dendrites in the thoracic neuromers which indicates that they receive many other inputs as well as that from the GDN's. The same is true for the TTMm's. (Courtesy N. Strausfeld)

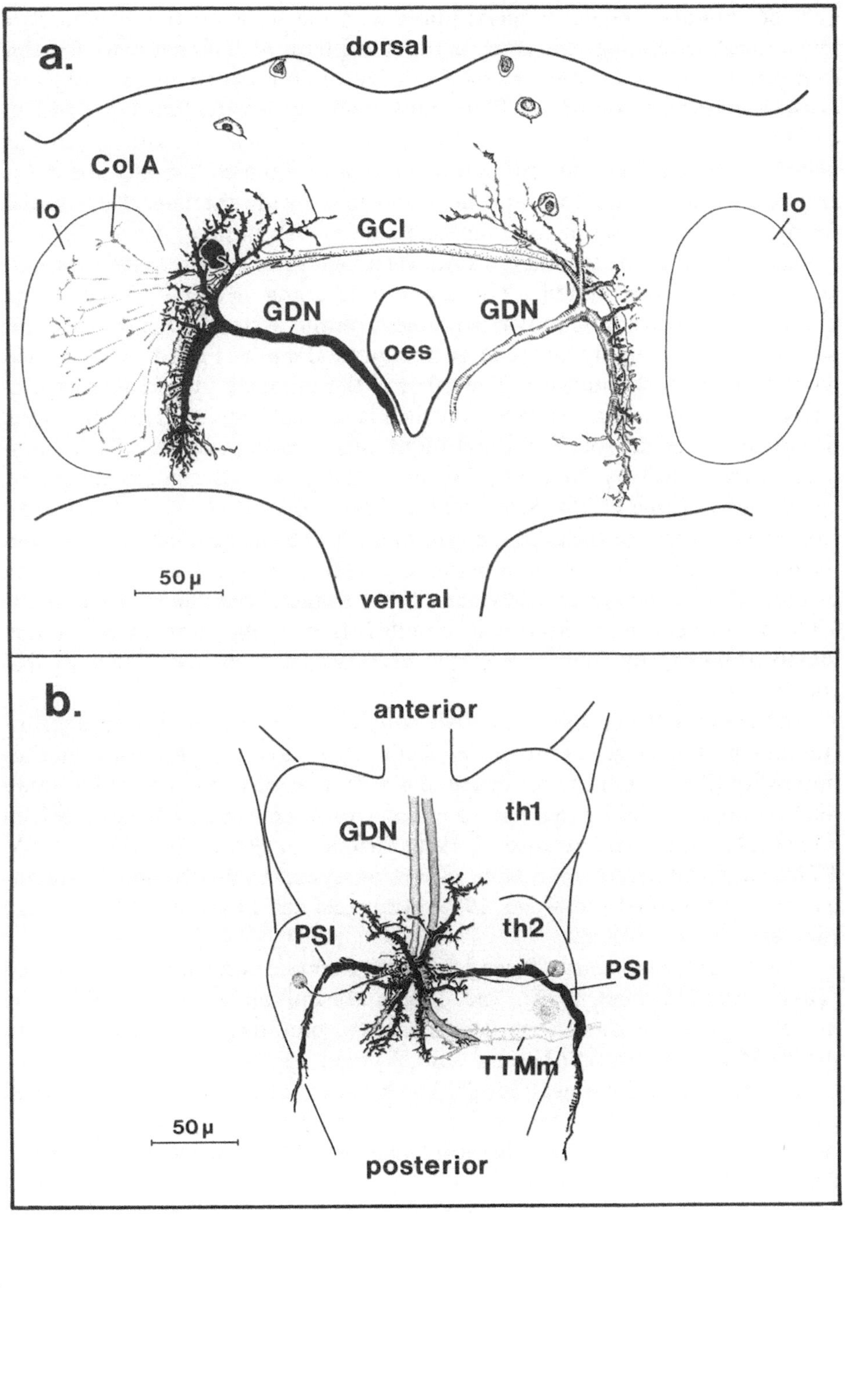
a.
dorsal
Col A
lo
GCI
lo
GDN
GDN
oes
50 μ
ventral
b.
anterior
th1
GDN
PSI
th2
PSI
TTMm
50 μ
posterior

light-off stimulus regularly elicits jumping if the fly is on the ground (but may stimulate landing if it is in flight) (see Chap. 3). The pathway for this response has recently been traced from the optic lobes to the mid-leg and dorso-longitudinal (indirect) flight muscles (Fig. 36). Its most prominent component, though certainly not the sole one, is a pair of giant descending neurons. These arise from cell bodies in the brain situated dorsally above the area known as the deutocerebrum. Dendrites invade the lateral brain and from there send an axon, one on each side of the midline, down the neck connective and to the second (mesothoracic) neuromer of the fused thoracic-abdominal nervous system. The giant descending neurons (GDN's) are known to be presynaptic to two prominent output systems. One is the motor neuron (TTMm) supplying the tergotrochanteral muscle (TTM, the so-called jump muscle) of the mid-leg. The other is the uniquely identifiable peripherally synapsing interneuron (PSI), which is subsequently presynaptic to the five axons of the dorsolongitudinal flight muscle motor neurons. These connections,described by King and Wyman (1980), are also substantiated by cobalt fills into the GDN (Strausfeld and Bassemir, 1983). Cobalt migrates from the terminals of the GDN into both the TTM and PSI. This finding and the measurement of delay times between a spike in the GDN and contraction of the TTM (Tanouye and Wyman, 1980) suggest that the GDN and the TTM motor neuron are electrically coupled. Interestingly, the jump is immediately followed by flight, as might be expected from the GDN to PSI connection.

In the brain, the GDN's are cobalt-coupled to a pair of large interneurons running in the large commissure. They also have contact with mechanosensory fibers from the antennae and with one class of isomorphic small field neurons of the lobula, the so-called Col A neurons, first described in large flies (Hausen and Strausfeld, 1980; Strausfeld, 1980). The GDN, Col A, TTMm and PSI neurons can all be filled transsynaptically through the antennal nerve (Strausfeld and Singh, 1980; Strausfeld and Bassemir, 1983; see also Fig. 36).

Single gene mutations affecting the giant fiber pathway have been isolated (Tanouye and Wyman, 1979). One of them, the mutant *bendless*, modifies the morphology of the giant fibers and blocks (or modifies) its main synapse to the TTM motor neuron (Thomas and Wyman, 1982).

A further visual pathway largely independent of the pathways for visual course control and the start reflex is that of the visually elicited landing behavior. In this case the behavioral investigation is ahead of the identification of the underlying substrate, but again the visual mutants and a special property of this behavior, its plasticity, may help to take the first step toward a structural identification.

The mutants with structural defects in the visual ganglia have the optomotor yaw torque and leg extension responses affected differentially. In the

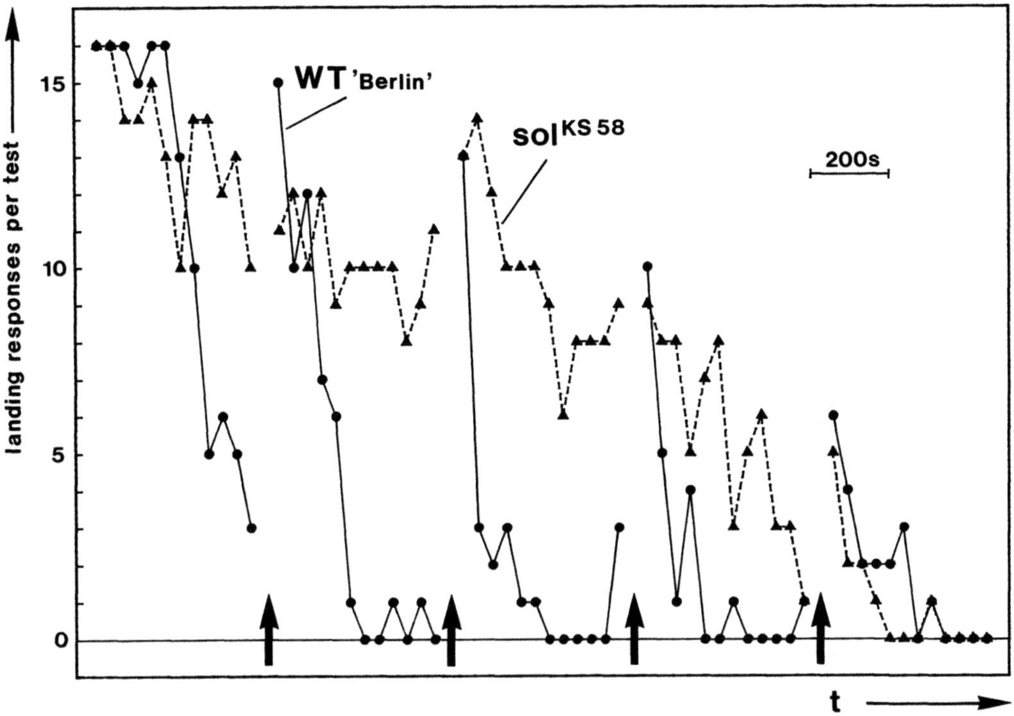

Fig. 37. Habituation and dishabituation of landing response of single WT (Berlin) and mutant *small optic lobes*^{KS58} flies. Stimulus consists of two vertical dark stripes (width $\delta = 7°$; height $H = \pm 12°$) on an evenly illuminated oscilloscope screen ($\Delta\psi_{disp} = \pm 37°$). Stripes suddenly start to move from front to back through frontal part of each visual half-field ($w = 60° s^{-1}$). For each point in the graph the stimulus is presented 16 times in 3 s intervals. Dishabituation is caused by landing, which is brought about by handing to the fly a tiny piece of paper (*arrows*). Note habituation of dishabituation and the distinctly slower habituation in *sol*^{KS58}. (Fischbach, 1983 b)

mutant *omb* unilateral front to back motion of vertical stripes elicits normal leg extension but only little yaw torque as compared to wild type. Inversely, in the mutant *sol* unilateral front-to-back motion of a stripe is very ineffective in eliciting leg extension, whereas the same stimulus causes a normal optomotor yaw torque response. This specific interference of mutations with either the landing behavior or the course control, even when both are guided by the same visual stimuli, suggests that visual processing for these two behavioral subunits proceeds largely in parallel (Fischbach, 1983 b).

The mutant *omb* strongly suggests that HS- and VS-cells are not crucial for landing to be elicited by front-to-back motion, a result supported by the contrast frequency dependence of these cells which matches that of course control, not that of the landing response. The identification of a neuronal substrate by exclusion of other neurons is, however, a lengthy procedure. Fortunately there is a more direct approach.

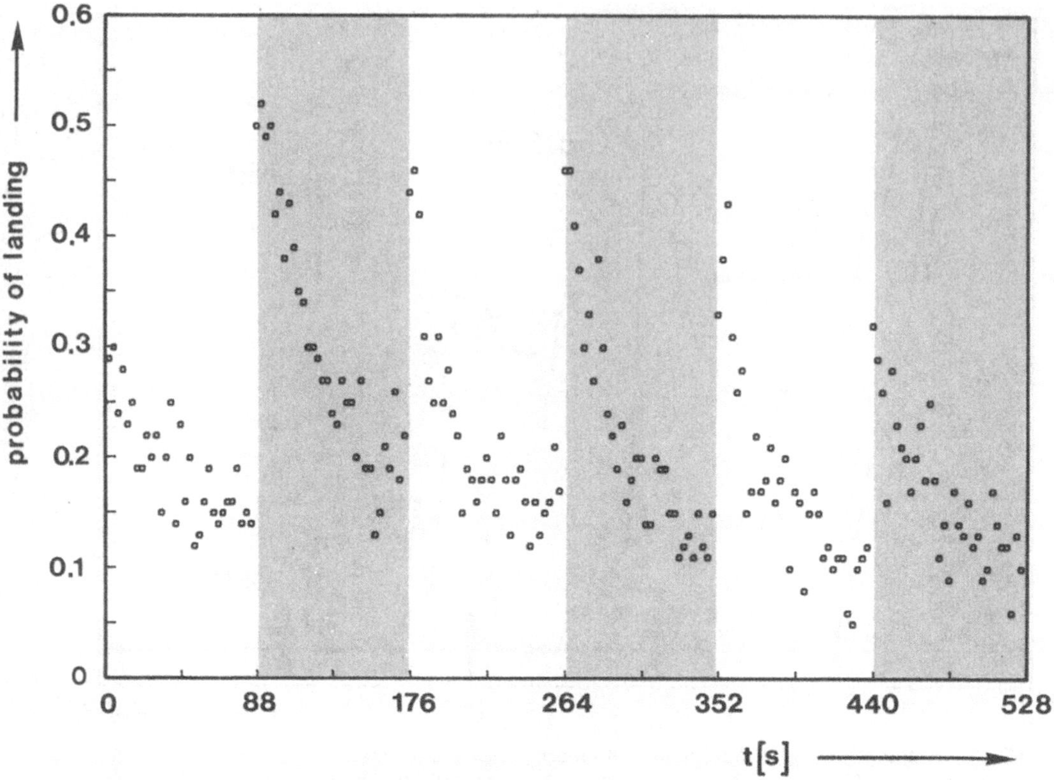

Fig. 38. Contralateral sensitization of landing response. One vertical dark stripe (width $\delta = 7°$; height $H = \pm 12°$) moves on an oscilloscope screen from $\psi = 18°$ to $\psi = 39°$ or from $\psi = -18°$ to $\psi = -39°$ with an angular velocity of about $w = 120° s^{-1}$. Eyes are stimulated alternately with volleys of 32 tests at a rate of 1 test per 2.75 s. Series starting with the left or with the right eye from altogether 116 flies are added up. *Shaded* and *nonshaded areas* mark alternation of stimulus between visual half-fields. *Ordinate* shows the probability of landing for the ensemble of flies. (Pooled data from WT (Berlin) and Canton S which did not differ significantly from each other; courtesy K. F. Fischbach)

Leg extension in response to a repeating visual stimulus gradually habituates. But a high response frequency can be restored by a dishabituating event such as real landing (Fig. 37; Fischbach, 1981). Habituation occurs at a peripheral level near the sensory input stage. Stimulation in one part of the visual field causes local habituation which does not spread to all other parts. This phenomenon provides the opportunity to determine the receptive fields of the habituating units involved and may lead eventually to the identification of the neurons and synapses undergoing the relevant modifications. So far it is clear that for vertical motion the habituating units are small. Their width has been determined to that of one visual element ($\Delta\varphi = 6°$). Their height is less well known, but it is certainly less than half the height of the total visual field. The receptive fields of the habituating units for horizontal

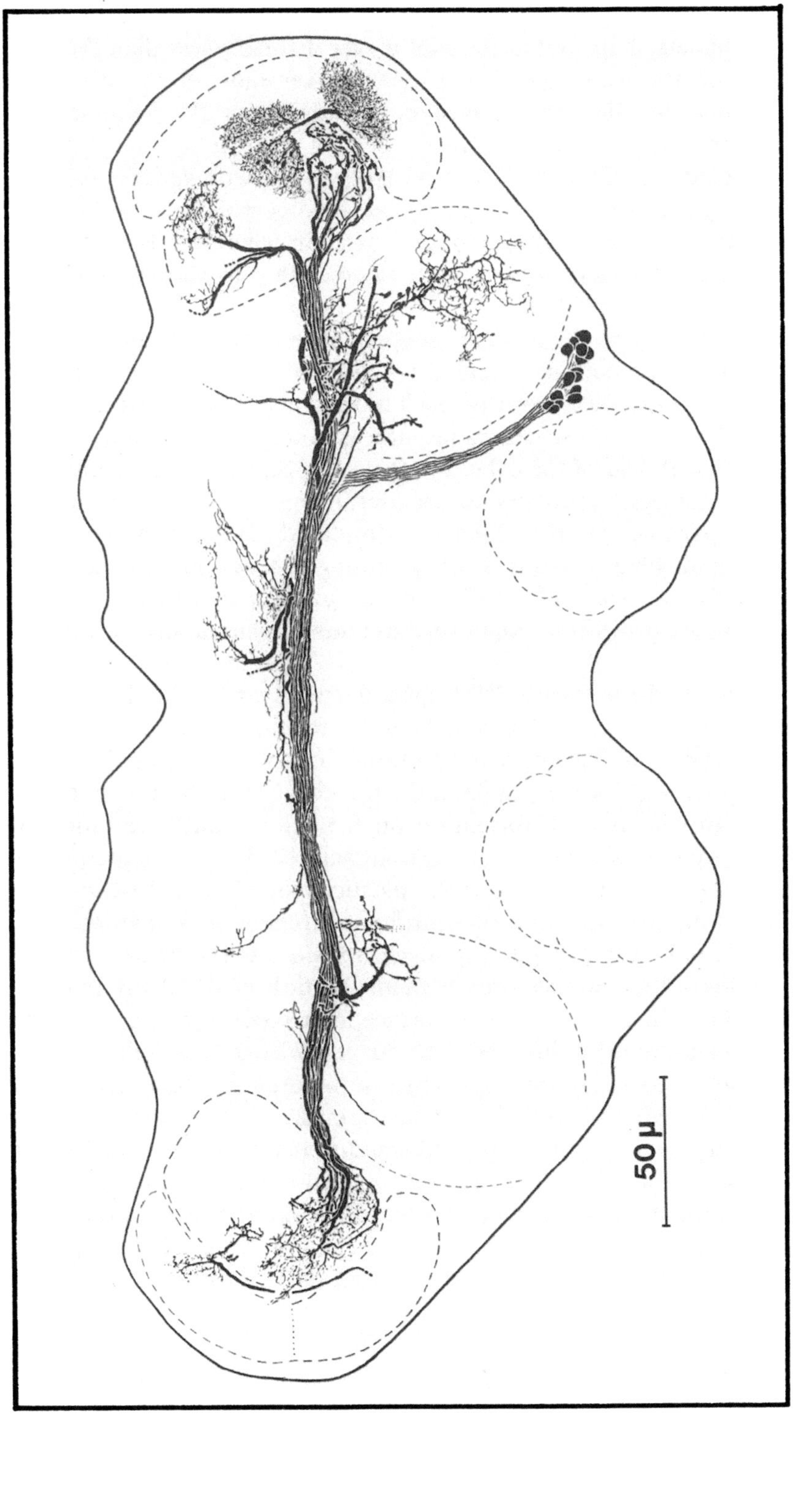

Fig. 39. Golgi-stained bundle of small field columnar neurons linking the two lobulae. Fibers stained occupy parts of lobulae corresponding to the dorso-frontal part of the visual field on both sides. Camera lucida drawing from 35 μ frontal sections of the mutant *sine oculis* (*so*). (Fischbach, 1983 a)

front-to-back motion extend vertically over two or three (adjacent) horizontal rows of visual elements and horizontally over distinctly less than the total sensitive area of the visual field (but clearly over more than one visual element). Vertical and horizontal motion as stimuli for the leg extension response are thus processed by different sets of neurons as the different receptive fields suggest. Cross-habituation experiments with vertical and horizontal motion in the same area of the visual field corroborate this conclusion. It is finally supported by the mutant *omb*, in which the leg extension response to horizontal motion is normal, whereas that to vertical motion is very feeble (Fischbach, 1983 b).

Equally interesting as the mutual independence of the habituating units are the sensitizing interactions revealed by stimulation in different parts of the visual field. For instance, front-to-back motion at a certain position in the frontal visual field of one eye causes an increased response to front-to-back motion in the visual field of the other eye (Fig. 38). Similar experiments with upward motion suggest that this contralateral sensitization is strongest if corresponding parts of the visual field are stimulated (data not shown). A set of columnar small field neurons directly linking the two lobulae has been described in *Musca* (Strausfeld, 1976) and *Drosophila* (Fischbach, 1983 a). Such a fine grained projection might mediate this functional mirror symmetry (Fig. 39).

Interestingly, in the mutant sol^{KS58} contralateral sensitization has a much shorter time constant. An interaction between the two hemispheres is still apparent, since bilateral front-to-back motion is a very strong stimulus while, as mentioned before (Chap. 3), unilateral front-to-back motion is much less effective than in wild type. Habituation on the other hand is significantly slower in *sol* than in wild type (Fig. 37; Fischbach, 1983 b). In Chapter 12 we will discuss another visual behavior, the plasticity of which is lost in *sol*.

Admittedly the identification of neural constituents of behavioral functions has made only a first beginning in *Drosophila*. At the present state of electrophysiological techniques only a small fraction of the neurons in the *Drosophila* brain will be accessible to intracellular recording and dye injection. (The two giant descending fibers, to our knowledge, are the only interneurons presently studied in this manner). On the other hand, activity staining with ^{3}H-deoxy-glucose and the genetic approach partially make up for this shortcoming. Studies with large and small flies will continue to support and depend upon each other. In *Drosophila* separate subunits of behavior will be identified and may even be traced in the CNS, but at a level coarser than the single neuron one. The identification of behavioral subunits will remain one of the main themes throughout the second part of this book.

Part B:
The Behavioral Structure of the Visual System

5 Flying Straight

Artificial Closed Loop Conditions

In our account of visual orientation we still discuss the first requirement, the stabilization of the animal. The last two chapters have dealt with optomotor behavior under open loop conditions. It is obvious from the properties of these responses that they serve to stabilize the trajectory of the fly against involuntary perturbations. This assertion, however, does not tell us how these course control functions operate when the fly moves freely. The simplest answer to this question would be that the mechanisms work permanently as long as the fly is confronted with visual motion (cf. Chap. 3; Optomotor Equilibrium). We will see in the second part of this book, however, that course control is more sophisticated. Even the basic function is modified according to more general requirements.

The first immediate requirement for an insect is to gain orientedness. *Drosophila* at the torque meter (see Fig. 14) with its head glued to the thorax has only few means left to acquire orientedness. It can still move its antennae, proboscis, wings, legs, halteres and abdomen, and presumably it can sense and often also see that these movements take place. It can probably hear the noise of its beating wings and it senses the air current generated by them. As mentioned before, good fliers even at the torque meter keep their front and mid legs folded underneath the body and their hind legs stretched backward as in free flight. Occasionally, they clean their legs and abdomen while flying. When flight stops they often try with their legs to free themselves from the hook and, lacking tarsal contact, they often perform righting maneuvres. Flies can see the distribution of light and darkness of their stationary surround. Whether patterns fade away like stabilized images for humans is a matter of debate. Certainly vertical stripes lose much of their attractiveness when they are not moving or at least changing their reflectance (for *Musca*: Reichardt, 1973; for *Drosophila*: unpublished observation). However, flies can move the optical axes of their visual elements slightly (Hengstenberg, 1971). This leads to perceptible brightness changes whenever the visual surround contains sufficiently contrasted edges.

In any case, with tethered flight most sensory experiences of free flight, the feedback stimuli of active changes in position, orientation, and also to a large extent of body posture are eliminated. Furthermore, the fly cannot explore its local environment. Thus a low level of orientedness must be the consequence.

It was the idea of Reichardt (1973; Reichardt and Wenking, 1969) to grant the fly (*Musca*) in this situation additional visual feedback stimulation by using the momentary yaw torque for driving the panorama, such as to simulate free flight far away from landmarks (in which case only rotation provides visual feedback). Reichardt calculated that for a fly to turn in flight most of the torque is used to overcome air friction and less than 1% to generate angular acceleration. Thus the fly's torque is nearly proportional to the angular velocity of the panorama. The next problem was to find a reasonable value for the relation between torque and angular velocity (coupling coefficient c; see Fig. 44). Reichardt solved this in two steps. First, he estimated the coupling coefficient by calculating the fly's air friction. Then he varied the coefficient during the closed loop experiment and adjusted it to the highest value which just would not cause oscillations in the loop.

We used Reichardt's experimental design for *Drosophila* (Fig. 40). The "optimal" coupling coefficient for *Drosophila* was determined accordingly. Later we found a way to determine the natural coupling coefficient which is about ten times larger than the one used here, but for reasons to be discussed later is not applicable to the artificial closed loop situation. We thus simulate unnaturally high air friction, which, however, does not seem to disturb the fly very much.

The next problem is the adjustment of zero torque, i.e., the torque at which the panorama is at rest. Without visual stimulation *Drosophila* modulates its torque in a nonrandom manner which would require long averaging before it would be usable for zero adjustment. (The torque compensator has a different zero for each fly and for a fly at rest and in flight due to imprecision of the alignment of the fly in the instrument). We therefore routinely use the open loop measurement of the optomotor response for this purpose. The panorama is rotated first clockwise, then counterclockwise, and zero torque is taken as the mean between the two response levels. Fortunately, with *Drosophila* the determination of zero torque is not critical for closed loop experiments, as will be seen in the following.

Optomotor Balance

In the preceding chapters we have discussed the optomotor responses under open loop conditions and have interpreted them as visual reflexes counteracting involuntary course deviations. Let us now investigate the yaw torque response in closed loop. When the closed loop apparatus with a cylinder of dark and bright vertical stripes as panorama and the fly in its center is switched on, the torque of the fly has an arbitrary value which in

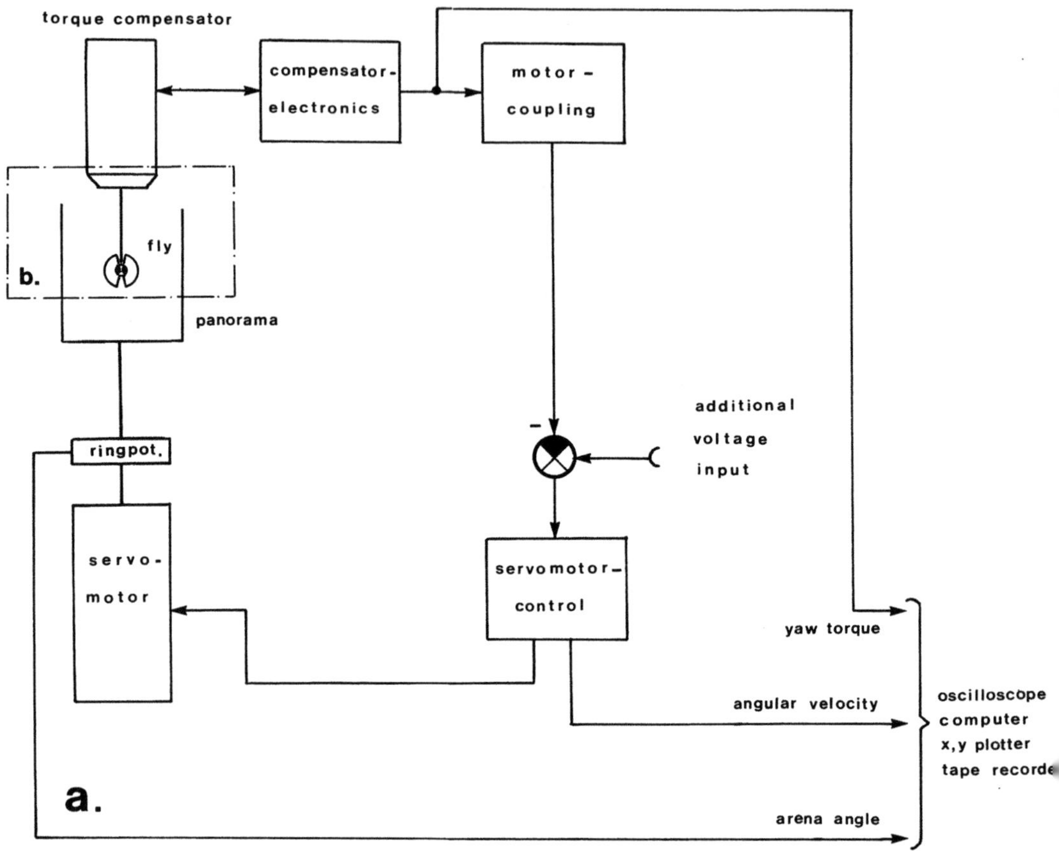

Fig. 40. Experimental set-up for studying yaw torque in stationary flight under visual closed loop conditions. Since in free flight the angular velocity of the fly is roughly proportional to its torque, the angular velocity of the panorama is made proportional to the torque compensator output. At ⊗ an additional voltage corresponding to an extra angular velocity (rotatory bias) can be "injected" into the closed loop. The ring potentiometer provides a voltage proportional to the angular position of the arena. (b) shows an arena with a random dot texture (Fig. 45c) and black stripe as panorama

most cases causes the panorama to rotate at a certain angular velocity; but the fly immediately stops the rotation and keeps the panorama at rest. This holds for any pattern which can be resolved by the fly. Applied to the free flight situation this result shows that the fly flies straight.

We now challenge this ability. We add a slowly increasing positive or negative D.C. voltage to the closed loop between compensator and motor control unit. Since angular velocity of the motor is proportional to the voltage driving it, this measure leads to a slowly increasing rotatory bias of the panorama or, what amounts to the same, can be considered as a slow drift of the zero setting of the torque signal. Such a situation is not totally artificial. One might imagine that one of the wings is progressively deteriorating

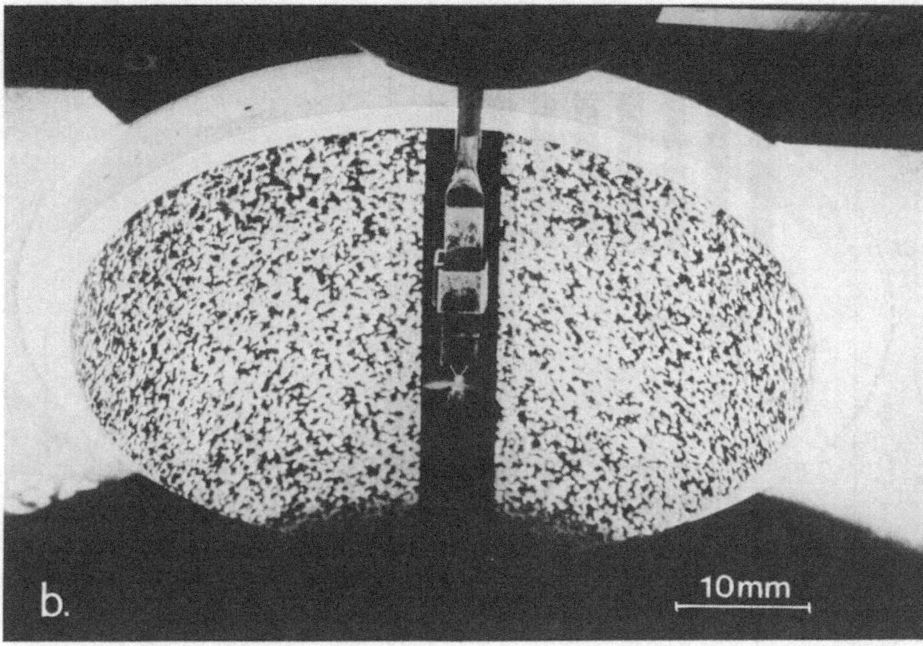

Fig. 40 b

during flight. In the experiment shown in Fig. 41 the rotatory bias is raised from $w_b = 0° s^{-1}$ to $w_b = \pm 1000° s^{-1}$. The fly is able to continuously stabilize the panorama with short intermissions up to a bias of $w_b = \pm 700° s^{-1}$, an achievement for which it has to generate steady torque of more than $T = \pm 6 \times 10^{-9}$ Nm. This is about twice the average torque measured in open loop experiments with the same pattern (Fig. 42). The fly maintains this very high yaw torque even if it does not fully compensate the bias when this is most extreme.

In the experiment just described the rotatory bias was increased gradually. Now we change it stepwise. At an arbitrary moment we add to the closed loop a constant D.C. voltage which is equivalent to a rotatory bias of $w_b = 110° s^{-1}$ or a change of zero torque by $T = 10^{-9}$ Nm (Fig. 43 a). The fly changes its average torque by precisely $T = 10^{-9}$ Nm, but it reaches this value only after several seconds during which the cylinder is often oscillating irregularly with large amplitude (Fig. 43 a). This result was obtained with a striped cylinder as panorama. It is typical for patterns uniformly covering a large fraction of the visual field. (With pronounced landmarks in the panorama, for instance in the case of a single vertical black stripe as in Fig. 43 b, the cylinder is stabilized in a fraction of a second. These different time constants suggest different mechanisms of stabilization at work; but for the discussion of responses to landmarks a later chapter is reserved).

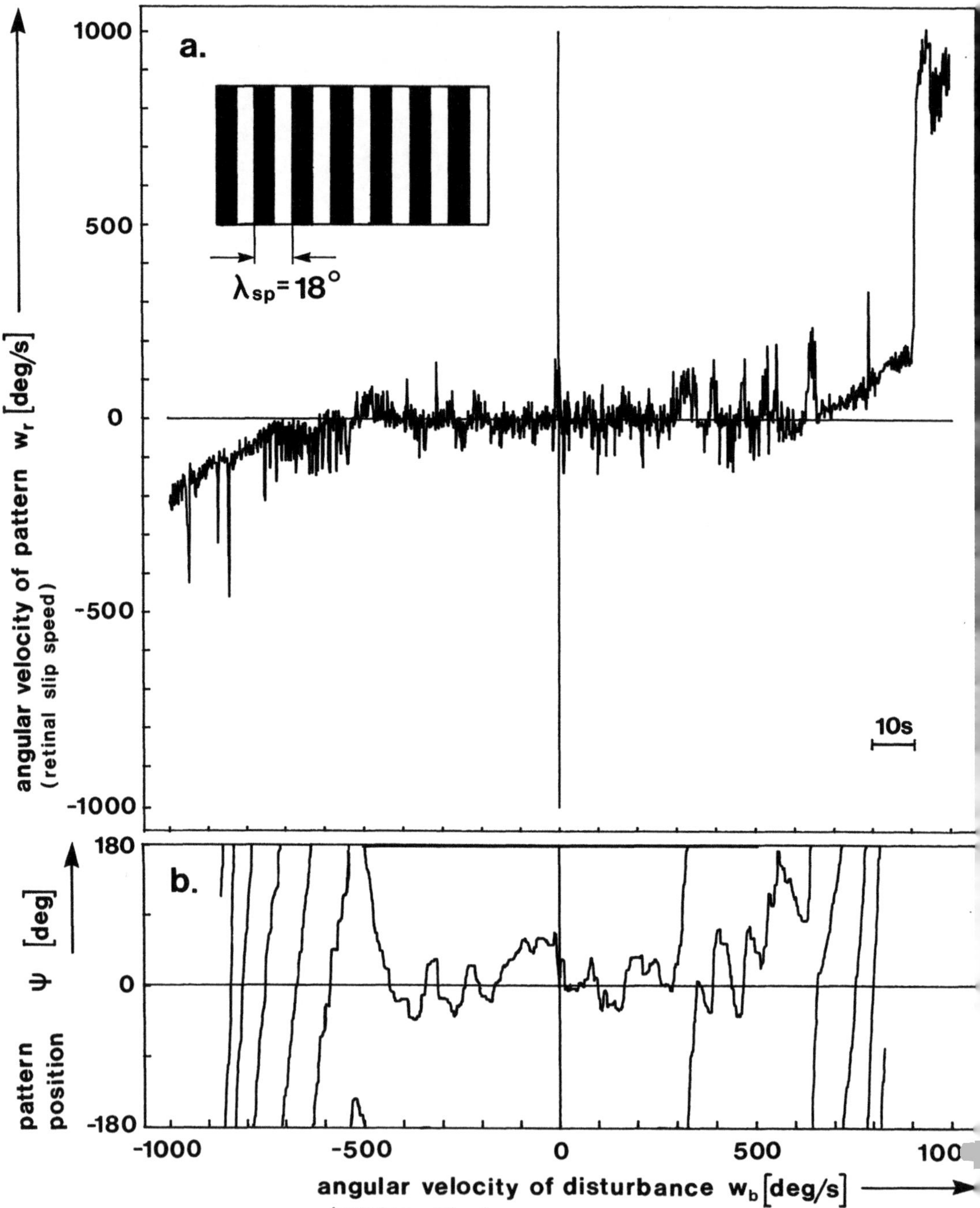

a.
b.
λ_{sp}=18°
10s
1000
500
0
-500
-1000
180
-180
100
angular velocity of pattern w_r [deg/s]
(retinal slip speed)
pattern position Ψ [deg]
angular velocity of disturbance w_b [deg/s]
(rotatory bias)

Now let us consider the question to which extent the known open loop properties of the optomotor yaw torque response discussed in Chapters 3 and 4 explain the closed loop behavior just described. We first regard the situation of the fly in the above experiment as a negative feedback loop disturbed by a rotatory bias $w_b(t)$ which leads to visual motion (retinal slip speed $w_r(t)$) stimulating a movement detection system (which might be the summed activity of an array of EMD's). In free flight the fly's torque output T(t) would lead to a rotation $w_f(t)$ of the fly counteracting the rotatory bias (Fig. 44). The closed loop gain G being defined as the relation between input $w_b(t)$ and output $w_f(t)$ in the steady state must be very close to $G = 1$, as can be seen from Fig. 41. In fact, even by extensive averaging, we are unable to detect any slip of the panorama for rotatory bias in the range between $w_b = -110° s^{-1}$ and $w_b = +110° s^{-1}$ (e.g., Fig. 43). For $G > 0.6$ an undamped negative feedback loop will cause the torque output to oscillate in response to even the slightest disturbance (as a computer simulation shows). Thus optomotor behavior cannot be as simple as that.

In a recent study of visual course control in the walking beetle *Tenebrio molitor* Varju (1975) also observed a closed loop gain $G = 1$ with a striped cylinder as panorama and a small rotatory bias introduced into the closed loop. The explanation he offered was that *Tenebrio* was actually keeping one of the edges of the stripes in frontal position, i.e., was responding not only to the movement but also to the position of the stripes (see *Fixation* below).

We tried to eliminate edge fixation in the *Drosophila* experiment by using patterns without vertical stripes as panorama as shown in Fig. 45. One has to look at these patterns with the eye of a fly. For instance, the squares of the checkerboard panorama (a) are seen on the average by about four visual elements. Thus the panorama is perceived by the fly as having high pattern contrast. The main pattern wavelength in the zebra stripe pattern (b) is about $\lambda_{sp} = 5.4°$. The pattern elements cover only part of the receptive fields of the individual photoreceptors. Thus apparent contrast must be quite low ($m_{eff} \approx 0.05$). Furthermore, it has to be a lower frequency component of the pattern not obvious to the human eye which dominates the response, since this has the same sign as the rotation of the cylinder, while the basic pattern wavelength lies in the range of geometrical interference ($2\Delta\varphi^* > \lambda_{sp} > \Delta\varphi^*$; see

Fig. 41. (a) Optomotor balance. A fly is flying stationarily under closed loop conditions as shown in Fig. 40. A slowly increasing positive or negative D.C. voltage equivalent to an angular velocity w_b is added to the loop. Angular velocity of panorama (retinal slip speed w_r) is shown as function of angular velocity the panorama would have under open loop conditions (rotatory bias w_b). (With a blind fly or at high angular velocities $w_b (w_b > 1000° s^{-1})$ the angular velocity trace would fluctuate around the diagonal (see for instance Fig. 94b). Left and right halves of the curve are measured starting at $w_b = 0° s^{-1}$. Maximal torque the fly exerts to keep panorama at rest is more than $T = \pm 6 \times 10^{-9}$ Nm. Trace from single experiment with WT (Berlin). The position trace of the panorama in (b) shows that retinal slip is small with respect to fluctuations in pattern position. For definition of ψ see Fig. 3a and Appendix 2

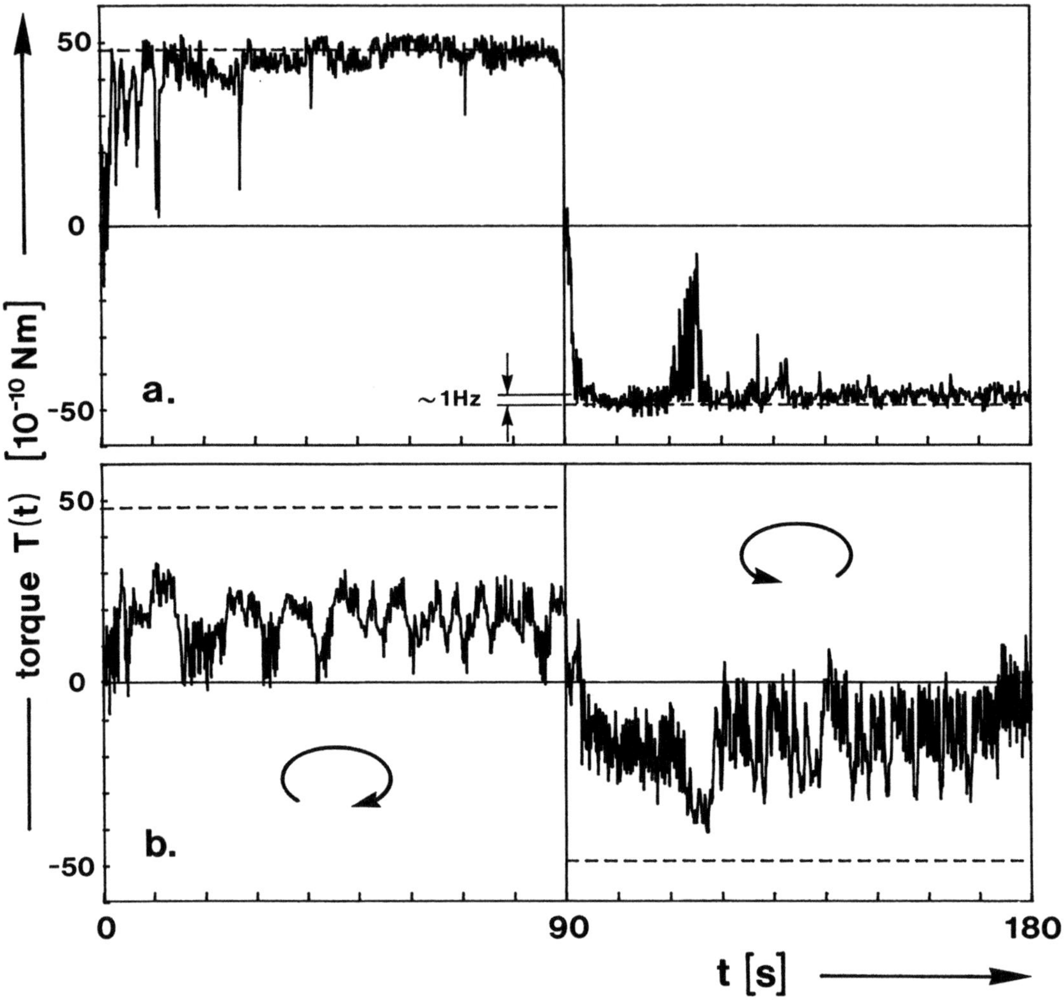

Fig. 42. Comparison between yaw torque response to a periodic grating with pattern wavelength $\lambda_{sp} = 18°$ under closed loop (a) and open loop (b) conditions with the same fly. In (a) a constant voltage corresponding to about T = + and $- 5 \times 10^{-9}$ Nm is added into the loop. *Dashed horizontal lines* indicate torque the fly would have to generate in order to keep panorama at rest. The fly happens not to stabilize the panorama completely causing the drum to rotate at an average angular slip speed of about $w_r = 18° s^{-1}$ ($w_r/\lambda_{sp} = 1$ Hz). This speed is applied in the experiment shown in (b). The fly exerts much larger torque when the feedback loop with the panorama is closed, than under open loop conditions. *Dashed horizontal lines* in (b) indicate the $T = \pm 5 \times 10^{-9}$ Nm torque level of (a) for comparison

Fig. 18) and should thus contribute negatively to the optomotor response. The grain of the "random dot" pattern (c) will not be resolved by the *Drosophila* eye. However, as one can see with half-closed eyes, the average gray will not be uniform for the fly but rather "cloudy". It will thus contain faint landmarks. The "random squares" pattern (d) obviously contains accu-

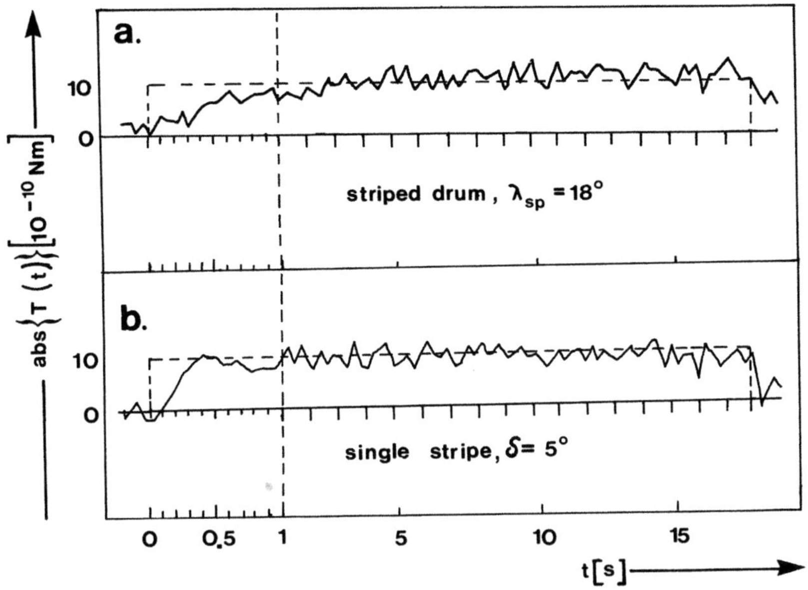

Fig. 43. Readjustment of optomotor balance in response to a sudden rotatory bias. At time t = 0 a constant voltage equivalent to torque $T = 10^{-9}$ Nm is added to the closed loop. Conditions as in Fig. 40. Note extended time scale for the 1. second. Torque traces are averages from n = 12 sweeps with the same WT (Kapelle) fly

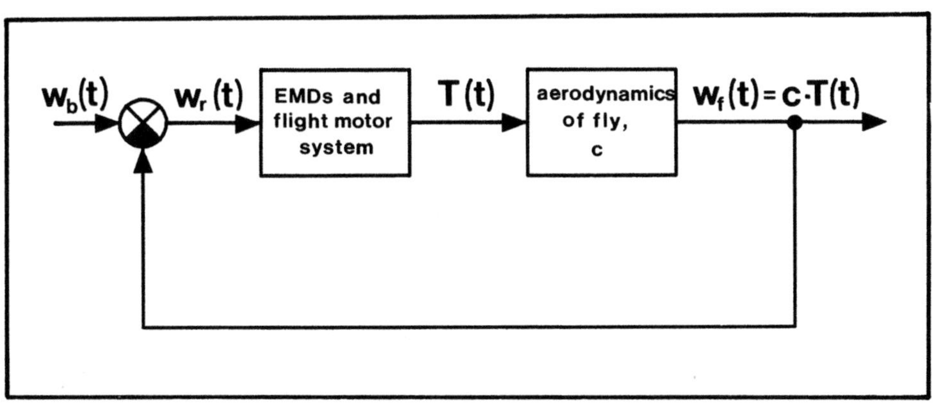

Fig. 44. Free flight in a rotating drum. Visual feedback loop consisting of a movement detection system (EMD's as shown in Fig. 17), flight motor system and aerodynamics of fly. w_b angular velocity of panorama (rotatory bias); w_r retinal slip speed; w_f angular velocity of fly; T flight torque

mulations of dark and bright squares which could serve as landmarks but has no extended vertical edges. Finally, in the "little stripes pattern" (e) the pattern elements are perceived by the fly with high subjective contrast and yet are distributed sufficiently evenly for no fixation landmarks to be apparent. None of these patterns may be perfect in the suppression of a possible edge

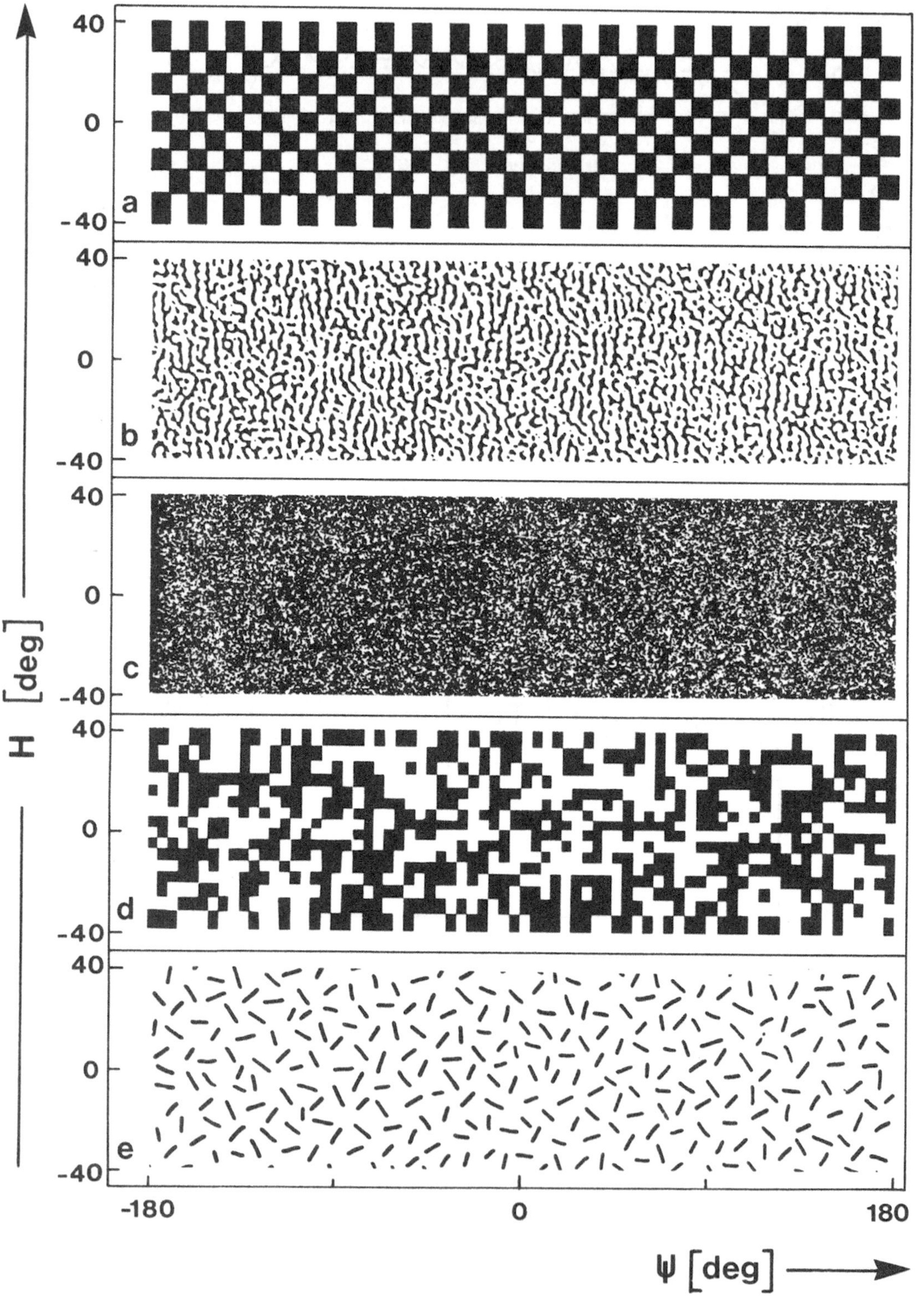
40
0
-40
a
b
c
d
e
H [deg]
-180
0
180
ψ [deg]

fixation response; but in all of them, edges have considerably less weight than in a periodic grating. *Drosophila* is able to stabilize the panorama with these patterns against a rotatory bias as well as with a periodic grating (Fig. 93). Thus edge fixation seems not a likely explanation for the high closed loop gain. We will discuss this matter again in the context of fixation.

The main argument against an edge fixation model comes from the fact that no particular angular positions of the panorama are maintained by the fly. Instead, the cylinder rotates stepwise without directional preference, as we will see in the next chapter.

A simple modification of the motion-dependent course control system would be one integrating the output of the movement detector and having the torque output proportional to the integral. This would make the closed loop gain dependent upon the length of time a particular direction of motion is maintained, but would allow for a steady state gain of $G = 1$ while generating a stable output. One now has three parameters to adjust, the influx into the integrator, its capacity and its leakiness. This time, the model can be made to fit the closed loop behavior, but it does not comply with the properties of the open loop optomotor response. When the movement stops, the optomotor torque response decays nearly as quickly as it builds up after movement starts (Fig. 22). This is not compatible with a simple integrator model. In open loop experiments, continuous net motion is needed to maintain a certain response level, while during closed loop it is not.

We have mentioned already another result which shows that the ability to fly straight is more than just the output of the optomotor machinery as we know it from the open loop experiments: The maximal average torque in closed loop conditions is about twice as much as during open loop experiments using the same visual pattern for stimulation (cf. Fig. 42a and b). For distinguishing between open and closed loop situations the fly uses reafferent information, as we will see later. We thus assign to the closed loop behavior at the torque compensator the new term "optomotor balance". It is the stationary equivalent of flying straight by visual means. [Note that optomotor balance has no immediate relation to Götz's (1975a) concept of "optomotor equilibrium" (see p. 43f).]

It may be surprising that *Drosophila* can adjust its zero torque over such a wide range. However, a fly needs a mechanism to balance the flight motors on either side. Any damage to or dirt on one of the wings will change the force of flight on that side. A mechanism providing a gliding zero torque reference

◄

Fig. 45. Various textures used throughout this study as panorama in closed loop and open loop experiments. Patterns cover the inner surface of a transparent cylinder. The fly is kept in the center. (a) Checkerboard panorama; (b) Zebra stripe pattern (LT 134, Letraset); (c) Random dot texture (Courtesy W. Reichardt); (d) Random squares pattern; (e) Little stripes pattern. (Magnified from LT 131, Letraset)

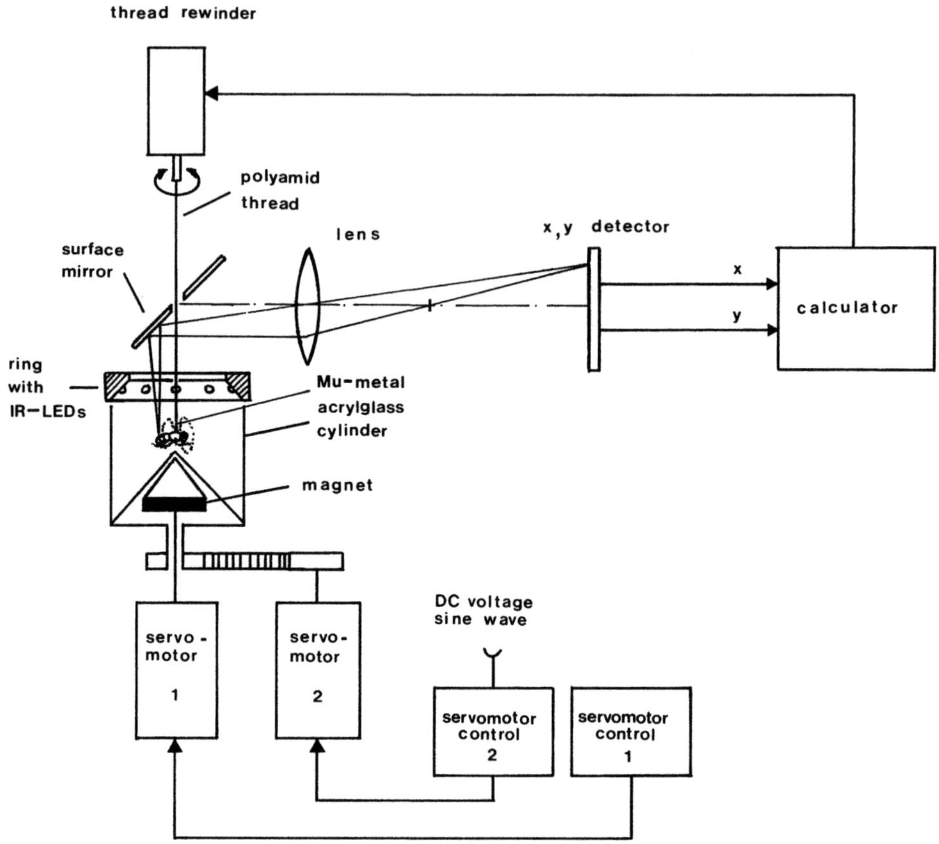

Fig. 46. Set-up for studying visual flight control of *Drosophila* which is free to rotate around its vertical body axis, but otherwise is kept stationary. Angular position of the fly is monitored by projecting its image onto an x-, y-detecting infrared sensitive photodiode. The suspended end of the polyamid thread can be turned with the rotations of the fly to avoid too much twisting. Fly is illuminated from above by 16 infrared emitting diodes which are shielded from direct sight by a circular infrared transparent filter. The magnet is mounted on a motor in order to eliminate any orientation cues from the magnetic field if necessary by high speed rotation of the magnet. (K. Vogtmann, in preparation)

can be slow, as it is in fact (Fig. 43), since no fast external changes in the efficiency of the two wings are to be expected.

Closed Loop Gain of Freely Rotating Flies

In free flight *Drosophila* is known to possess additional nonvisual mechanisms to stabilize its course. This is demonstrated in an experiment in which optomotor balance is tested with flies which are free to rotate around their vertical body axis but otherwise are kept stationary.

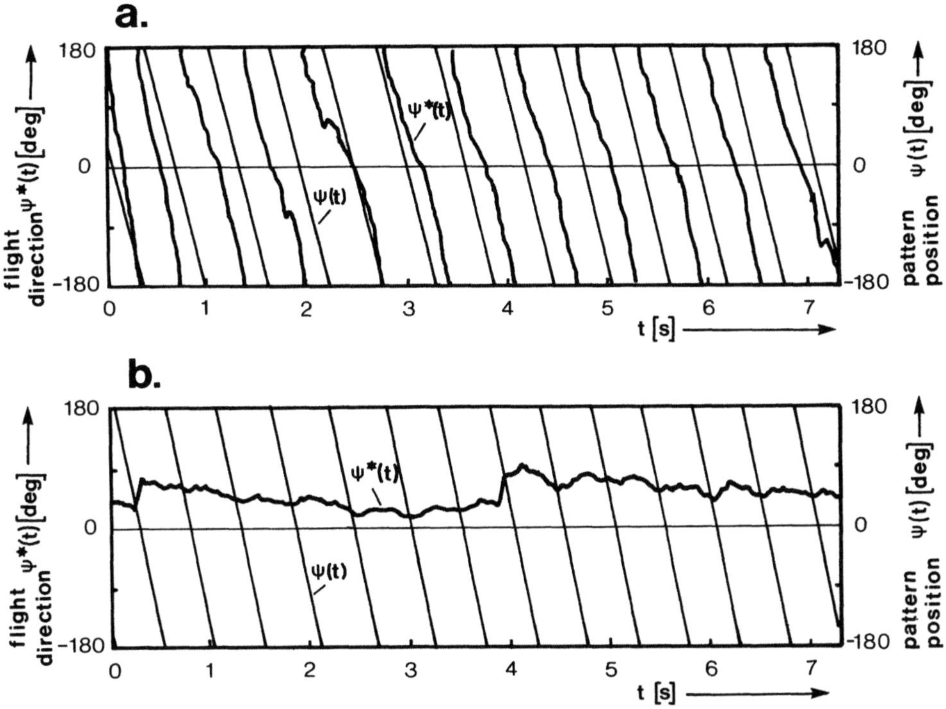

Fig. 47. *Drosophila* in stationary flight, but free to rotate around its vertical body axis (for experimental conditions see Fig. 46) may either stabilize its orientation with respect to its visual surround (a) or with respect to external space (b). The fly is surrounded by the "little stripes pattern" (Fig. 45e) which in this special case rotates at angular velocities of $w_b = 610° s^{-1}$ (a) or $w_b = 640° s^{-1}$ (b). ψ, the angular position of a certain point in the panorama and ψ^*, the angular position of the fly are measured with respect to the same point of reference in external space. [During flight at the torque meter this reference coincides with the direction of flight ($\psi^* = 0°$).] In (a) the fly rotates nearly as fast as the panorama. Mean retinal slip speed for this recording period happens to be $w_r \approx 24° s^{-1}$. But as in experiments at the torque meter (Fig. 41) the mean slip speed is small compared to the slip speed fluctuations. In (b) mean retinal ship equals angular velocity of panorama ($w_r = w_b$). The fly is able to fly straight in this set-up at any pattern velocity. In this case the contrast frequency of the small spatial wavelength components may be beyond flicker fusion of the eye but as the angular position trace of the fly (ψ^*) shows the fly responds to a $\lambda_{sp} = 360°$ component in the pattern with a transient course deviation

The fly is glued to a small piece of Mu-metal suspended at a 10 μ polyamid thread. With a pointed magnet underneath, the fly can be kept at a defined position but is still able to rotate freely (Fig. 46; Vogtmann, in preparation; E. Buchner, unpublished; Heisenberg and Wolf, 1979). Surrounded by a stationary striped cylinder or by one carrying the pattern of Fig. 45e *Drosophila* is able to keep stable flight orientations. If now the cylinder is rotated, the fly is in a conflicting situation. The visual system and the nonvisual course control mechanism will differ in what they report as straight flight. But the fly can still follow the rotation without measurable

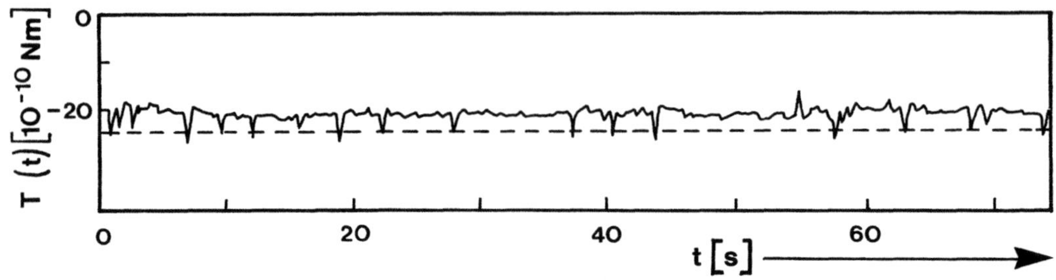

Fig. 48. Constant slip during closed loop with striped drum ($\lambda_{sp} = 18°$). A strong rotatory bias amounting to $T = 25 \times 10^{-10}$ Nm has been introduced into the loop. The fly generates a steady torque of about $T = 20 \times 10^{-10}$ Nm and arrests the panorama only for brief moments; Canton S; closed loop coupling: 10^{-9} Nm $\triangleq$ $55° s^{-1}$. (Heisenberg and Wolf, 1979)

slip. Even at an angular velocity of $w_b = 1000° s^{-1}$, the fly is able to fully compensate the motion. The closed loop gain in this case is about $G = 1$.

Part of the time, however, the fly does not follow the movement of the panorama. It may fly straight in relation to external space or may take turns in either direction. The fly distinctly switches between following and ignoring the visual cues. This result shows that proprioceptive signals monitoring the fly's rotation provide sufficient information for straight flight at the polyamid thread. It seems that when this information is contradictory to the signals from the visual system the fly takes its choice (Fig. 47; Vogtmann and Heisenberg, in preparation).

In free flight, where coherent motion of the panorama in most cases is due to the fly's own rotations, the signals are not in conflict with one another but cooperate in stabilizing the fly's orientation. For freely flying hoverflies, *Syritta pipiens*, a closed loop gain $G < 1$ ($G = 0.8$) in the steady state was found (Collett, 1980 b). It may be that *Syritta* had some stationary visual cues in addition to the rotating ones.

In complete darkness *Drosophila* on the polyamid thread can stabilize its flight against fast rotations, but does perform slow turns. It is certainly not completely disoriented. On the other hand, with both halteres amputated it is very much impaired even when visual information is available. Removal of the halteres interferes with optomotor balance when the fly is free to rotate but not at the torque compensator. A possible explanation for this effect will be discussed below.

Not always will *Drosophila* at the torque compensator stabilize the panorama completely. Occasionally it maintains a certain slip velocity over considerable time (Fig. 48), particularly after long closed loop flights. When the bias is strong the slip will be in the direction of the bias. Since with a strong bias the fly bends its abdomen and hind legs toward the side to which it tries to turn, one cannot help getting the impression that the fly has to make an effort to generate this much torque and that it is just a bit lazy when it allows for the slip.

6 Endogenous Behavior in Yaw Torque Fluctuations

For decades optomotor responses were treated as time averages in recordings and modeling. What matters for the fly, however, is the momentary torque it exerts and the momentary movement stimuli it receives. Therefore we will now consider the fine structure of yaw torque at the torque compensator in open and closed loop situations without applying extra movement stimuli. We want to watch the endogenous flight behavior that *Drosophila* performs while attached to its hook and to the recording device. We hope to identify some of it in our yaw torque recordings. If it is true that the primary goal of an animal is to acquire orientedness and that behavioral activity is part of this process, one might be able to observe it.

Saccades and Torque Spikes

We start with the torque during optomotor balance. As mentioned in the previous chapter, *Drosophila*, while maintaining optomotor balance, does not keep the panorama immobilized but moves it stepwise with equal probability in either direction (Fig. 49 a). These steps are due to short bursts of torque – so called torque spikes – which have a stereotype time course and under given conditions often have uniform amplitude. For a torque spike the fly reduces by about 12° the wing beat amplitude on the side to which it is going to turn (wing hitch; Götz et al., 1979; Götz, 1983). The equivalent in free flight can easily be observed when *Drosophila* is hovering. In this flight mode, in which the fly is oriented rear down by 60° and the wings are beating in a nearly horizontal plane, the vector of the force of flight is pointing upward. The fly keeps a certain altitude and can move in any horizontal direction, normally at low speed. In this type of flight one can see stable orientations interrupted by abrupt turns in the horizontal plane.

Such sudden reorientations which have a much faster time course than most other voluntary turns are ubiquitous among the insects. They were called body saccades by Land (1973, 1975; Land and Collett, 1974; Collett and Land, 1975 a) in analogy to eye movements of mammals. In *Drosophila* body saccades of freely flying flies were filmed by Bülthoff et al. (1980). The histogram of angular velocity in the horizontal plane (Fig. 49 d) obtained from selected flight episodes by frame to frame analysis has a trimodal

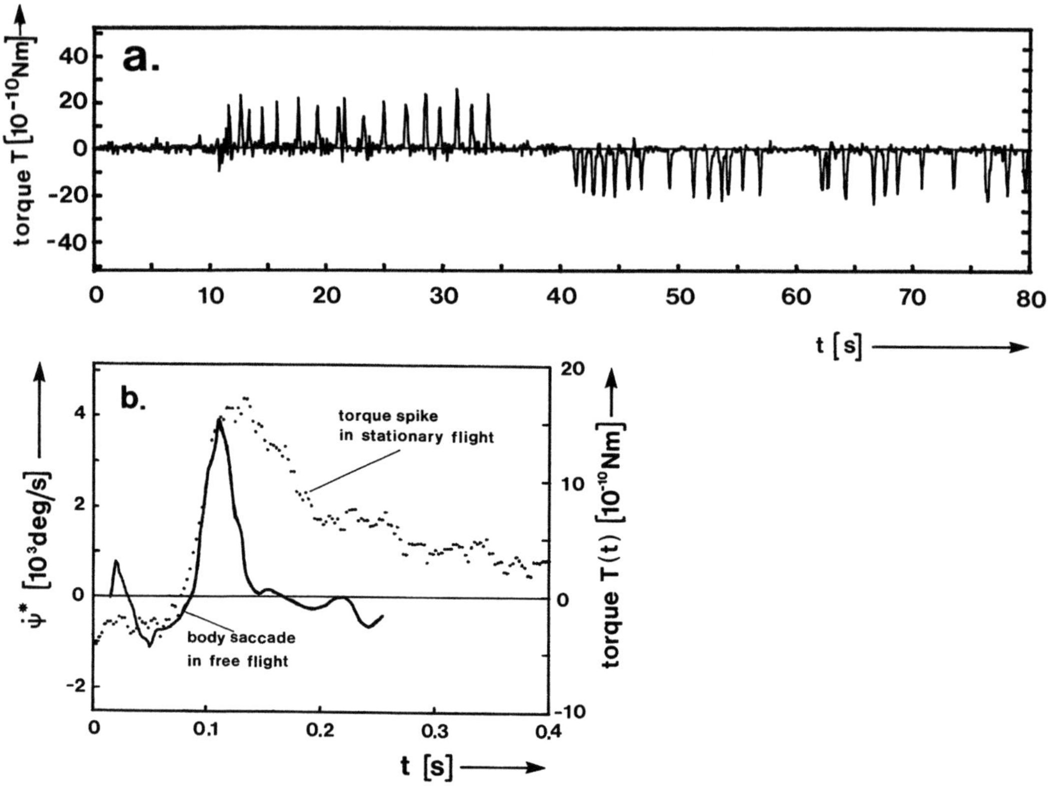

Fig. 49. (a) In stationary flight *Drosophila* generates short pulses of torque (torque spikes) which under closed loop conditions lead to stepwise rotation of the panorama. Similar torque pulses in free flight lead to abrupt turns of the fly (body saccades). (b) Time course of angular velocity during body saccade reconstructed from high speed cinematographs and of torque spike in stationary flight. Duration of saccades is shorter than that of torque spikes and during saccades flies rotate much faster than the panorama does during torque spikes, given the coupling conditions in our closed loop experiments. (Free flight saccade redrawn from Bülthoff et al., 1980). (c) Body saccades at the polyamid thread as illustrated by the fly's angular position (upper diagram) and velocity (lower diagram). Although flight maneuvres are restricted nearly as much as at the torque meter, saccades are about as fast as in free flight. At the thread as well as in free flight the fly receives mechanosensory stimuli from its real angular velocity and acceleration. (d) Histogram of angular velocities during selected episodes of free flight. The two flanking peaks of angular velocities between $\dot{\psi}^*_{abs} = 1500$ and $4000° \, s^{-1}$ are due to body saccades. (After Bülthoff et al., 1980). *Shaded area* indicates retinal slip speed range obtained at the torque compensator with closed loop coupling of 10^{-9} Nm $\hat{=}$ $110° \, s^{-1}$

distribution with the flanking peaks of velocities between 1500 and $4000° \, s^{-1}$ belonging to saccades of either polarity. Figure 49b shows the time course of one of them, which takes between 30 and 50 ms. Comparing saccades with torque spikes in stationary flight, one notices that in free flight saccades are five to ten times shorter. Also, on the polyamid thread, when *Drosophila* is free to rotate around its vertical body axis, it performs saccades of the short type as in free flight (Fig. 49c). Obviously the real angular

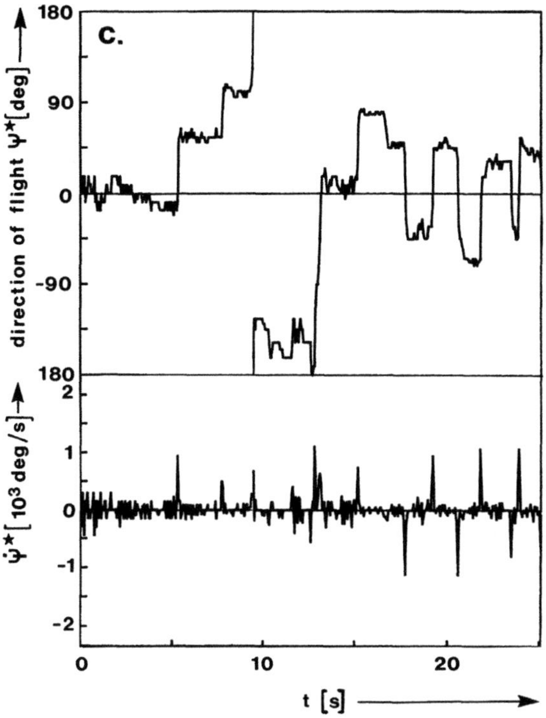

Fig. 49c

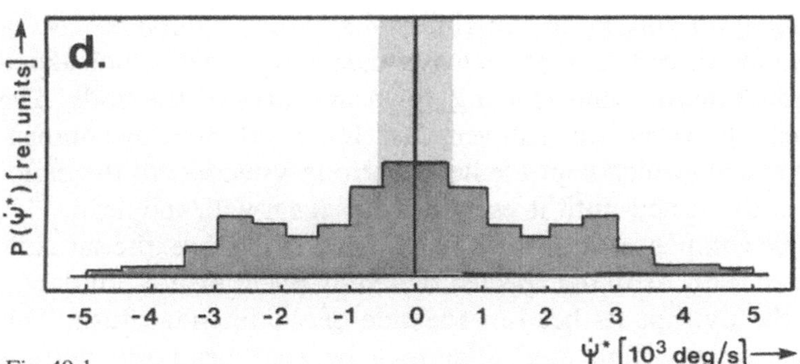

Fig. 49d

motion of the fly shortens the burst of torque. It is conceivable that signals from the halteres interfere with the torque spike motor program.

The equivalence between saccades and torque spikes allows the experimenter to adjust, in closed loop experiments with tethered flies, the coupling constant between torque and the angular velocity of the panorama. We chose the relation 10^{-9} Nm $\triangleq$ $110° \, s^{-1}$ ($55° \, s^{-1}$ in earlier experiments), since in this way an average-sized torque spike displaces the panorama by $\Delta\psi = 60°$

which is only slightly less than the average-sized saccade on the polyamid thread ($\Delta\psi = 80°$). One might be tempted, after knowing the yaw torque the fly generates during torque spikes and the angular velocity it reaches during free saccades, to adjust the coupling constant accordingly. However, this is not feasable for reasons of stability, and it would not be a good approximation to the free flight situation either. Due to the long duration of the torque spikes the angular displacement of the arena would be about ten times larger than what would correspond to a saccade in free flight. The instability encountered with the simulated natural coupling conditions may simply reflect the missing course control through the presumed angular acceleration and/or velocity proprioreceptors counteracting the turning commands with a slight delay as is apparent on the polyamid thread during saccades.

Torque Spikes as Fixed Action Patterns

Short as they are, saccades are complex sensory-motor patterns. In addition to the torque generated by the two wings and to the proprioceptive influence just described, one observes, in stationary flight, bending of the head, hind legs and abdomen to the side of the intended turn. In freely rotating flies the motor program is different. This is revealed in motion pictures of saccades at the polyamid thread. Abdomen, hindlegs, and head stay well aligned with the thorax during the whole maneuvre (although they are poorly coupled mechanically).

In his first studies of saccades in flies Land (1973, 1975) reported head saccades of *Calliphora* flying in an experimental situation similar to *Drosophila* on the polyamid thread. *Calliphora* was attached to a pivot which added a fivefold rotational inertia, thus slowing down the turns of the body. The turning of the head, however, was still very fast. It was followed by compensatory head movements which kept the head stationary relative to the panorama while the body turned until it was lined up again with the head.

If at the torque compensator the *Drosophila* head is left free, the saccadic turning of the head leads to an unexpected switching phenomenon. Initiating a torque spike, the fly flips its head to the side of the intended turn. This causes the torque to stay high until alignment between head and body is achieved. But now the thorax, being hooked to the torque meter, does not follow. Without visual stimuli this situation seems only to be terminated by a torque spike into the opposite direction which gives the head free until the next torque spike occurs (Fig. 50). (The records of torque and stripe position in real time shown in this figure will be used throughout. They are difficult and tedious to read, but they provide direct insight into the behavior of the fly. We would like to encourage the reader to get acquainted with them right away). The data indicate that misalignment of head and thorax during torque spikes prevents yaw torque from declining to base line.

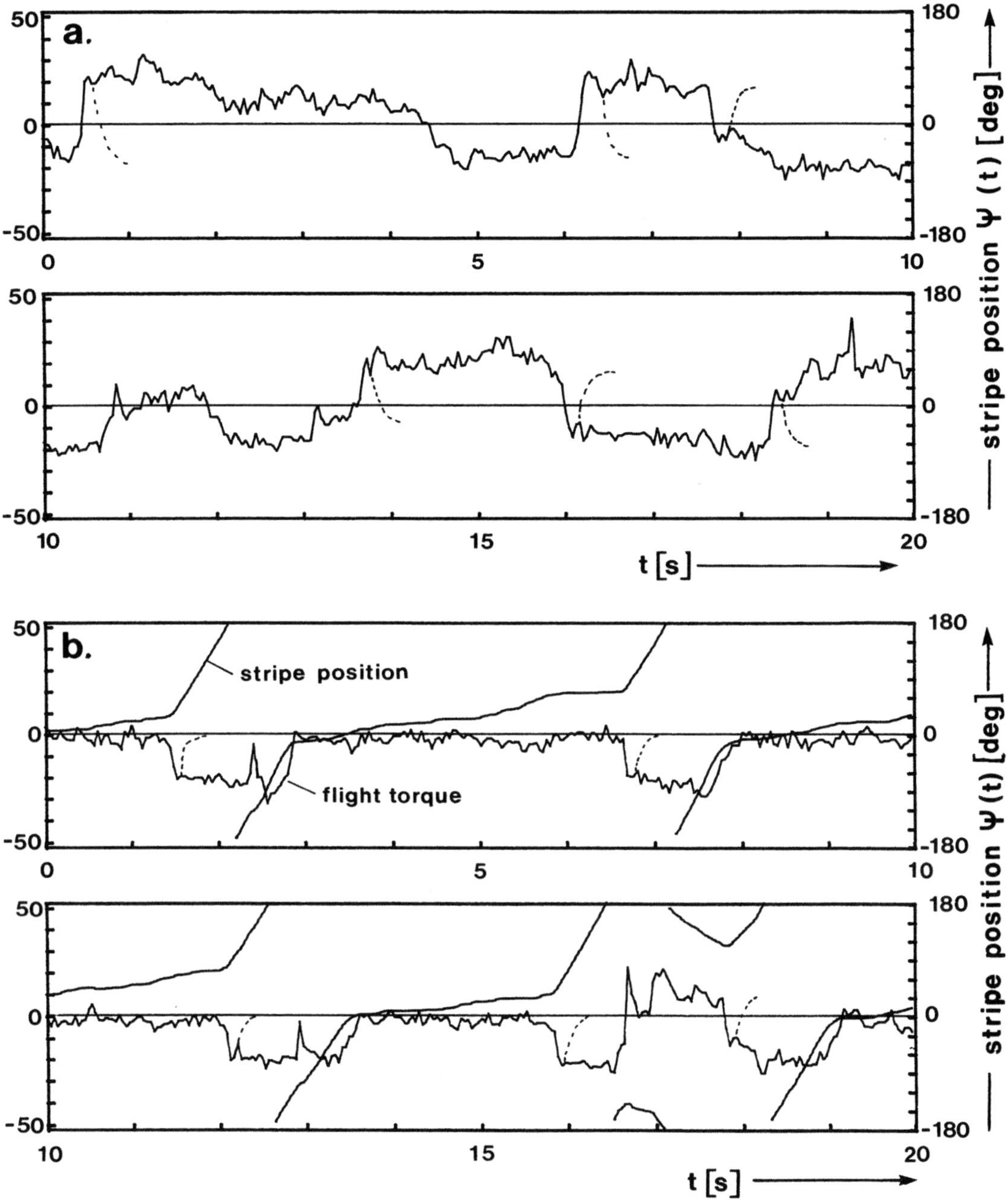

Fig. 50. Stationary flight with free head. The fly is attached to the torque compensator by the thorax only. (a) No moving patterns are presented. In the entire record (of which only 20 s are shown) normal torque spikes are missing. However, abrupt changes in torque are observed which are as fast and of about the same size as the rising phase of torque spikes. (b) Under closed loop conditions with one vertical dark stripe ($\delta = 5°$) (*smooth trace* shows position of stripe) such flies have difficulties in keeping optomotor balance. The sudden torque changes do not decay as do torque spikes and keep the stripe rotating. Torque returns to baseline when the stripe approaches the frontal position. *Dashed* part of presumed torque spikes hidden by subsequent torque activity

As shown by Liske (1977), even passive turns of the head evoke yaw torque responses of the body mediated by the sensory hairs of the prosternal organ. It is thus debatable whether in Land's experiment mentioned above the thorax receives independent turning commands during the head saccades at all, or whether the slightly delayed turns of the body are induced exclusively by the head movements. The latter interpretation is unlikely, however, since even with the head glued to the thorax at least in *Drosophila* and *Musca* (unpublished), torque spikes as the equivalent of body saccades are found.

Rather, what all these results tell us is that body saccades are due to a sensory motor program which assures that the whole fly turns in a coordinated fashion. Head, thorax, abdomen, and hind legs receive independent turning commands, but in addition their turning movements appear to be coupled by posture reflexes.

Reafference Control

The visual system even participates in the saccadic program. This is apparent in experiments with the torque compensator: The time course of torque spikes during closed loop and open loop experiments shows hardly any difference. The reafferent visual feedback stimulation does not shorten the torque spike in a manner similar to the mechanosensory feedback from real rotations.

This raises an old problem which had been discussed 30 years ago (von Holst and Mittelstaedt, 1950; Mittelstaedt, 1951). Do stabilizing reflexes interfere with voluntary movements in the same way as they do with involuntary ones? Mittelstaedt had observed that the fly *Eristalis*, walking inside a rotating striped drum, appears to be forced to turn with the pattern movement, whereas in a stationary drum it seems not to be limited in its own turns. Mittelstaedt rejected the idea that the fly might just be blind during voluntary turns, since twisting the fly's head by 180° and thus inverting the sign of the visual feedback caused continuous rotations of the fly. With the *Principle of Reafference* v. Holst and Mittelstaedt (1950) gave a very elegant explanation by proposing that a copy of the command ("efference copy") might be passed on to the visual system where it compensates for just that part of the visual afference which is due to the organism's voluntary movement (reafference). This allows the organism to respond to all movement due to involuntary turns and to motion occurring in the visual environment (exafference) without interference from its own voluntary motion. There are several ways in which a turning command can interact with a feedback loop, stabilizing the animal against involuntary turning. The special virtue of the efference copy scheme is to make the effect of the command independent of the magnitude of the reafferent stimulus. But the scheme requires that the

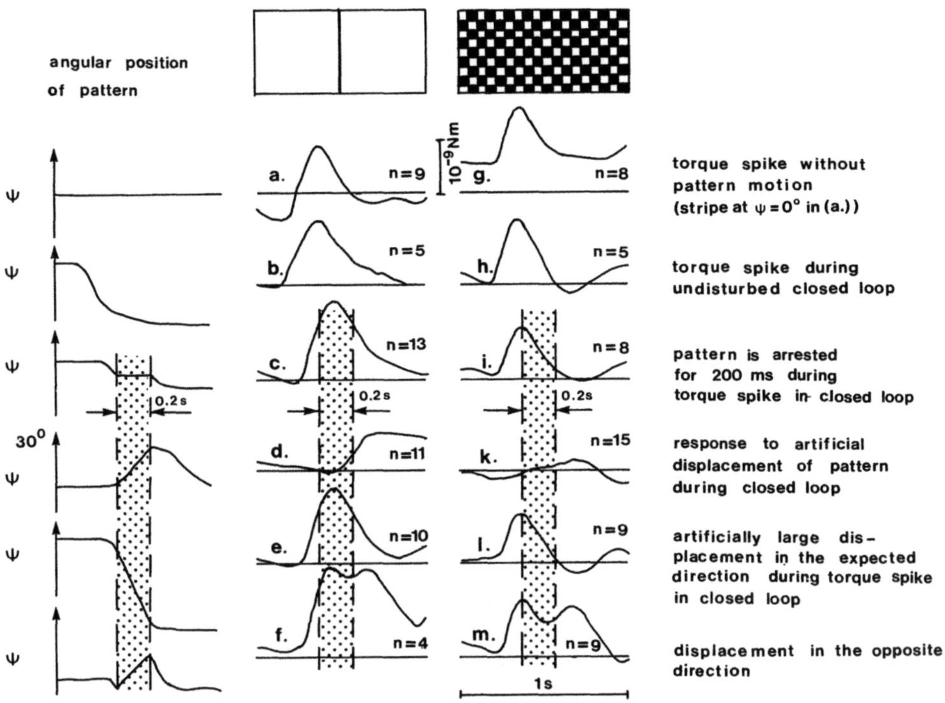

Fig. 51. Responsiveness of a fly is modified during torque spikes. Comparison of (a) or (c) with (b) or (e) shows that the amount of visual feedback stimulation does not alter the time course of torque spikes as long as the pattern moves in the "expected" direction. Motion in the "unexpected" direction (f, m) stimulates a correction response during torque spikes. This is particularly noteworthy in (m) since the displacement of the checkerboard panorama during straight closed loop flight with the same angular velocity normally does not elicit a correction response (k). *n* no. of sweeps. Data in a–f and g–m are from two different flies. (After Heisenberg and Wolf, 1979 where experimental details are explained)

reafferent stimulus can be quantitatively predicted, which only rarely can be the case. If all visual input were inhibited during the saccade, the command were independent of the magnitude of the reafference as well. Since this mechanism does not require predicting the reafferent stimulus, it is error-proof, but obviously implies other dangers.

The similar time course of torque spikes with and without visual feedback suggests that *Drosophila* distinguishes between reafferent and exafferent stimulation. In stationary flight at the torque meter the mechanism now can be studied.

First we have to make sure that the same visual stimuli occurring reafferently during a torque spike do elicit a response if applied exafferently during times when the fly does not generate torque spikes (Fig. 51). To begin with the fly is in closed loop with a single vertical black stripe which happens to be held in an about frontal position. We now open the visual feedback loop for 200 ms and displace the stripe during this time by applying a constant voltage to the motor turning the panorama. Such experiments will be extensively discussed in Chapters 8, 9 and 11. Here it may suffice to say that if we displace the stripe with about the same angular velocity and by the same amount as the fly does during a torque spike, the fly in most instances responds to the displacement with a phasic burst of torque to counteract the disturbance (Fig. 51 d). Thus the visual stimuli occurring reafferently during torque spikes are in an angular velocity range in which the visual system is responsive. Yet no responses to the reafferences are observed during torque spikes.

As a next step we interfere with the closed loop during a torque spike. This is not simple since we cannot tell in advance when the fly will generate one. Fortunately, many flies keep a very smooth torque base line in between torque spikes. This enabled us to design an electronic detector which signals torque spikes a few milliseconds after their onset and allows us to interfere with the visual feedback during most of a spike. The results show that *Drosophila* is neither blind during a torque spike, nor does it possess an efference copy quantitatively matching the reafferent stimulus. It suppresses responses to pattern movement into the direction into which the panorama should rotate during the maneuvre. For movement into the opposite direction, however, the fly is fully responsive (Fig. 51 f). It even responds to certain types of pattern movement into the "unexpected" direction to which it would respond only rarely or not at all during straight flight. Artificial displacements of a stripe from back to front normally lead to responses only in 10–15% of the trials. During torque spikes the probability goes up to over 90% if the displacement is in the "unexpected" direction. Also, patterns subtending a large part of the visual field, if displaced under our experimental conditions, do not elicit phasic torque responses. They do, however, during torque spikes if displaced into the wrong direction (Fig. 51 m; Heisenberg and Wolf, 1979). Thus, for the body saccades of *Drosophila* the problem envisaged by the *Principle of Reafference* has a solution which goes beyond the theoretical predictions in that the responsiveness during torque spikes seems to be profoundly modified in many respects. Vision during fast, active turns appears to be an important matter.

→

Fig. 52. Block diagram summarizing unidirectional saccadic suppression and channeling. During an active (saccadic) turn in clockwise (*cw*) direction the response to the concomitant counterclockwise (*ccw*) reafferent motion is suppressed. At the same time in the channels for clockwise motion gates (filters) are opened which when closed abolish certain motion responses. The "saccade generator" (*SG*) is under the control of proprioreceptors which influence the time course of the saccade. The central brain (under the influence of the periphery) determines their mean frequency and polarity. The "gates" (filters) are also under the control of the central brain. On the motor side not only wing beat amplitude is modulated. Head, hind legs, and abdomen are turned independently into the direction of the intended rotation

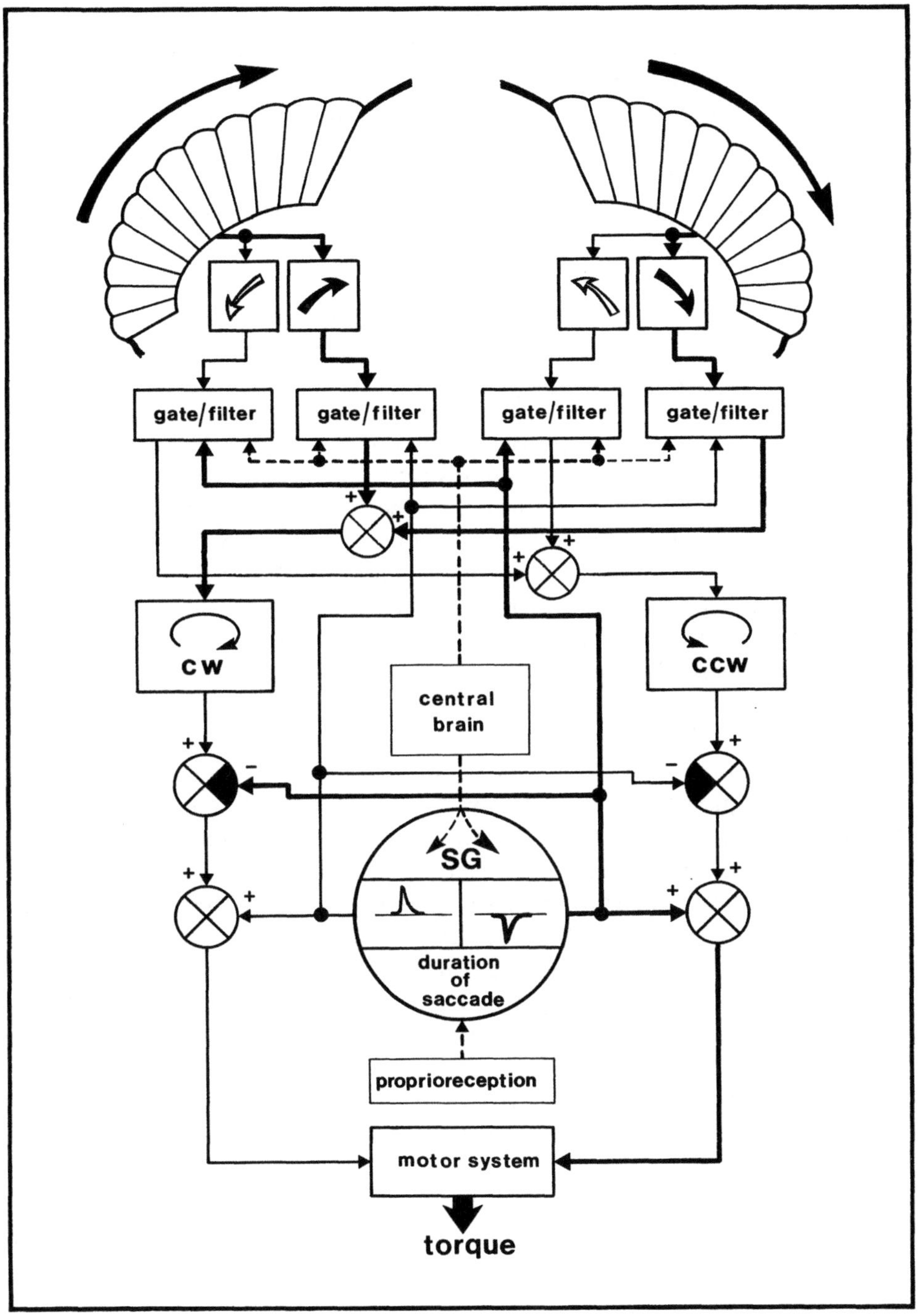
gate/filter
gate/filter
gate/filter
gate/filter
cw
ccw
central brain
SG
duration of saccade
proprioreception
motor system
torque

A further result of these experiments is noteworthy. We have seen above that saccades in free flight are much faster and shorter than at the torque meter, probably due to additional proprioceptive course control mechanisms. The experiment discussed here thus shows that the timing of the modification in the visual system is coupled to that of the motor pattern. This coupling is also apparent in the closed loop experiments with free head where the fly remains caught in the saccade because of the misalignment between head and body. There, too, the fly stays unresponsive to pattern movement into the expected direction for a much longer time than the duration of a torque spike. These functional properties are summarized in Fig. 52.

Spontaneous Occurrence of Torque Spikes

Torque spikes are not responses. As mentioned above, they occur spontaneously. Spontaneous behavior of unrestrained animals in the natural habitat always remains somewhat questionable, since it is nearly impossible to keep track of all possible variables. For behavior at the torque meter we can continuously monitor possible variables such as local turbulences, humidity, electrostatic forces, noise, vibration, odors, visual signals, etc. We have not been able to find stimuli eliciting the torque spikes. In averages of many torque traces aligned in respect to an external event (i.e., a stimulus) torque spikes are normally not found. In Fig. 51 the averages are obtained by pre-trigger analysis, using the spike detector for triggering. The responses in Fig. 51 d, f and m (or, e.g., in Figs. 64 and 86) do not, in our interpretation, qualify as torque spikes since they vary in size and time course, depending upon velocity and duration of the moving stripe. Admittedly this is not an invincible argument, since some responses to displacements of stripes are indistinguishable from torque spikes. Relying only on operational definitions, one would have to retreat to the statement that torque spikes *can* be spontaneous. But we like to assume that torque spikes and responses to the displacement of stripes mostly result from different dispositions and may even involve quite different motor programs, even if the dynamics of torque happens to be the same.

We have proposed regarding body saccades as active, voluntary behavior (Heisenberg and Wolf, 1979) based upon the initiating activity of the brain. Independence of releasing stimuli is regarded as a defining property of voluntariness, whereas reafference control is a characteristic elaboration of this mode of behavior (see the Chap. 13 for discussion of voluntary behavior).

Torque Spike Rules

Actions are not independent of sensory information; but this relation is less direct than in stimulus-response chains. Sensory data modify the proba-

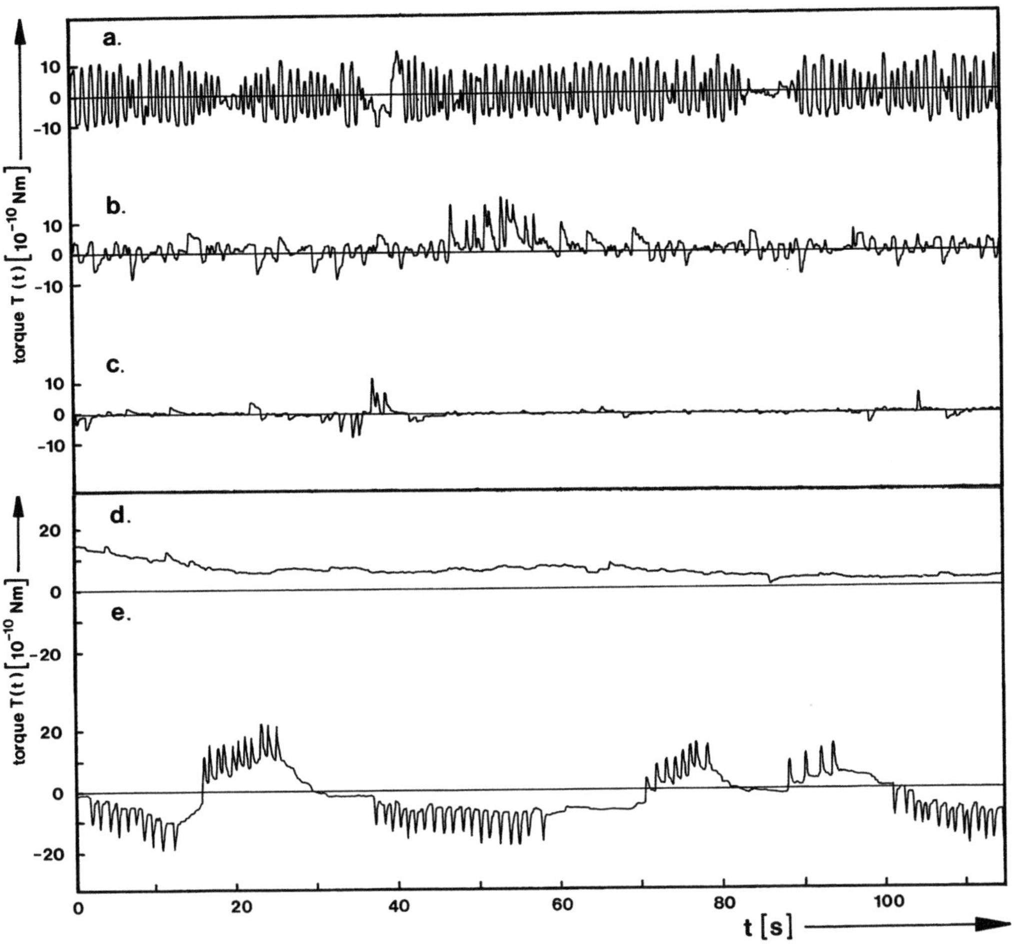

Fig. 53. Typical endogenous patterns of flight torque. (a)–(c): Undisturbed closed loop with a single vertical black stripe. (a) Oscillating mode. (b) Noise mode. (c) Quiet mode; closed loop coupling: 10^{-9} Nm ≙ 55° s^{-1}. (d) In complete darkness torque spikes are particularly rare. (e) In a brightly illuminated arena without moving patterns flies perform what has been interpreted as searching maneuvres. They generate large, slow torque fluctuations superimposed by volleys of torque spikes with the same polarity as the slow modulations (compare torque histogram in Fig. 57b). Note that traces are drawn by a pen recorder introducing a mechanical low-pass filter (cut-off frequency at about f = 3 Hz). For unfiltered torque trace during "quiet mode" see Fig. 81. (After Heisenberg and Wolf, 1979)

bility of an action occurring. Thus also the probability for torque spikes varies with the situation. In total darkness, for instance, torque spikes are rare (Fig. 53d). Under stationary illumination without visual feedback or moving patterns, flies often generate volleys of spikes with the same polarity and similar interspike intervals. They are supported by slow torque modulations of the same polarity as the spikes. We interpret this flight pattern as search for visual stimuli (Fig. 53e).

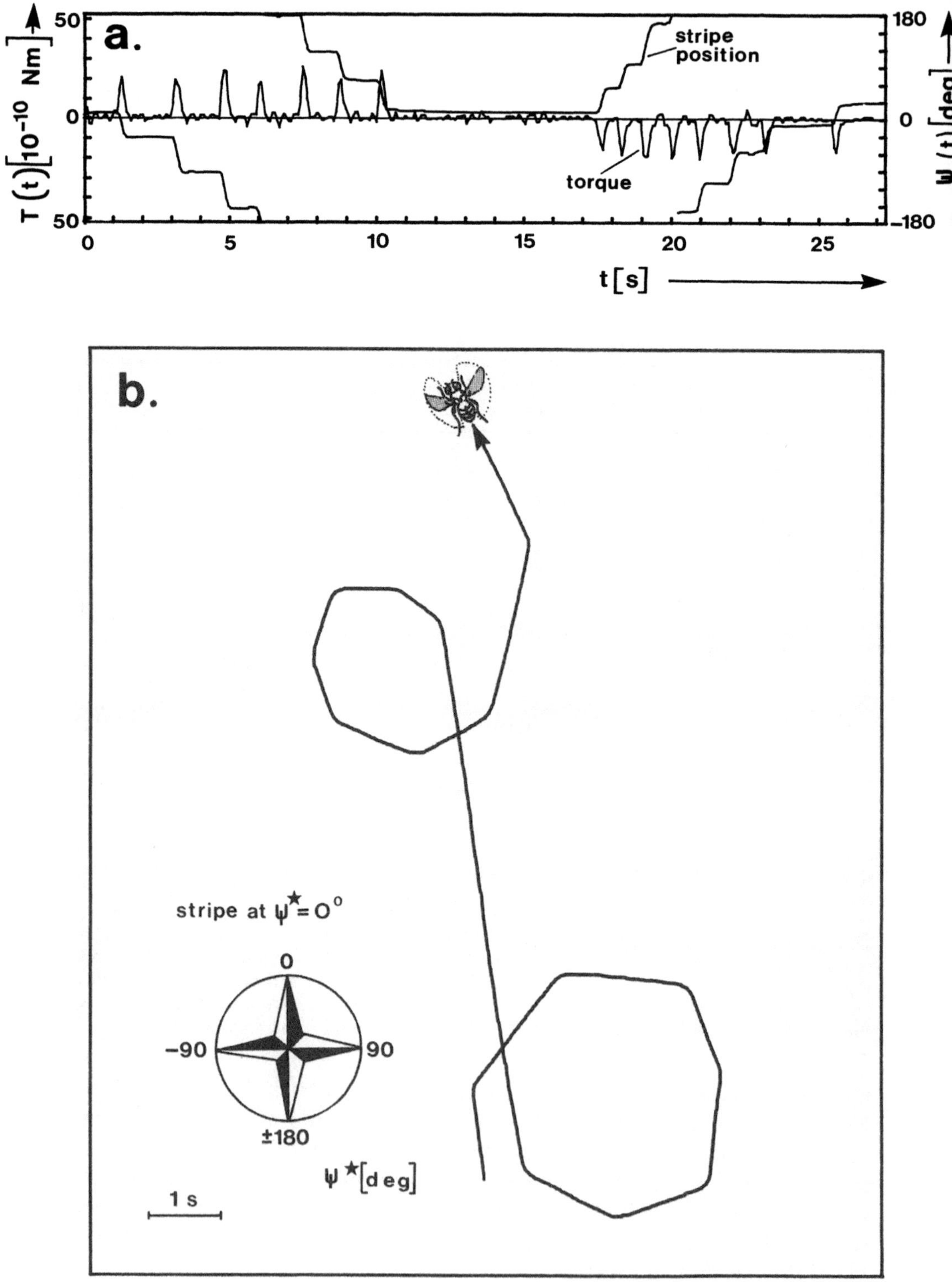
a.
50
0
50
T (t) [10⁻¹⁰ Nm]
180
0
–180
ψ (t) [deg]
stripe position
torque
0
5
10
15
20
25
t [s]
b.
stripe at ψ* = 0°
0
–90
90
±180
ψ* [deg]
1 s

Under normal closed loop conditions, for instance with a single vertical stripe, similar volleys of torque spikes can be observed. The fly generates them for flying 360°-loops (Fig. 54). Here the torque spikes are part of a more complex behavior. We wondered how the loops are programmed, whether the fly possesses a motor program in which the angular shift per torque spike and the number of torque spikes are determined in advance to give a full circle, or whether the fly generates torque spikes until the loop is complete. The latter alternative holds for *Drosophila*. If we reduce the coupling coefficient during such a maneuvre, the fly adds a few torque spikes until the circle is complete. Thus, apparently the visual pattern at the beginning of the loop is stored and must be matched for the volley of torque spikes to be terminated.

Under open loop conditions, the frequency and polarity of torque spikes is strongly influenced by movement stimuli. In a rotating drum flies tend to generate torque spikes against the movement of the panorama (nystagmus). *Drosophila* also has a tendency to direct torque spikes toward dark, vertical stripes. The coincidence of these two conditions, e.g., during back-to-front movement of a stripe, has a particularly strong effect (Fig. 68 on p. 130). It seems that torque spike frequency is particularly high if motion and object position favor the same spike polarity (torque spike rule; Fig. 55).

Chromatic adaptation has a strong effect on torque spike frequency in the eye color mutant *white*. Brief preadaptation with red light drastically increases torque spike frequency, irradiation with white light restores the normal level (Fig. 77b). Finally, it should be mentioned that overall spike frequency and size are characteristic properties of certain mutant strains and to a lesser degree of individual flies.

Flight Modes

Freely flying *Drosophila* has two readily distinguishable flight modes: cruising flight and hover flight. During the former the fly keeps its body axis roughly parallel to the direction of flight and may thus acquire a high forward velocity. During the latter the body axis is tilted rear down by 60° (see p. 95; David, 1978; Heisenberg and Wolf, 1979), and the fly has only small translatory velocities in arbitrary directions. As David (1978) has pointed out (p. 48), cruising flight and hover flight may be considered the two

Fig. 54. (a) Flies often perform 360°-loops by generating volleys of torque spikes. No. of torque spikes is adjusted by the fly to make full loops. (b) Flight trajectory is calculated from the data in (a) assuming forward velocity to be constant. The pattern, a single vertical dark stripe ($\delta = 5°$) in an otherwise homogeneously illuminated background, is assumed to be at infinite distance from the fly (i.e., angular position of stripe does not change during straight flight)

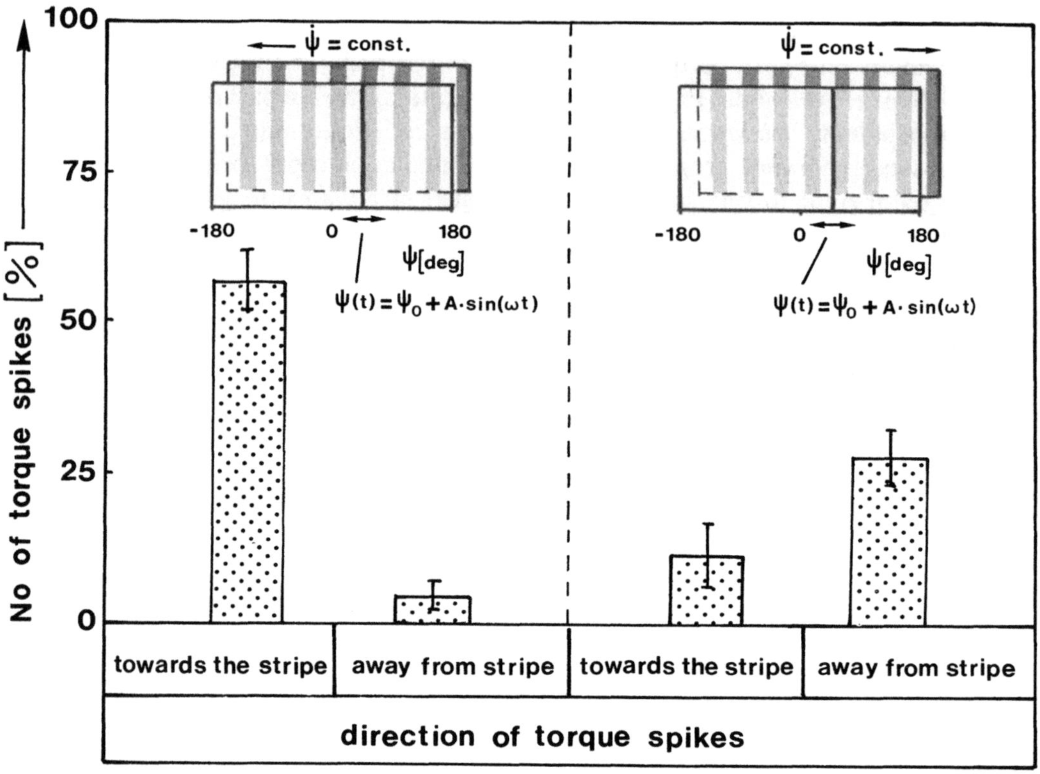

Fig. 55. Frequency of torque spikes depends upon stimulus situation. The fly is flying in the center of a rotating striped drum ($\lambda_{sp} = 45°$; pattern contrast m = 0.1; $w/\lambda_{sp} = 0.4$ Hz). In front of the grating a vertical dark stripe ($\delta = 5°$) is oscillating sinusoidally around the position $\psi = 45°$ or $-45°$ (frequency f = 2.5 Hz; amplitude $A_{p\text{-}p} = 5°$). Data from 48 min of flight time with six flies (WT Kapelle) are pooled for mirror-symmetrical stimulus conditions. *Vertical lines* indicate standard errors of the means among flies. Note the strong effect of the oscillating stripe which without the background grating would have only a marginal influence

extremes of a continuum used for over-ground speed control. How this behavioral variable relates to the different "flight modes" observed with the fly at the torque compensator, where the fly is always kept horizontal and we have nothing to observe but the modulation of torque, remains an open question.

A smooth torque base line in closed loop ("quiet mode", Fig. 53c) is often found after several minutes of the experiment. At the start some flies generate more torque "noise" (Fig. 53b) and fewer torque spikes. Occasionally, flies show the so-called "oscillating mode" (Figs. 53a and 74). They generate large torque fluctuations which keep the stripe oscillating around the frontal position. This torque modulation is due to the visual feedback, since opening the loop arrests the torque at its momentary value (Fig. 74a II, a III). Such flies may suddenly switch to the quiet flight mode, but just as abruptly they may fall back into the oscillating mode after an artificial disturbance. We assume that the oscillating mode reflects an incompatibility between the conditions of our closed loop and certain parameters of the fly's visual system.

It must indeed be only a small fraction of the fly's behavioral repertoire that we can comprehend from observing the torque traces. Most flight behavior will be suppressed with tethered flight and lack of feedback stimulation. (For instance, freely flying flies do not fly straight all the time). Probably other parts of endogenous behavior which occur at the torque meter are not yet understood.

7 Orientation Toward Objects

We have started to discuss the ability of the fly to stabilize its flight course as revealed by its yaw torque behavior at the torque compensator. The fly was surrounded by patterns covering large parts of the visual field such as precisely spaced vertical stripes, random dot textures, zebra stripes or a checkerboard panorama (Fig. 45, p. 90). Since we wanted to study only this first level of orientation behavior, the pattern elements were more or less evenly distributed in order to provide little incentive for the fly to choose a particular direction. The closed loop situation gave the fly the opportunity to gain a small but important amount of orientedness, i.e., "knowledge" about its own presumed angular velocity.

We now try to approach the second level of the investigation asking about the orientation of the fly relative to a singularity in the pattern. In this chapter experiments will be reviewed in which flies head toward such singularities. The question will be raised: How does a fly assume a certain orientation with respect to a landmark? One of the simplest landmarks (at least for scientists) is a single dark vertical stripe in an otherwise homogeneously white arena. Flies have the tendency to fly or walk toward vertical, dark stripes. We will briefly discuss work on large flies showing that this behavior may be brought about simply by the mirror-symmetry of the animal and the properties of the local channels in the lattice of visual elements.

Orientation toward objects in *Drosophila*, however, turns out to be a multi-faceted problem. This is not unexpected. The angular position of an object is represented in the brain with all the detail provided by the eye and, as we have already discussed, orientation behavior can not be understood without considering activity and orientedness. Thus turning toward an object is a matter of needs and choices. (Nevertheless, it will be shown in this chapter that the orientation mechanism worked out in large flies does apply to *Drosophila* under certain conditions as well.)

Let us start with *Drosophila* at the torque compensator confronted with a single stripe under closed loop conditions (see Fig. 40a). The time average of the relative orientations between stripe and fly is depicted as orientation histogram, a plot introduced by Reichardt and Wenking (1969) for large flies. Remember that the fly's orientation is fixed, but the angular position of the stripe is variable. The fly's yaw torque may cause the stripe to rotate horizontally around its vertical body axis. The angle between the fly's longitudinal body axis and the stripe is called ψ (see Fig. 3a, in experiments with a rotating fly it will be called ψ^* as in Figs. 49b–d and 47). The averaged orientation histogram of 4-min flight periods of ten *Drosophila* flies (stripe width $\delta = 5°$, height $H = \pm 40°$; Fig. 56a) shows that the flies prefer to fly toward the stripe $[p_{max} = p(\psi = 0°)]$, however, in this experiment more than 21% of the time they keep the stripe outside the frontal ($-25° < \psi < 25°$) part of the visual field. While the stripe is in lateral positions, its angular velocity is distributed as closely around zero as always during optomotor balance. This is shown in the torque histograms of Fig. 56b recorded in the same flight episodes evaluated in Fig. 56a.

If one now looks at these torque and angular stripe position traces in more detail, one notices that at the onset of the experiment a fly normally keeps the stripe in a frontal position using one of the torque modes described in Fig. 53a–c. After a while, most flies acquire mode c and start shifting the stripe to lateral and caudal positions by means of torque spikes. In between these sudden changes of orientation, the stripe position is generally very stable. The mechanisms involved in this stabilization will have to be discussed in view of the problem of how optomotor balance is maintained. Even during this phase of the experiment the frontal stripe positions are preferred. In many experiments one finds a second slightly preferred position in the back of the fly.

Stabilization of the stripe in frontal position was first discovered with the fly *Musca* by Reichardt (1973), who called it fixation. Taking up this nomenclature for *Drosophila* we originally called the behavioral state in which the stripe is stabilized in other parts of the visual field nonfixation. We will use the term "menotaxis" here (see Chap. 8), since this behavior does not necessarily exclude all kinds of fixation, as will be shown below.

Menotaxis is not found with *Musca* at the torque compensator (Reichardt, 1973, and personal communication). The reasons for this difference between the two species of dipterans are not clear and we can only speculate about them. Certainly free flight behavior of *Musca* and *Drosophila* appears to be very different. Of free-flying *Musca* we mostly notice a behavioral state in which short darts interconnect various landing sites. As mentioned earlier, hover flight is commonly observed in *Drosophila* but only rarely in *Musca*. On the other hand, it is obvious that in free flight neither *Drosophila* nor *Musca* are always heading toward dark vertical objects. Possibly in *Musca* this discrepancy between free and stationary flight is due to the detailed properties of the visual stimulus. However, it is also possible that the special conditions of tethered flight at the torque compensator keep *Musca* in a certain behavioral state and *Drosophila* in another one. [Recently Götz (1983) has begun to explore visual orientation of *Drosophila* by recording wing beat amplitudes (see p. 48f). Using the difference between the amplitudes on either side to control the angular velocity of the panorama, he found fixation

Fig. 56. (a) Orientation histogram; average of 10 4-min flight episodes from 5 flies (for details see text). (b) Torque histograms obtained from the same data of which the orientation histogram was derived. *Bottom* histogram shows torque distribution for total measuring period. *Above* data are split according to stripe position. Histogram in the *middle* indicates torque distribution for the 21% of the time while stripe was outside the (arbitrarily chosen) area $-25° < \psi < 25°$. (Top histogram shows complementary data). Closed loop coupling, 10^{-9} Nm ≙ $55°\ s^{-1}$

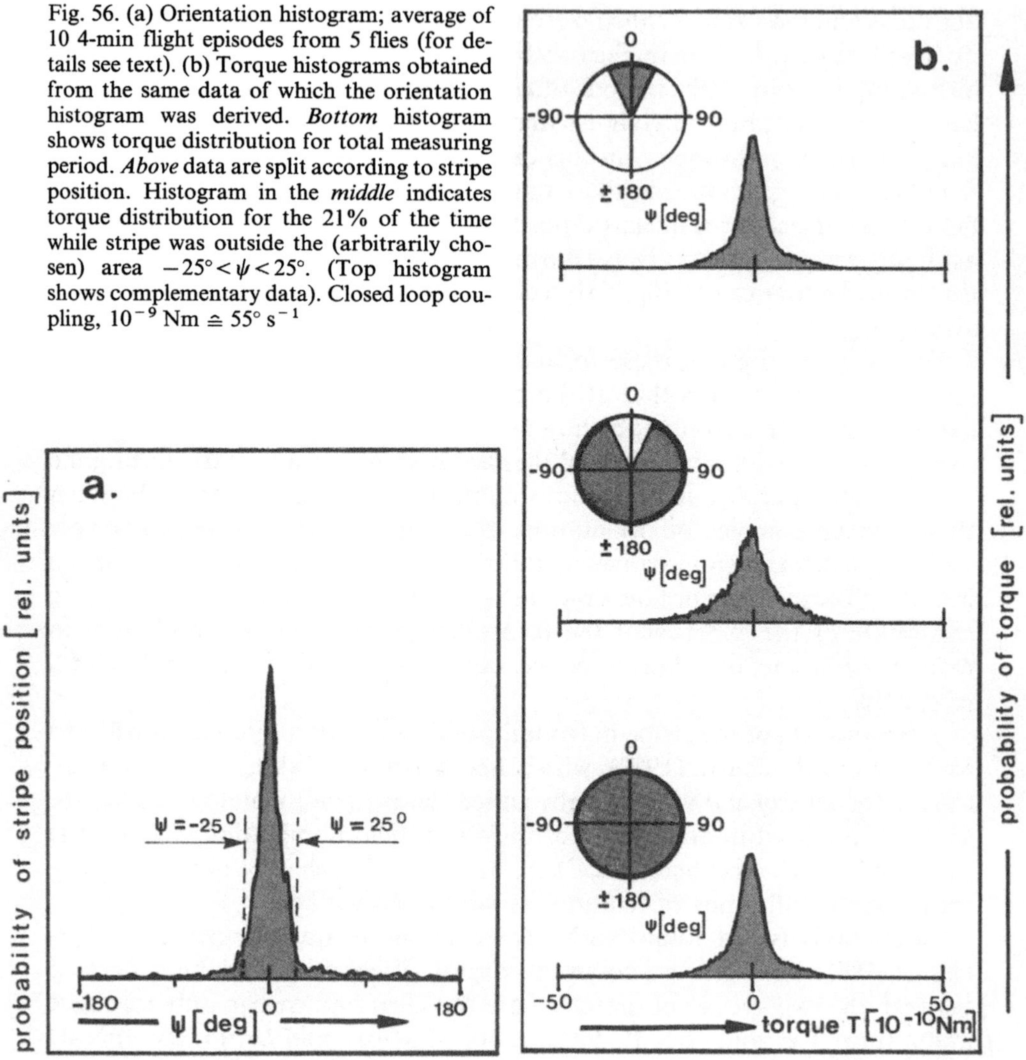

behavior most of the time and only rarely menotaxis. Although these discrepancies are not yet understood they emphasize the importance of the experimental procedure.] The torque histograms of *Musca* and *Drosophila*, measured without any visual motion or flicker stimuli present, also suggest different behavioral states of the two species at the torque compensator (Fig. 57). Part of the torque trace of *Drosophila* leading to the two-peaked distribution is shown in Fig. 53e and is tentatively explained as search behavior.

We will now very briefly review fixation behavior of *Musca* which has been studied over many years in most detail by W. Reichardt's group (see,

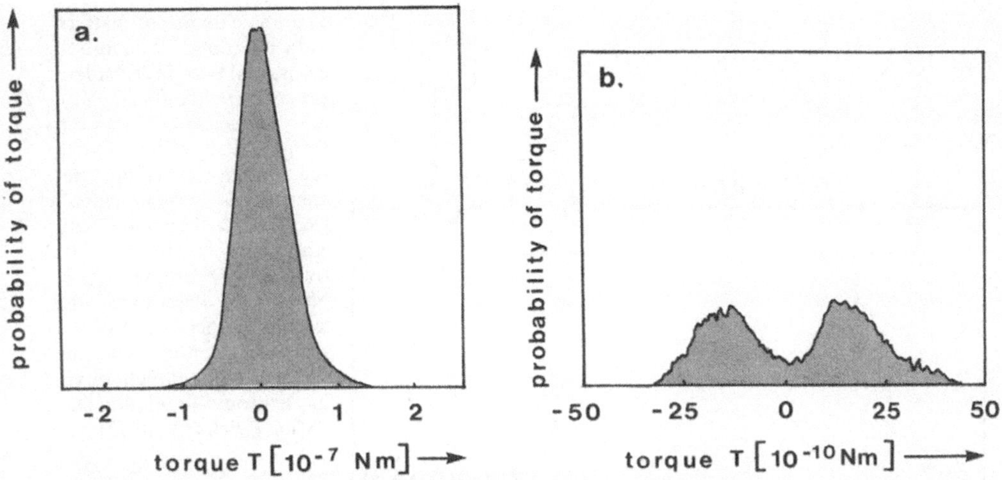

Fig. 57. Torque histograms of *Musca* (a) and *Drosophila* (b) measured in a homogeneously illuminated stationary arena without any pattern. Each histogram is derived from a 2-min flight period of a single fly (*Musca* data from Reichardt, 1973)

e.g., reviews by Reichardt and Poggio, 1976; 1981; Poggio and Reichardt, 1976 b; 1981 a). Subsequently these results will be compared with fixation behavior in *Drosophila*. A comprehensive review on movement and position computation in the fly's visual system is given by E. Buchner (1984).

Fixation Behavior in *Musca*, a Theory

In *Musca* fixation behavior with one vertical dark stripe is well described by the interplay of two otherwise independent processes: "random" torque fluctuations generated by the fly and visual responses of the fly to the resultant movements of the stripe (Fig. 58). For many reasons torque fluctuations are an important part of this process. For instance, in a stationary fly a stationary object does not elicit a significant torque response (Reichardt, 1973).

The visual response essentially consists of torque *toward* the *position* of the object. A mechanical model for this behavior would be a ball rolling in a trough being agitated continuously by a randomly acting force (e.g., Brownian motion).

The two constituents of fixation behavior can be measured in open loop experiments. In an arena without any moving or flickering visual patterns *Musca* modulates its yaw torque spontaneously (Fig. 57). For a theoretical treatment these flight maneuvres can be approximated by a stochastic process N(t) with well-defined properties (Poggio and Reichardt, 1973 a; Reichardt and Poggio, 1975).

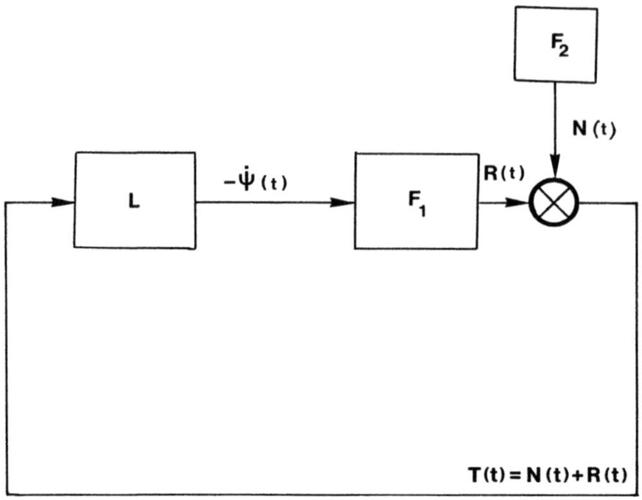

Fig. 58. Theory of pattern induced flight orientation envisages two independent processes in the fly (F_1, F_2), one generating random torque fluctuations [$N(t)$], the other providing responses to visual stimuli [$R(t)$]. In a surround with stationary landmarks the total torque output $T(t) = N(t) + R(t)$ determines the angular velocity $\dot{\psi}(t)$ of the panorama either via the dynamics of real flight or via its electronic simulation (L). (After Reichardt, 1973)

The visual responses involved can be measured in many ways, for instance by slowly rotating a stripe around the fly and recording the torque as a function of its angular position ψ (Fig. 59). Adding and substracting for each value of ψ the response functions for clockwise and couterclockwise rotation, one obtains two new functions which can be visualized in the mechanical model just mentioned. The summation of the local torque values gives the "position function" [$D(\psi)$-function], which would constitute the force driving the ball back toward the bottom of the trough. Integration of the position function leads to the shape of the trough potential $U(\psi)$, indicating the potential energy of the ball at the position ψ (Fig. 59e). The difference of the local torque values [difference function $r(\psi)\dot{\psi}$] is comparable to the friction between the ball and the trough (Fig. 59d). For brevity we will write $r(\psi)\dot{\psi} = r^*(\psi)$.

Independence of the visual responses as to whether the loop is open or closed is a crucial part of the analysis of fixation behavior in *Musca*. The following short summary of the theoretical model describing this behavior (Reichardt and Poggio, 1976) shows the importance of this assumption.

In free flight yaw torque $T(t)$ leads to changes in orientation of the fly [$\psi^*(t)$] with respect to space-fixed zero in as much as it overcomes the fly's rotatory inertia $\Theta\ddot{\psi}^*(t)$ and aerodynamic friction $k\dot{\psi}^*(t)$. This can be written as

$$\Theta\ddot{\psi}^*(t) + k\dot{\psi}^*(t) = T(t). \qquad (1)$$

As mentioned earlier, estimates of both parameters show that for a fly $\Theta\ddot{\psi}^*(t)$ is negligibly small compared to $k\dot{\psi}^*(t)$.

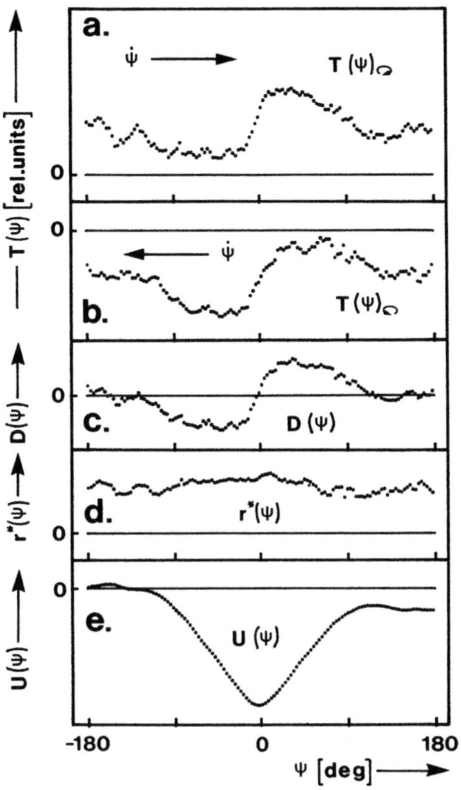

Fig. 59. Musca yaw torque response to a slowly rotating stripe segment (width $\delta = 5°$, height $H = 22.5°$), presented in lower part of visual field. In (a) the stripe moves clockwise in (b) counterclockwise around the fly. Rotation starts at $\psi = \pm 180°$. By adding and subtracting the functions in (a) and (b) for each ψ a "position function" ($D(\psi)$-function) (c) and a "difference function" ($r(\psi)\dot{\psi}$)-function to be written $r^*(\psi)$ for brevity) (d) are obtained. In (e) the integral of the position function, the potential $U(\psi)$ is shown. (c) and (d) are redrawn from Poggio and Reichardt (1976 a), (a) and (b) are not the original recordings. They were reconstructed from (c) and (d) in order to make this figure comparable to Fig. 62

The spontaneous torque fluctuations N(t) and the position function $[D(\psi)]$ as well as the difference function $[r^*(\psi)]$, are taken to contribute linearly to the total torque output:

$$T(t) = D[\psi(t)] + r^* [\psi(t)] + N(t). \tag{2}$$

Equations (1) and (2) lead to

$$k\dot{\psi}(t) = - D[\psi(t)] - r^* [\psi(t)] + N(t) \tag{3}$$

for the closed loop situation. This equation contains several further simplifications as compared to the most general description. First, the second and higher derivatives of $\psi(t)$ are not taken into account. Secondly, the delay between visual stimulus and motor output is neglected. Thirdly, the response of the fly is considered time-invariant. Nevertheless, the validity of Eq. (3) can be proven under various experimental conditions at the torque compensator and the biological significance of the position function being antisymmetrical with respect to $\psi = 0°$ has been shown independently in free flight experiments with tracking flies (see below).

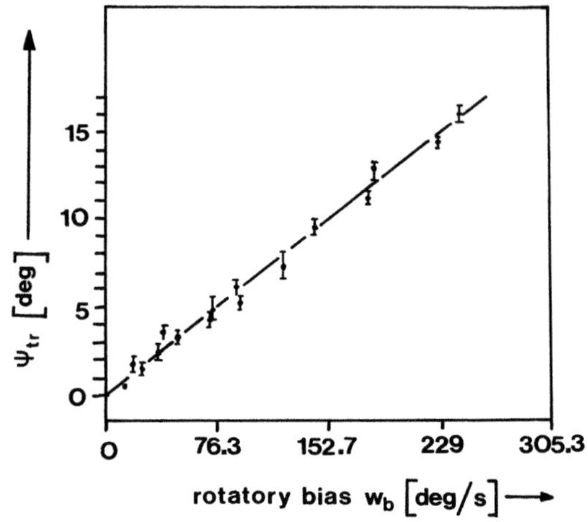

Fig. 60. Error angle ψ_{tr} between direction of flight and stripe position in closed loop as a function of the rotatory bias w_b of the stripe under closed loop conditions for *Musca* (see Fig. 41 for further explanation of w_b). *Straight line* shows the predicted relation between error angle and rotatory bias. (After Virsik and Reichardt, 1976)

It is not essential for the theory (and in fact is equivalent on a coarse time scale) whether the position function is due to flicker detectors monitoring the position of the stripe or whether it is mediated by movement detectors responding more strongly to front-to-back motion than to motion in the opposite direction (Poggio and Reichardt, 1981 b).

Originally, Reichardt (1973) proposed that the yaw torque response to front-to-back motion had a larger open loop gain than the reponse to back-to-front motion. This conjecture was based on the observation that neither a stationary stripe nor a stripe flickering with a certain frequency and pattern contrast elicited an appropriate response. Subsequently, however, Pick (1974) showed that sizable yaw torque responses could be evoked by a flickering stripe without any motion being involved. Thus, the position function was called "direction-insensitive". However, Pick did not determine the dynamic range of the flicker effect. For real flight tracking it seems to be too slow to contribute much to the behavior. The original conjecture appears now to be strongly supported by recent experiments of Wehrhahn and Hausen (1980), and Wehrhahn (1981).

Fig. 61. (a) *Musca* position functions [D(ψ)] and respective potentials [U(ψ)] for different angular separations $\Delta\psi_{sep}$ of two stripes. One or two vertical black stripes (width $\delta = 5°$, height $H = \pm 45°$) are slowly rotated around the fly as in the experiment shown in Fig. 59, but here in addition the pattern is randomly oscillated by maximally $\pm 5°$. The line of symmetry of the patterns is chosen as position $\psi = 0°$. (b) Orientation histograms with one and two stripes separated by the same angular distances ($\Delta\psi_{sep}$) as used in (a). *Left column* shows computer simulation of histograms using the position functions shown in (a). In the *right column* histograms of closed loop measurements are plotted. Note that two maxima in the orientation histogram may be obtained even if the associated potential profile has only one minimum ($\Delta\psi_{sep} = 60°$). For $\Delta\psi_{sep} = 60°$ the angular distance of the two maxima is smaller than that of the stripes. (After Reichardt and Poggio, 1976; for *Drosophila* cf. Fig. 75)

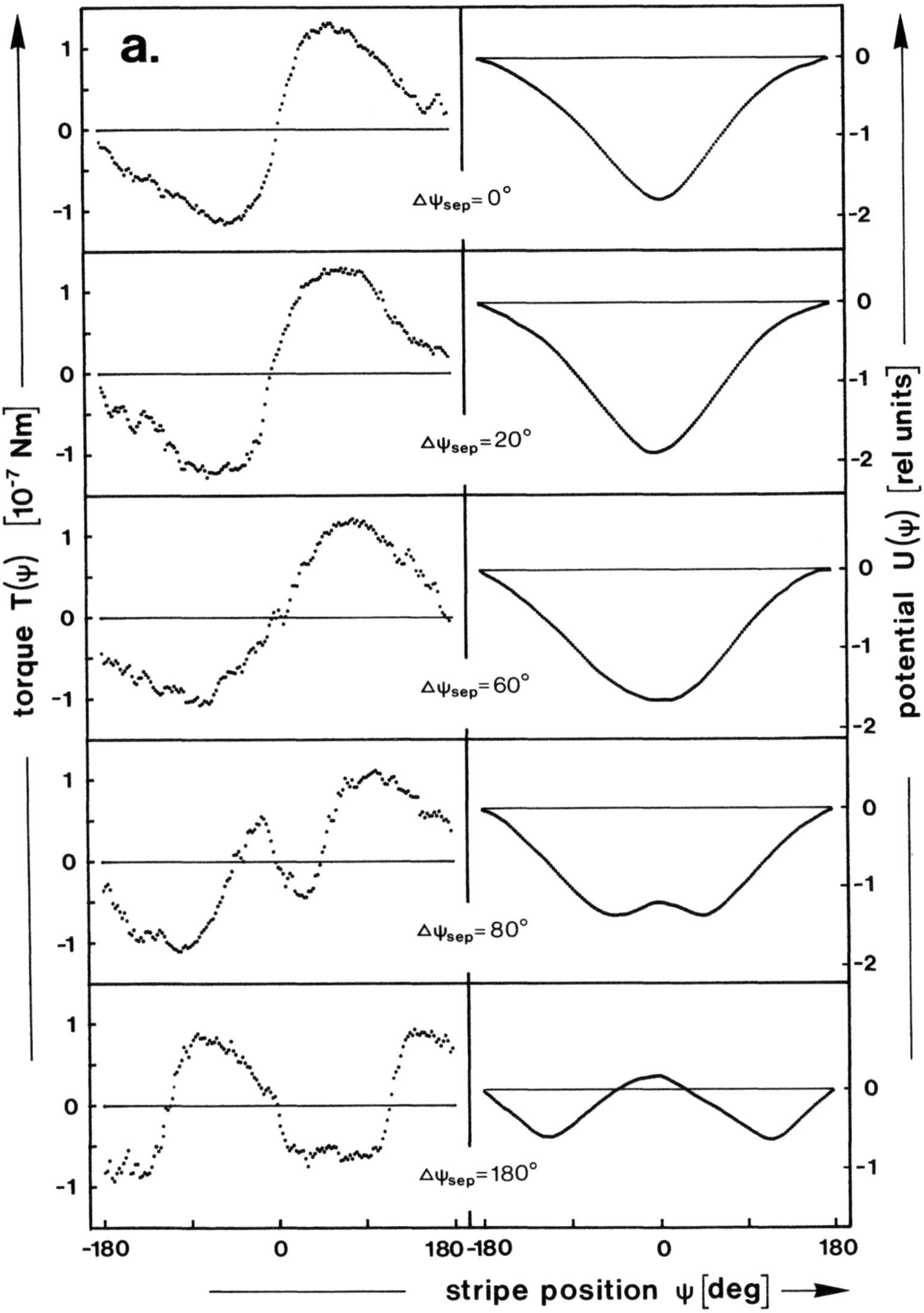
a.
Δψsep= 0°
Δψsep=20°
Δψsep=60°
Δψsep=80°
Δψsep=180°
torque T(ψ) [10⁻⁷ Nm]
potential U(ψ) [rel units]
stripe position ψ [deg]
1
0
-1
-2
-180
180

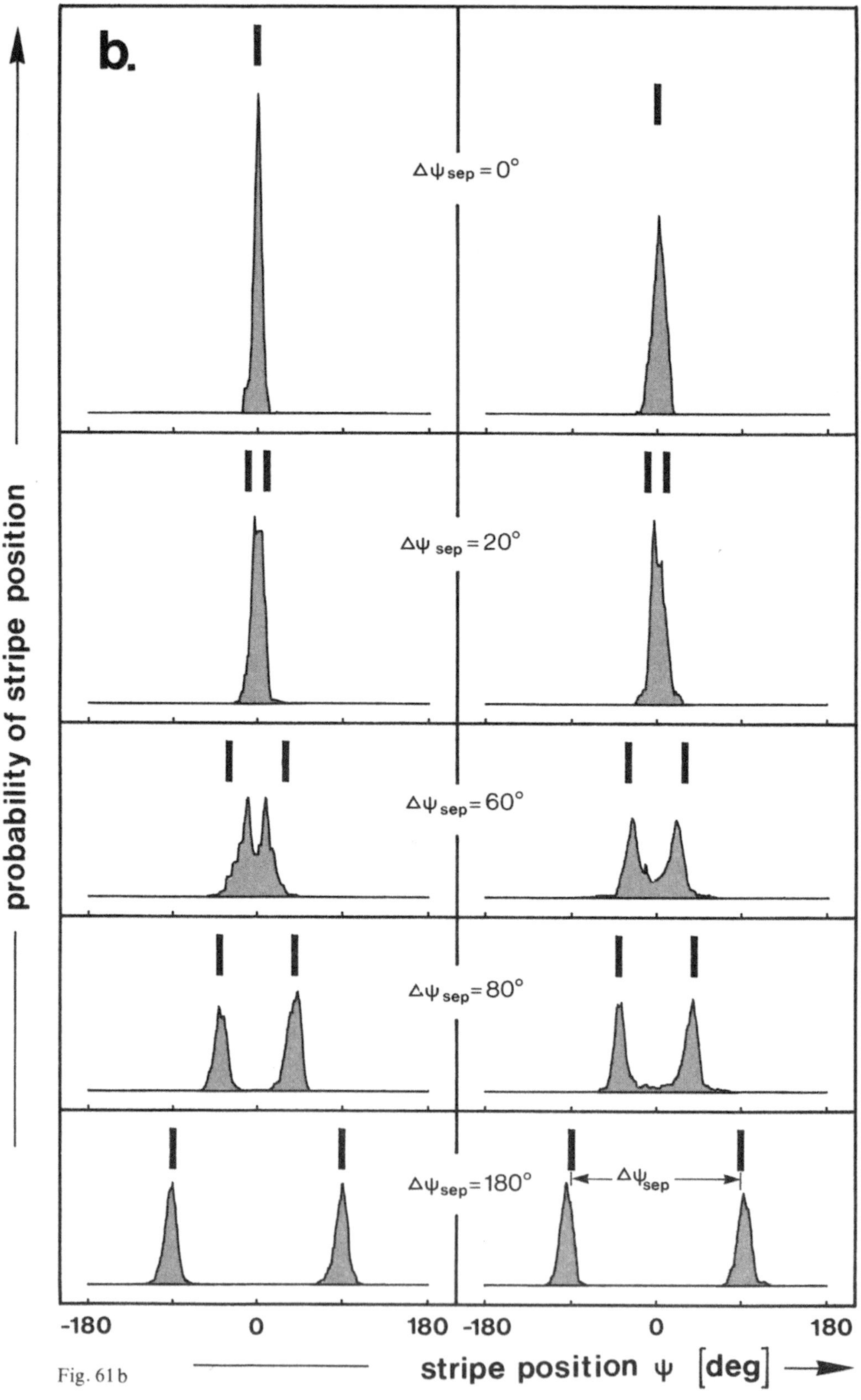

b.
probability of stripe position
Δψsep = 0°
Δψsep = 20°
Δψsep = 60°
Δψsep = 80°
Δψsep = 180°
Δψsep
-180
0
180
-180
0
180
stripe position ψ [deg]

Fig. 61 b

Whatever the cause for the position function, the theory quantitatively predicts the outcome of several experiments at the torque compensator. We will discuss two of them which can be directly compared to equivalent experiments in *Drosophila*. If a D.C. voltage which would give the stripe a constant angular velocity w_b (rotatory bias; cf. Figs. 41, 42) is added into the closed loop, *Musca* fixates the stripe at positions shifted with respect to $\psi = 0°$ into the direction the external signal would turn it (error angle: ψ_{tr}). Applied to the free-flight situation this implies that in pursuing a moving object (stripe) the fly lags behind (Fig. 60; Virsik and Reichardt, 1976). In the model [Eq.(3)] this situation can be well approximated by adding into the closed loop a term for the external disturbance S(t).

$$k\dot{\psi}(t) = -D[\psi(t)] - r^*[\psi(t)] + N(t) + S(t). \tag{4}$$

In the present case $[S(t) = kw_b]$ the error angle ψ_{tr} is proportional to the rotatory bias, as long as ψ_{tr} stays in the linear range of the position function around $\psi = 0°$ (Fig. 59c). In Fig. 60 the good agreement between the theoretical prediction (curve) and the experimental results (dots) is shown.

Another prediction of the theory, a less obvious one though, is that in a fixation experiment with two stripes the maxima in the stripe position histogram should be closer together than the two stripes (for $0° < \Delta\psi_{sep} < 80°$), if one assumes that the torque responses induced by each of the stripes alone add linearly to produce the response to the compound pattern. The reader may convince himself of this property of the theory by superimposing two potential troughs as shown in Fig. 59 with an (angular) distance of $\Delta\psi_{sep} = 60°$. The effect is indeed found experimentally (Fig. 61; Reichardt and Poggio, 1976).

The Object Response in *Drosophila*

Let us now return to the orientation behavior of *Drosophila* under these conditions. As we have already seen, two of the central premises in the model, the invariance in time of the behavior at the torque compensator and the equivalence of open loop and closed loop experiments, are not realized in *Drosophila*. However, it is not obvious off hand how deeply these differences influence fixation behavior. The theory might still provide a good approximation. We will in a later chapter go through the constituents of the model and see to which degree they are realized in the small fly; but our analysis in *Drosophila* will have to proceed along different lines. We will go back and forth between open and closed loop experiments in the hope of learning from one type of behavior about the other, even if equivalence cannot be assumed. As in the case of course control, we start with the discussion of open loop situations and of data which are averages of many identical experiments.

In this way, observations are strictly confined to stimulus–response chains. The biological significance of such responses is difficult to grasp, since open loop behavior hardly ever occurs in real life. On the other hand, if under such adverse laboratory conditions an animal reliably responds again and again to a stimulus, one can be fairly confident that this behavior must have an important function in real life.

Special Role of Front-to-Back Motion

We first consider the open loop response to a single rotating stripe by recording yaw torque as a function of position and direction of motion of the stripe. As mentioned above, this experiment has been used in large flies to measure the position function (Fig. 59), showing that equidirectional torque (i.e., torque which would turn the fly into the same direction as the direction of the stimulus motion) is larger while the stripe moves from front to back than during motion in the opposite direction. Although the experimental conditions used are slightly different, the important features of the position function in *Drosophila* are quite similar to those of *Musca*. The sign of the function differs for the two visual half-fields. By definition it is negative in the left and positive in the right visual half-field, the slope of the curve being positive at $\psi = 0°$ (Fig. 62). *Musca* and *Drosophila* may differ in the relative response amplitude for stripe positions in the back of the animal (see Heisenberg et al., 1978). However, at high angular velocity and when no rest periods are interspersed between clockwise and counterclockwise rotation motion, after effects have to be taken into account. For instance, in Fig. 62c the additional increase of torque for $\psi_{abs} > 140°$ is due to such after effects, as control experiments have shown.

For stationarily walking *Drosophila* Bülthoff (1980; 1982 b) has tried to measure the same function (Fig. 72). The difficulty with such experiments is that the fly decides when to walk and when to rest. Thus the stimulus sequence has to be repeated over and over again to obtain a reliable response average which normally reflects very long measuring periods. Using as a stimulus a small black square ($5° \times 5°$) rotating at $200° \, s^{-1}$ Bülthoff found a position function which had no positive slope at $\psi = 0°$ and seemed, judging from its shape, to be dominated by motion after effects. At slower angular velocities, the position functions were altogether very shallow.

So far, no clear-cut evidence is available showing that with a vertical stripe a position function similar to the one recorded in flight can be obtained with stationarily walking flies. In an earlier study with *Drosophila* walking on a tread compensator and with moving gratings in restricted parts of the visual field as stimuli, Götz and Wenking (1973) had found dynamic differences in the turning responses to front-to-back and back-to-front motion. At low

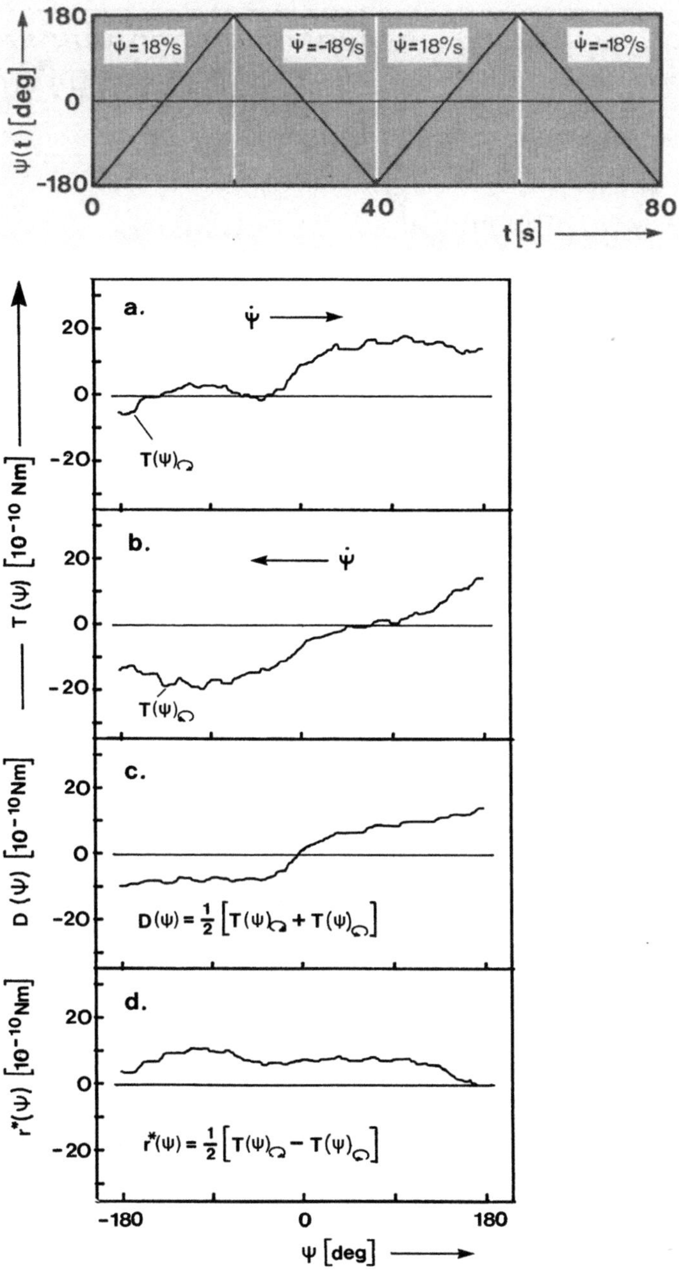

Fig. 62. Position function (c) and difference function (d) of *Drosophila* (for *Musca* see Fig. 59). A vertical dark stripe (width $\delta = 5°$, height $H = \pm 40°$) is rotated (angular velocity $w = 18° s^{-1}$) around the fly. As shown in the upper diagram one stimulus sequence consists of four full circles, two of clockwise (a), two of counterclockwise (b) rotation. Traces in (a) and (b) show for each direction of motion local torque averages from 40 such sequences as functions of stripe position ψ measured with 13 flies (WT Berlin). Functions in (c) and (d) are obtained by adding and substracting functions (a) and (b) for each value of ψ

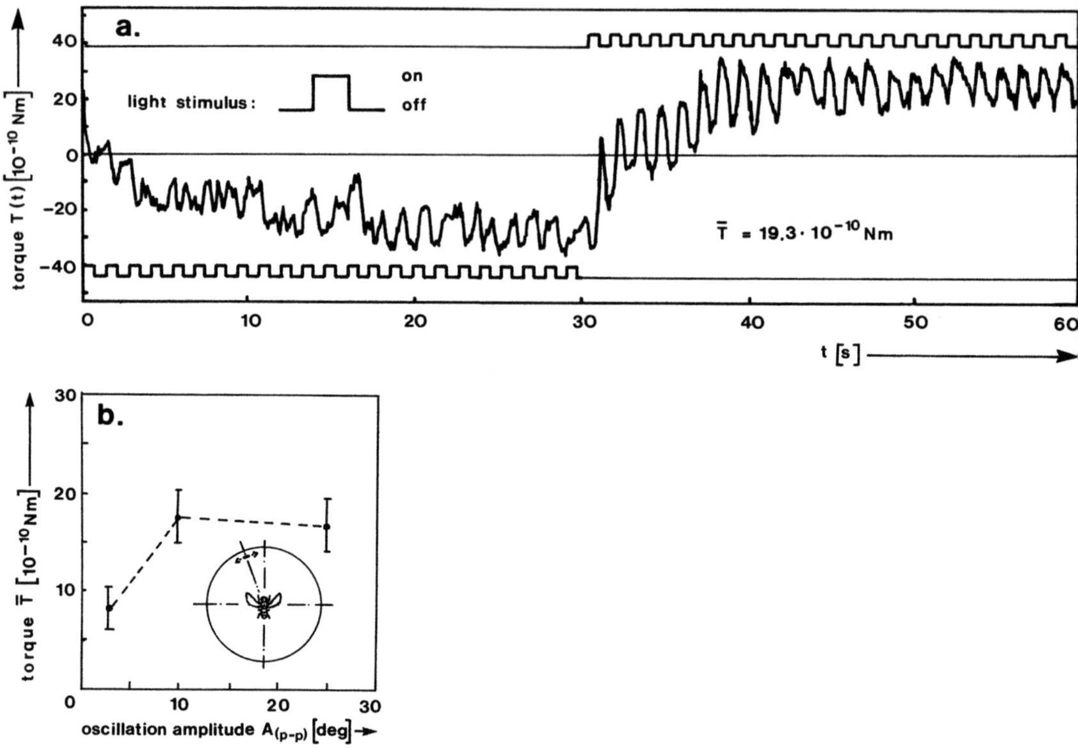

Fig. 63. (a) Yaw torque responses toward bright flashes of light in complete darkness. Light from a 2.2 V, 0.25 A tungsten bulb reaches the fly through one of two optical wave guides (D = 10°) directed toward the fly at the angular position $\vartheta = +45°$ and $\psi = +45°$ (or $\psi = -45°$). The fly shifts the base line of torque toward the side from where the flashes come and in addition responds to nearly each flash by trying to turn further into the same direction. Single record from one fly (WT Berlin). (b) Yaw torque toward a vertical black stripe ($\delta = 3°$; $H = \pm 45°$) oscillating (with small amplitude and frequency f = 1 Hz) alternatingly at position $\psi = 45°$ and $\psi = -45°$. Data show half the difference between torque responses at the two stripe positions and are averages from ten 2-min flight periods at each position and for each oscillation amplitude [6 WT (Berlin) flies]

angular velocities of the grating back-to-front motion and at high angular velocities front-to-back motion elicited more turning. This result suggests that the responses to the two directions of motion are mediated by differently tuned unidirectional EMD's. How it relates to the position function recorded in flight remains to be investigated. We will return to this problem in the context of the closed loop situation.

Since the work of Pick (1974; 1976) local flicker detectors (i.e., detectors for changes in light intensity) are known to contribute to the position function in *Musca* at least in a certain range of angular velocities of a moving stripe. In *Drosophila* yaw torque responses to flickering light have been little investigated. A bright flash in the dark under certain conditions induces a large torque response toward the light source (Fig. 63 a). Whether this effect is mediated by a lattice of flicker detectors is questionable. Bülthoff and Götz (1979) have elegantly demonstrated that flickering intensity gradients cause an illusion of motion for the fly. Apparently in the input channels to the movement detectors signals are distorted in such a way that (in this case) local intensity differences are interpreted as phase shifts. In any event, with moving stripes only small flicker effects are observed (Götz et al., 1979). A stripe oscillating at a lateral position by less than the divergence angle $\Delta\varphi$ induces but little torque toward that side (Fig. 63 b). As in *Musca* the position function of *Drosophila* seems to be mediated in large parts of the dynamic range primarily by unidirectional movement detectors.

An experiment relevant in this respect was performed many years ago by Zimmermann (1973). As a stimulus, a bright field (height: $H = 30°$; width: $\delta = 36°$) moves over an oscilloscope screen which is imaged by wave guide optics onto an arena surrounding the fly. After some time of complete darkness the luminous field appears at $\psi = -135°$ and moves forward, where it disappears at about $\psi = -20°$. Next it appears at $\psi = +20°$ and moves front-to-back disappearing at $\psi = +135°$. Finally the back-to-front stimulation is repeated. To each event *Drosophila* wild type responds with a short burst of equidirectional torque and, in the case of front-to-back motion, with an additional long-lasting change in the torque base line which takes more than a minute to return to its original value (Fig. 64 a). We have looked for this after effect with the set-up used in most of the experiments described here. When a dark or gray stripe rotating around the fly is suddenly stopped, a similar long-lasting shift in the torque base line can be observed (Fig. 64 b). This effect is less pronounced with a striped drum. The Zimmermann experiment gives no indication of a flicker response. What it suggests is the existence of two types of direction-sensitive movement responses, one with a fast decay time, being stimulated equally strongly by motion from back to front and motion from front to back, the other with a slow decay time, being sensitive only to motion from front to back. The two kinds of responses seem to interact additively.

Range of Stimulus Parameters for the Position Function

In flight experiments with a single stripe essentially similar position functions can be recorded for a broad range of conditions. Irrespective of the particular angular velocity, irrespective of whether the stripe moves smoothly in one direction or is oscillating at random, the function is antisymmetric for the two visual half-fields and has a positive slope at $\psi = 0°$.

Even with gratings subtending certain fractions of the visual field position functions can be obtained (for *Musca*: Wehrhahn and Hausen, 1980; for *Dro-*

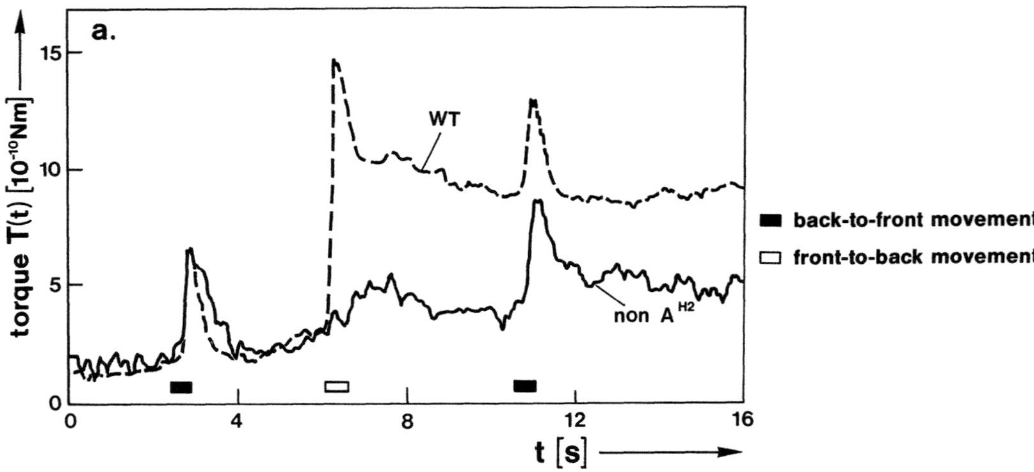

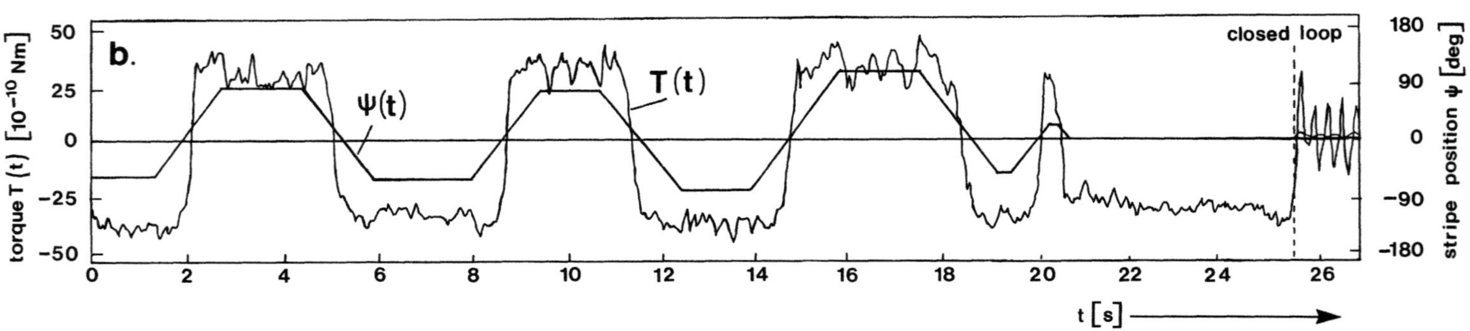

Fig. 64. (a) Yaw torque in response to a luminous moving field (30° × 36°) consists of a fast-decaying and a long-lasting component. The long after effect is stimulated only by front-to-back motion, whereas the fast decaying one is stimulated by front-to-back as well as back-to-front movement. (Zimmermann, 1973). In the mutant *no on-transient* A^{H2} (*non A*) specifically the response to front-to-back movement is impaired. (Heisenberg, 1972). (b) A similar long-lasting after effect is obtained by front-to-back motion of a vertical gray stripe ($\delta = 5°$; $H = \pm 40°$) in a brightly lit arena if the stripe is abruptly arrested (WT Berlin)

sophila: unpublished). In such experiments the differences of the responses to front-to-back and back-to-front motion are determined at various positions (ψ). In *Drosophila* a grating measuring 90° × 90° and extending unilaterally from the frontal midplane to the side elicits about twice as large a response to motion from front to back than to motion in the opposite direction (Götz, 1975 b). Interestingly, in recent studies under similar stimulus conditions no difference in response amplitude was found (Götz et al., 1979). We will see below that the ambient illumination is an important factor influencing the position function.[4] This may have been one of the uncontrolled variables distinguishing the two experiments. These experiments indicate that the position function is not a fixed property of the large field course control system. Unidirectional motion detectors with adjustable gain must be involved.

Two Responses Contribute to Yaw Torque

Let us consider the following two possibilities: The fly might possess independent control systems for motion from front to back and for motion from back to front, each with adjustable gain depending upon the specific properties of the stimulus and the state of the fly. Alternatively, one could assume that yaw torque consists of a symmetrical direction-sensitive response, for instance the large field course control response, and in addition of a unidirectional response to motion from front to back. For the second model to be a real alternative to the first, we have to assume different stimulus characteristics for the symmetrical and the unidirectional response. To the symmetrical response, for instance, any moving contrast in the visual field, be it texture or object, might contribute, whereas for the unidirectional response to be effectively stimulated a well-discernible landmark might be required.

Direct and indirect evidence has accumulated in favor of the second model. The position function seems to be due to a response which can be experimentally separated from the large field course control response. We assume it to be directed toward isolated objects and thus call it object response. The term "position function" used so far refers to the experiment of Fig. 59 or similar ones. The term "object response" alludes to the biological significance of the response causing the position function. This will be dis-

---

[4] The effects of ambient illumination on torque and turning responses have been little investigated. Pick and E. Buchner (1979) noticed in *Musca* movement-sensitive wide-angle interactions of negative polarity which were abolished by illumination of the intervening visual elements. They proposed lateral inhibition (Arnett, 1972; Zettler and Jaervilehto, 1972) distal to the lattice of EMD's as an explanation. Wolf et al. (1980) discovered that the turning tendency of walking *Drosophila* in response to motion in a small fraction of the visual field depends upon the E-vector orientation of the stimulating light. This response modulation is abolished if the rest of the visual field is kept dark.

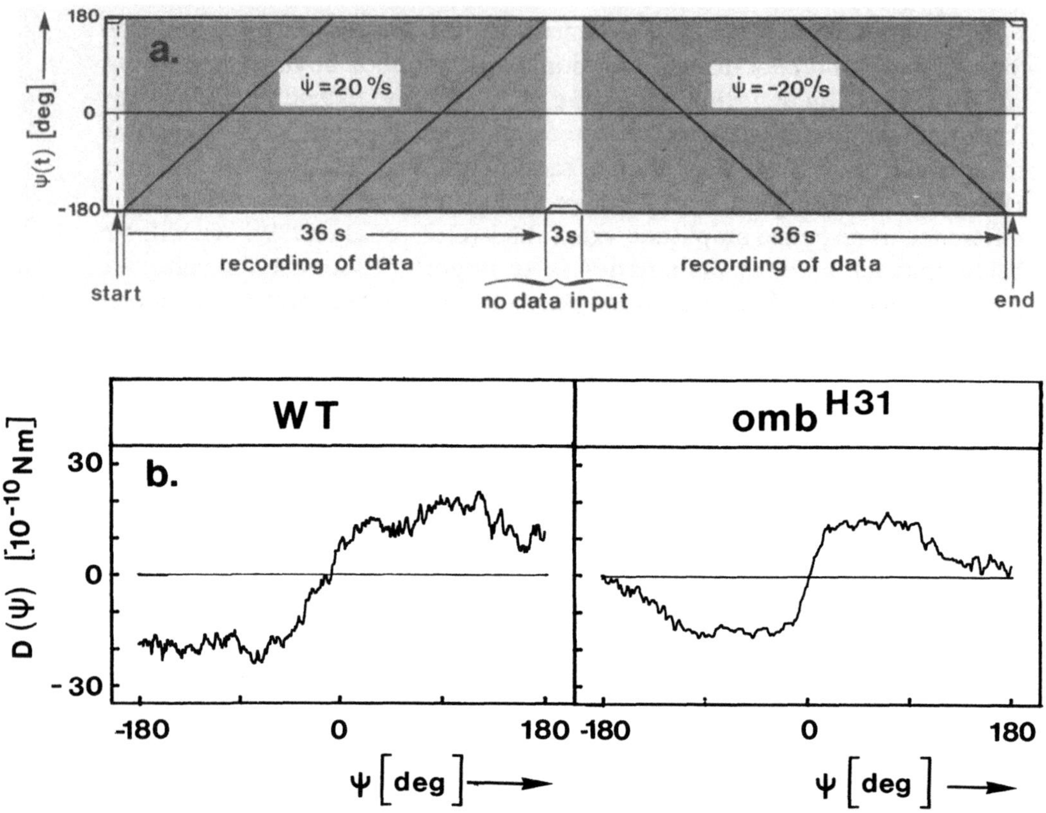

Fig. 65. Position functions (b) of WT (Berlin) and *omb*. Stimulus sequence (a) is slightly different from that used in the experiment of Fig. 62. Traces are averages of 7 (WT) and 9 (*omb*) experiments with 7 WT and 5 *omb* flies; width of stripe $\delta = 5°$; height $H = \pm 40°$; angular velocity $w = 20° s^{-1}$

cussed at the end of Chapter 9. In the same manner we use "optomotor response" as an operational term referring to particular experimental situations. The underlying "large field course control response" again refers to the biological function of this response and suggests that it is due to a distinct neural unit in the brain.

The best indication of two separate direction sensitive movement responses comes from neurogenetics. Many flies of the mutant strain *omb* in which the giant fibers of the lobula plate (HS- and VS-cells) are missing (see Fig. 34b) display a normal-sized position function (Fig. 65), while only about 30% of the torque response to a periodic grating is left (Fig. 33). Most of this remaining torque can be accounted for by the object response to the stripes in the grating. This is shown in the following experiments. A stationary screen of $\delta = 120°$ in width reaching from the frontal position to the side on which stripes are moving front-to-back completely abolishes the response in *omb* but not in the wild type (Fig. 66). Thus *omb* does not respond signifi-

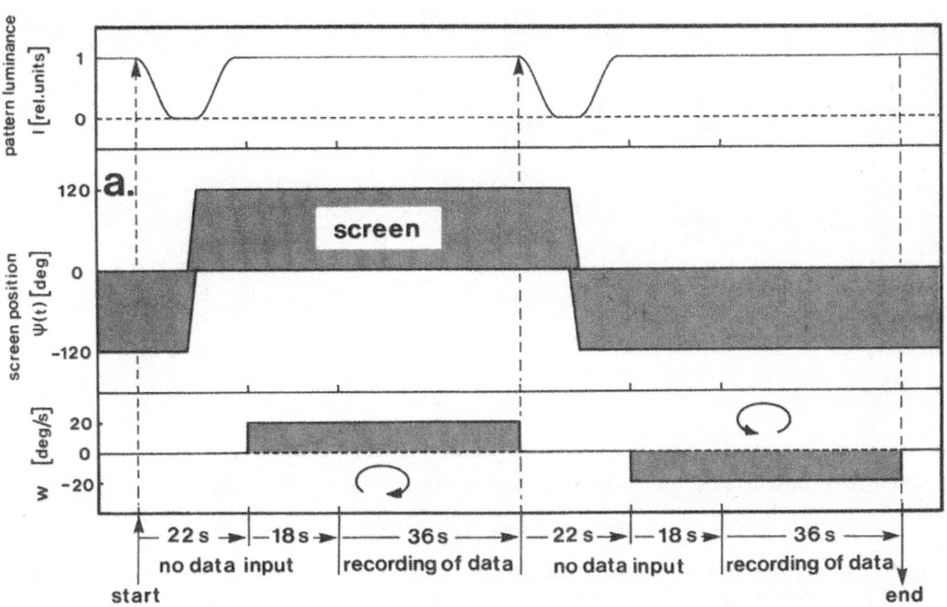

Fig. 66. Optomotor torque responses of WT (Berlin) and *omb* to a rotating striped drum ($\lambda_{sp} = 22.5°$; $m = 0.3$; $w = 20° s^{-1}$). *Upper diagram* (a) explaining the experimental procedure refers to *1st* and *2nd* row of (b). The screen is moved to the desired position in the dark in order to avoid motion after effects. Since switching the light on and off causes torque responses as well (see Fig. 63) pattern luminance is changed gradually by means of two polarizing filters. In this sample of flies *omb* shows 60% of WT response [*3rd* row in (b)]. Experiments with a stationary white screen [*1st* and *2nd* row in (b)] show that mutant response is entirely due to front-to-back motion in frontal part of visual field. The small negative response of *omb* to back-to-front motion (1st row) seems to be elicited by motion in the back part of the visual field. (K.G. Götz, unpublished). *Numbers* are time averages of torque responses (as the inset numbers in previous Figs., e.g. Fig. 22) and are averages of 10(14) measurements with 5 WT and 7 *omb* flies

b.	$\bar{T}$ [10^{-10}Nm]	
	WT	omb^{H31}
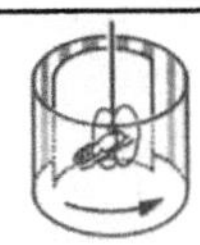	15.4±1.1	-5.9±1.4
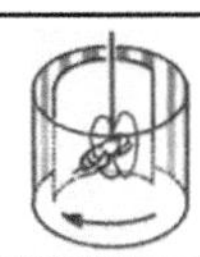	20.8±1.5	10.4±1.9
	20.9±1.6	12.5±1.1

cantly to back-to-front movement, whereas large field course control in the wild type is sensitive to both directions (see Chaps. 3 and 4).

Secondly, experiments of Bausenwein (1984) indicate that in *omb* the yaw torque response is mediated by EMD's with a sampling base larger than that of nearest neighbors along the x- and y-axes in the lattice of visual elements.

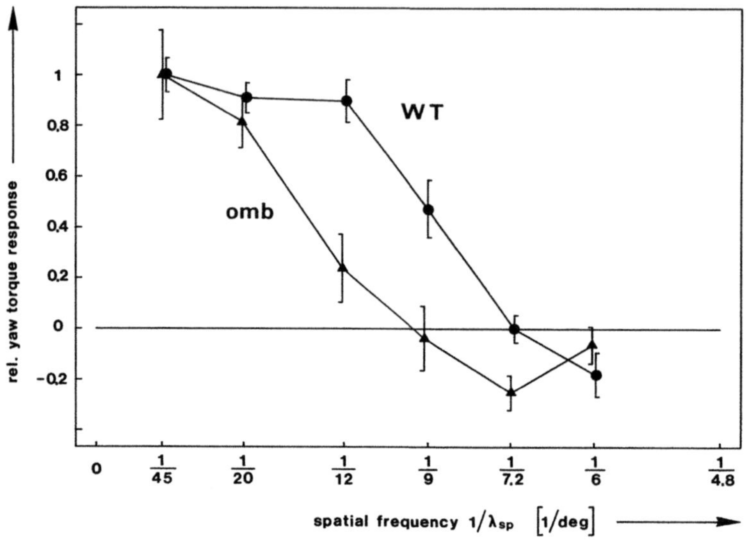

Fig. 67. Optomotor yaw torque responses (open loop) of WT (Berlin) and *omb* as functions of spatial frequency of grating. Stimulated area of visual field (H = ± 40°) extends from $\psi = 15°$ to $\psi = 60°$ (or alternatively from $\psi = -15°$ to $\psi = -60°$). Each response value is derived from 450 s measurements: At each of the two positions of the grating torque is recorded separately for front-to-back and back-to-front movement. Values are differences of the responses to the two directions of motion. Position of grating is shifted in the dark. Average luminance of stimulated area is slightly above that of the unpatterned screen. Data are means of 10 experiments with 10 WT and 10 *omb* flies (15 flies at $\lambda_{sp} = 7.2°$). Vertical bars: standard errors of the means. (After Bausenwein, 1984)

In our present stock the mean sampling base in the fronto-lateral part of the visual field is about 1.5 times that found for wildtype yaw torque responses (Fig. 67). However, in the mutant flies visual acuity for yaw torque seems to undergo a slow drift. Identical measurements with the textures shown in Fig. 45b and e in 1982 and 1984 revealed a small decrease of the mean sampling base (see also p. 131).

Nevertheless, the missing response to back-to-front movement and the reduced visual acuity for yaw torque in *omb* tell us that large field course control and the object response are two distinct control units and that the object response neither requires the large field course control response nor the HS-neurons of the lobula plate to operate (for further discussion see p. 170 ff). We will see below that also in wild-type flies independence between the two responses can be shown in behavioral experiments.

We have stated above (p. 120) that the object response has not been observed so far in stationarily walking flies. This nicely explains the apparent discrepancy in the optomotor yaw torque and turning responses of *omb* flies in flight and on the locomotion recorder (Figs. 32 and 33). If the remaining yaw torque in *omb* is due to the object response alone, an equivalent turning tendency should be missing in the walking mode.

Geiger and Poggio (1975) discovered that the position function in *Musca* is not present in newly hatched adult flies, but needs 11 days to develop fully. No such development is observed for the difference function in the same experiment and it is known from independent pilot studies (Reichardt and Götz, unpublished) that the optomotor yaw torque response is well developed in very young flies. As mentioned earlier (pp. 67, 69), Geiger and Nässel (1981, 1982) have been able to eliminate in *Calliphora* the giant fibers of the lobula plate (HS- and VS-cells) on one side of the brain by ablating a certain larval neuroblast with a laser microbeam. The behavioral effect was a reduction in the response to stimulation with a periodic grating on the side of the ablation. On the other hand, a normal full-sized response on this side to front-to-back motion of a single stripe is observed. Thus, again, the object response seems to be independent of the large field course control response and independent of the giant fibers of the lobula plate.

Analysis of Individual Torque Traces

If one analyzes the individual torque traces averaged in Fig. 62 one notices that the fly exhibits distinct behavior patterns which get lost by averaging (see Fig. 68). When the stripe moves back-to-front, frequent torque spikes into the direction of the stripe are observed. When passing through the frontal position, the stripe elicits a vigorous sustained torque response into the same direction as the movement. Note that this following response which normally is considered as a response to front-to-back motion can start as much as 45° on the contralateral side (Fig. 68b), i.e., while the stripe still moves back-to-front and has by far not yet reached the zone of binocular overlap. The precise position at which this behavior occurs may vary from sweep to sweep.

A similar pattern is found when the stripe is slowly (e.g. $f = 0.3$ Hz and $A_{p\text{-}p} = 30°$) oscillating in the fronto-lateral part of the visual field. Again torque spikes into the direction of the stripe during back-to-front movement and an equidirectional sustained response during front-to-back motion alternate over long periods of time (see Chap. 11: Selective Attention). The onset of the sustained response does not coincide precisely with the inversion of the direction of motion, but can anticipate it, and varies from sweep to sweep (Fig. 68c).

As mentioned in Chapter 6, the occurrence and polarity of torque spikes obeys the torque spike rule in that the torque spikes are directed toward the

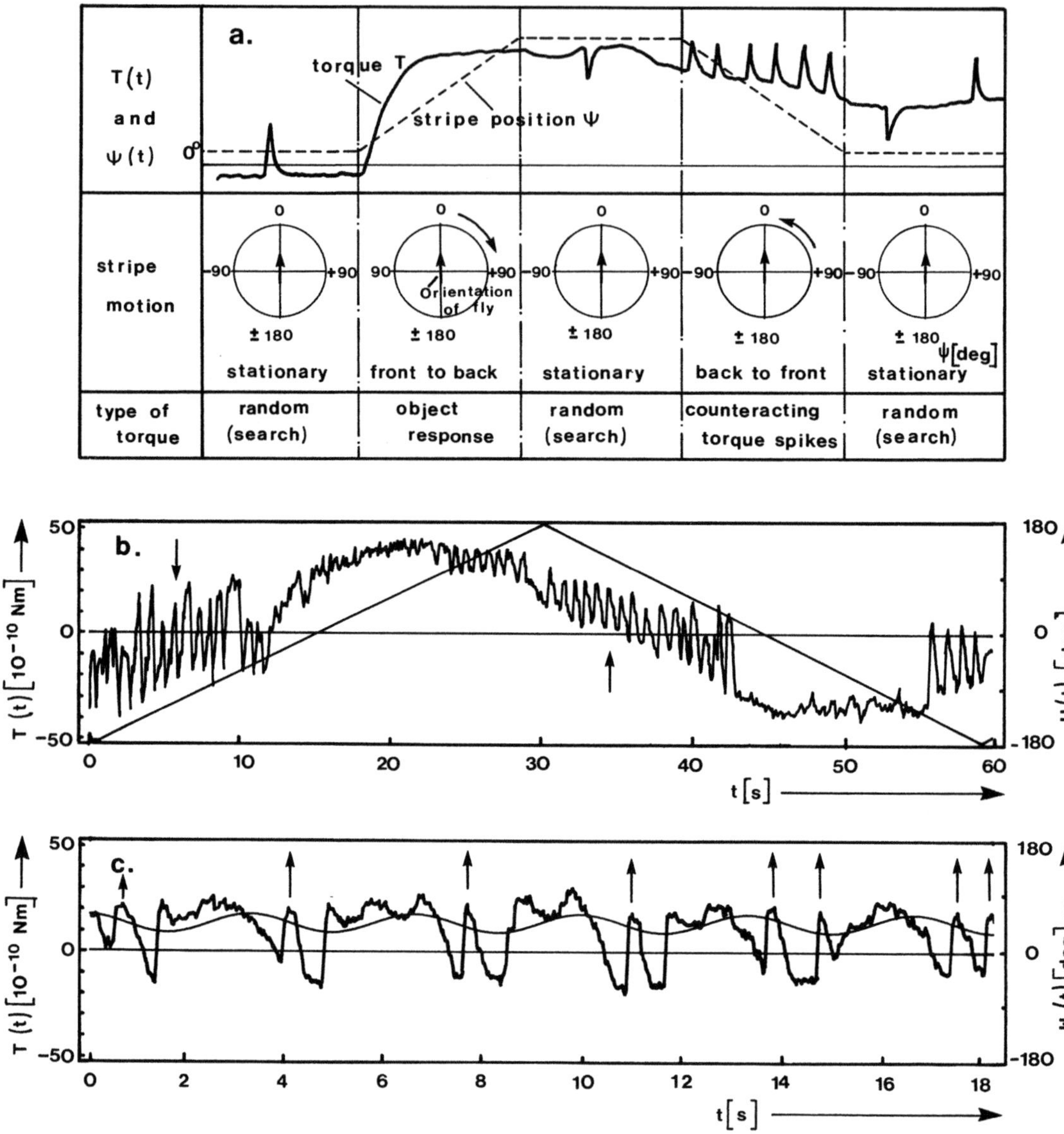

Fig. 68. (a) Schematic representation of yaw torque in response to front-to-back and back-to-front movement of a single vertical dark stripe. Front-to-back movement causes an equidirectional sustained response. During back-to-front movement the base line of torque seems undetermined (but for the mean a regression towards "zero" torque is observed). The fly generates torque spikes (identified by their steep rising phase) towards the side of the stripe and against its direction of motion (torque spike rule). The peaks of the spikes normally just reach the plateau held during front-to-back movement. If the torque base line still happens to be at that level, torque spikes may be very small. Occasionally in such situations the fly will abruptly reduce its torque just prior to the spike, after which the full sized torque spike is visible (example marked * in Fig. 101). (b) Position function (Fig. 62c) is due to typical torque pattern which is lost by averaging, but can still be recognized in the original recordings. The fly starts the period of sustained torque (to the front-to-back motion) somewhere in the vicinity of $\psi = 0°$. (c) A similar torque pattern is observed when the stripe oscillates in the fronto-lateral part of the visual field. *Smooth curves* in (b) and (c) show stripe position (right hand scale); *arrows* indicate polarity of torque spikes

position of the pattern and the fly generates them during back-to-front motion of the pattern. Their occurrence is not precisely timed by the stimulus. This leads to the elimination of torque spikes from averages of many traces.

Flies show remarkable individuality in the shape of the torque traces in response to the rotating stripe. Some flies exhibit no position function and occasionally one even finds a fly with a completely inverted one. It may be in part the visual experience prior to the test which is responsible for this variability. This effect is documented in an experiment in which flies are tested for their position function and are subsequently allowed to fly in closed loop with the stripe for 7 min. After this exercise the position function is measured again. Surprisingly it is greatly reduced (Fig. 69) as judged by the slope of the curve at $\psi = 0°$. The increasing amplitude at large ψ values may again be attributed to motion after effects. The optomotor response to a $\lambda_{sp} = 6°$ grating is hardly affected by this course of events excluding the uninteresting explanation of general fatigue (data not shown). The position function partially recovers if the flies are put back into their cages for several hours. We have not been able so far to find specific visual stimuli which reestablish it.

Plasticity of the position function in *Musca* is suggested from a fixation experiment performed by Geiger (1975), who showed that orientation toward a small stripe segment is improved if the fly is allowed to fixate a long vertical stripe first. A considerable interindividual variability of the position function in *Musca* has been observed by Reichardt (1973).

Mutants with Altered Orientation Behavior

Mutations specifically eliminating the position function have not been described so far. However, mutants with altered orientation behavior are quite frequent (Heisenberg, unpublished; Bülthoff, 1980; 1982 a, b). We have argued above that flying straight is the first level of orientation. Thus for instance, in *omb* visual orientation behavior is severely affected in flight (see Fig. 95) due to the missing large field course control function. If one considers only orientation histograms *omb* flies perform as well as, or even better than wild type in keeping the stripe in a frontal position. Yet the procedure which the *omb* flies use for this task is quite abnormal: if to begin with the stripe happens to be in the back part of the visual field or during the time when it moves forward, it appears to have no influence upon the behavior of the fly. As soon as it starts moving front-to-back in the frontal visual half-field, however, the fly applies the object response and keeps the stripe from moving backward (see Fig. 95). Fixation in *omb* seems to be generated by the object response. The majority of *omb* flies in the present stock show a normal

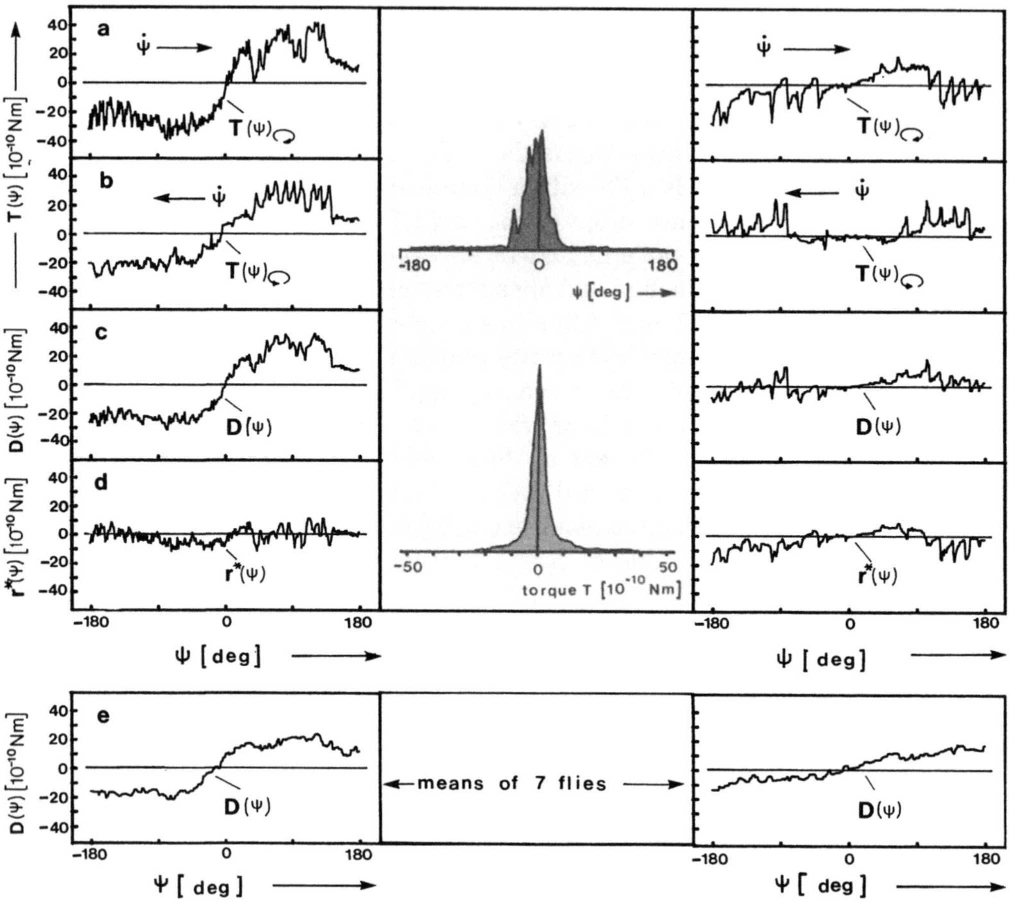

Fig. 69. Position function (c,e) and difference function (d) before (*left*) and after (*right*) a 7-min closed loop period (for which torque and stripe position histograms are shown in the middle column). Stimulus sequence as in Fig. 65a. A vertical dark stripe ($\delta = 5°$; $H = \pm 40°$) is used throughout (for calculation of position and difference functions see Fig. 62). (a–d) Experiment with a single fly (WT Berlin). (e) Means of seven experiments (7 flies). The 7-min closed loop period reduces (or abolishes) the position function

position function with a slowly rotating stripe under open loop conditions (Fig. 65).

In the original publication of visual behavior of *omb* a reduced position function was described (Heisenberg et al., 1978). The reasons for this discrepancy are not entirely clear. It is conceivable that it is due to the details of the experimental procedure. In the old experiments the position function was recorded, together with closed loop behavior which we now know reduces the subsequently recorded position function. Possibly, however, the flies have slightly changed their properties in the years between the first and second measurement (see p. 127).[5] Since the position function is not due to a fixed reflex the overall state of the animals may have a pronounced effect. Even in our wild-type stocks a small amplitude of the position function is recorded occasionally (see Fig. 71). It should be kept in mind that the absence of a behavioral response in a mutant can only tentatively be attributed to the concomitant absence of a neural structure. It is a much stronger argument to conclude that if a behavioral response is observed in the absence of a neural structure, this structure is not necessary for the response (see p. 65).

A second mutant with different orientation behavior, but in this case with an apparently normal large field course control response, is *no object fixation* A^{S100} (*nof A*) (mentioned as S100 in Heisenberg and Wolf (1979)). Although again closed loop behavior in this mutant is severely affected (Fig. 70; Bülthoff, 1980; Götz, 1980), the position function as measured by a slowly rotating stripe is not reduced (Fig. 71 b), indicating that one has to be very cautious in equating open loop and closed loop behavior. However, for *nof A* a defect in the position function can be found if one records it in a different manner. The traces in Fig. 71 c are obtained by the following procedure: First the torque of a wild-type *Drosophila* during a closed loop experiment with a single stripe is recorded in real time. Subsequently the stripe is rotated in open loop with the voltage from the tape to stimulate another wild-type or mutant fly. The curves then show the torque as function of the position of the stripe. With the special dynamics of *Drosophila* torque from a closed loop experiment, a moving stripe does not elicit in open loop a normal position function in *nof A*. This defect of *nof A* is as yet not fully understood.

Two interesting mutant strains in which freely walking flies show abnormal visual orientation behavior have been found by Bülthoff (1980; 1982 a, b). On the locomotion recorder the flies of these stocks show abnormalities in the large field course control response as well as in the position function. In both mutants the position function has a negative slope around $\psi = 0°$. The inverted position function is most pronounced in the mutant *nof* E^{B12}. As discussed above, the wild-type function measured by Bülthoff has slope 0 at $\psi = 0°$ and might be due to motion after effects. The deviations in the mutant traces (Fig. 72) are difficult to interpret.

A further orientation mutant *abnormal pattern orientation*^{S129} (*apo*); (Heisenberg, 1979), as well as two others not mentioned in publications as yet (*apo* B^{S71}, *apo* C^{S131}), have not been tested for their position functions in flight. All these strains should be investigated since they may constitute valuable tools for further analysis of orientation behavior.

[5] The visual behavior of *omb* has undergone certain changes before (Heisenberg and Götz, 1975; Heisenberg and E. Buchner, 1977; Heisenberg et al., 1978). It can now be studied in more standardized flies since a deletion has been found uncovering the *omb* gene in female flies heterozygous for the deletion and the chromosome carrying the *omb* allele. These flies should be heterozygous for most other genes affecting the phenotype.

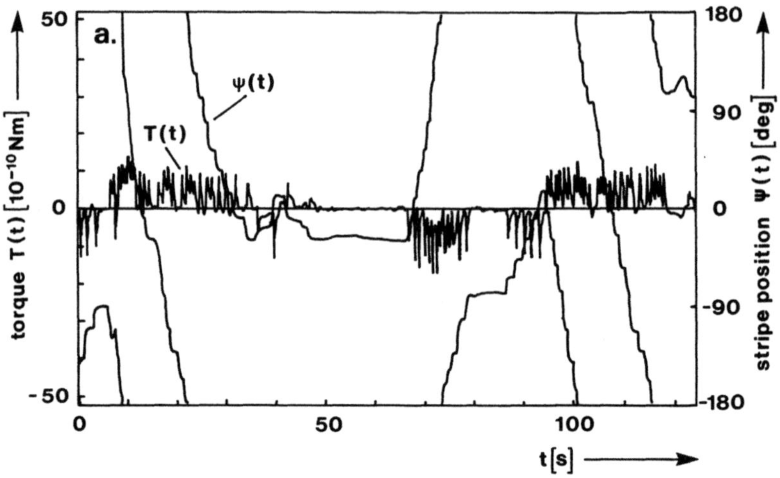

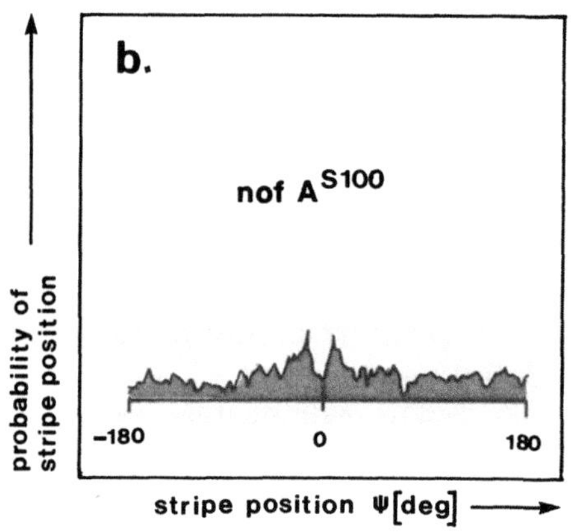

Fig. 70. (a) Records of stripe position and yaw torque of the mutant *nof* A^{S100} in a closed loop single stripe experiment. Optomotor balance is well established but frequency of torque spikes is somewhat increased. Due to slow speed of pen recorder torque trace does not return to base line of torque in between torque spikes. (b) Orientation histogram of ten 4-min flight periods of ten flies (cf. Fig. 56 for WT and experimental details)

In summary, we note that two distinct responses contribute to yaw torque, the large field course control response and the object response. Each one can operate without the other. In stationarily walking flies the object response has not been conclusively demonstrated so far.

Fixation in Flight

Fixation in free flight is difficult to observe. However, flies that escape during an experiment often head toward high contrast edges in the room where they are most difficult to find. On the polyamid thread (for experimental details see Fig. 46) orientation histograms in a white arena with a single stripe look quite similar to stripe position histograms obtained from flies at

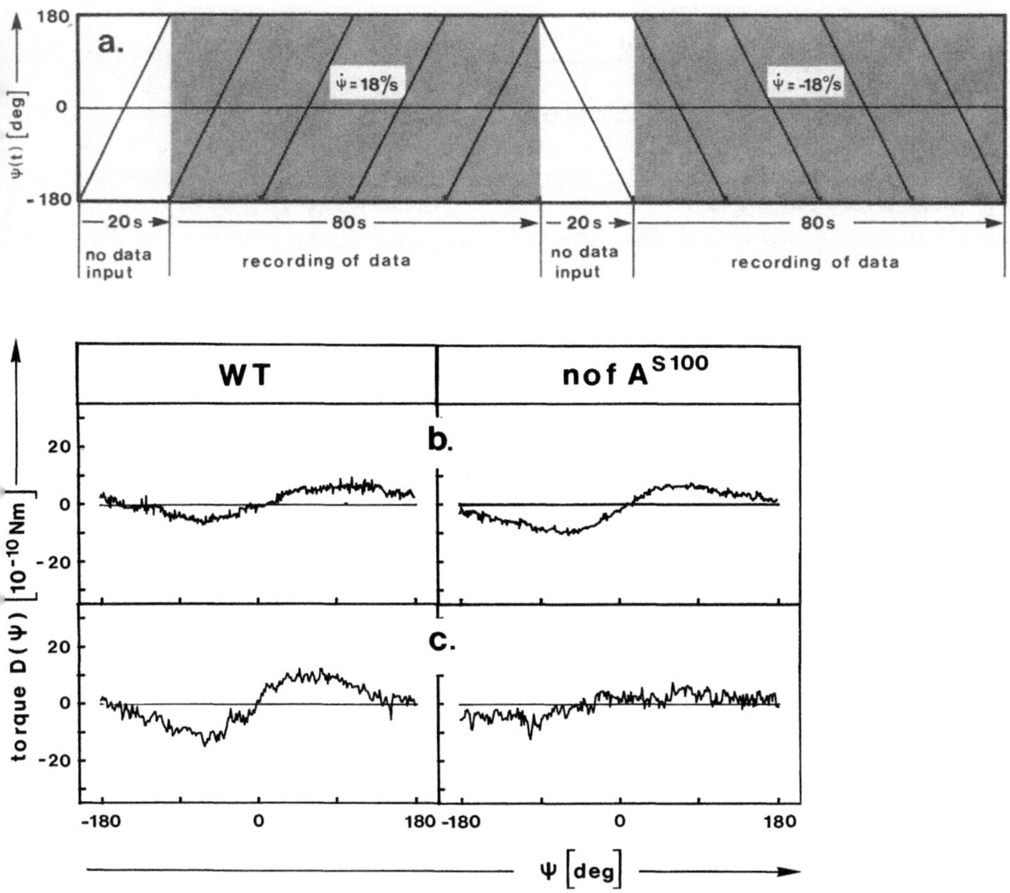

Fig. 71. Postion functions (b, c) of WT (Berlin) and *nof* A^{S100}. (a) Recording procedure in (b) is similar to that used in the experiments of Fig. 62 and 65. In (c) stripe does not rotate smoothly as in (b) but is driven by a closed loop experiment of another fly. For this purpose torque of WT (Berlin) during an undisturbed closed loop experiment with a vertical dark stripe is recorded on magnetic tape. This torque trace is made to drive the panorama of the test fly of which (open loop) torque at positions during front-to-back and back-to-front movement of the stripe is recorded separately. Traces in (b) are averages from 6(10) experiments with 4 WT and 5 *nof* A^{S100} flies. Position functions in (c) are from 22 5-min periods of flight each,measured with 7 WT and 5 *nof* A^{S100} flies. Note rather small amplitude of WT position function in (b) as compared to those in Fig. 62 and 65. This variability reflects long term changes in the stocks

the torque compensator (Fig. 73; cf. Fig. 56a). However, fixation seems to be slightly favored on the thread on account of menotaxis. The difficulty with this experimental design seems to be that the axis of rotation is slightly anterior to the insertion points of the wings.The resulting artificial instability of orientation may disfavor menotaxis, as we will see below. Nevertheless, menotaxis is readily observed on the polyamid thread as well (see Fig. 49c). So far only few data are available for stationarily flying and freely rotating

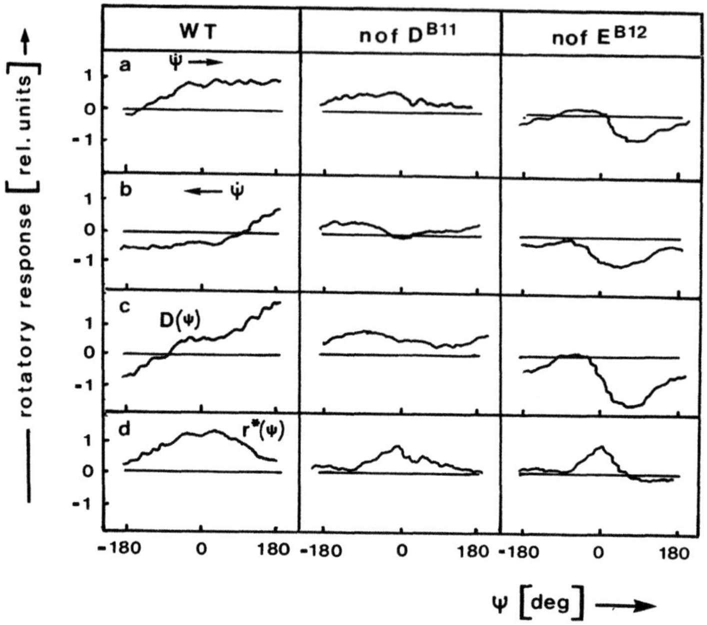

Fig. 72. Position function (c) and difference function (d) of walking flies on the locomotion recorder as measured with a 5° × 5° square rotating at $200° s^{-1}$. Note the flat position function of WT for $-60° < \psi < 60°$. (The positive slope outside this region may possibly be attributed to motion after effects.) One of the orientation mutants *no object fixation* E^{B12} (*nof E*) has an inverted position function with a negative slope at $\psi = 0°$ (redrawn from Bülthoff, 1982 b). Traces in (c) and (d) are marked D(ψ) and r*(ψ) to indicate that they have been obtained from (a) and (b) by the same procedure as the corresponding functions in flight experiments (see Figs. 59, 62)

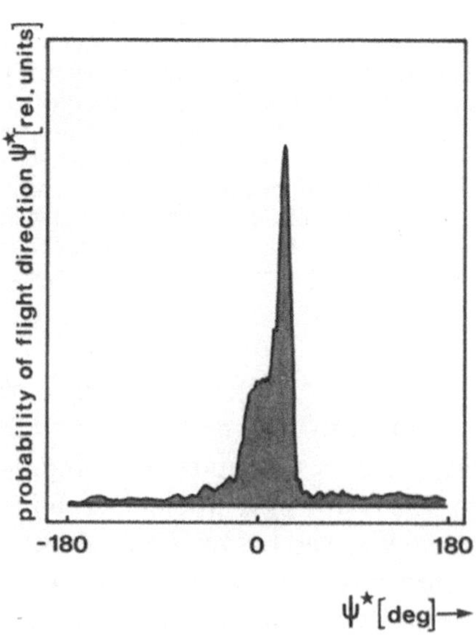

Fig. 73. Orientations of WT (Berlin) during a 4-min flight period in the apparatus shown in Fig. 46 (polyamid thread; fly is free to rotate in the center of the arena). ψ^* indicates the fly's orientation with respect to space-fixed zero. *Vertical black stripe* ($\delta = 5°$; $H = \pm 45°$) is positioned at $\psi = 0°$. (K. Vogtmann, unpublished)

Drosophila flies. We still rely primarily upon experiments with the torque compensator.

It has been stressed (i.e., Reichardt and Poggio, 1976) that the position function (open loop) and fixation behavior (closed loop) are equivalent in the sense that a fly keeping the stripe in frontal position must show a position function with a positive slope at $\psi = 0°$. This statement holds for the theory outlined above. As an alternative, however, they discuss pseudo-fixation in which a symmetrical movement response alone $[D(\psi) = 0]$ can lead to a preferred orientation toward the object if the local gain for movement in the frontal part of the visual field is larger than that for movement elsewhere (Poggio and Reichardt, 1976 a). [In the mechanical model of the closed loop situation a disproportionately large direction-sensitive movement response (difference function) for certain parts of the visual field is equivalent to large local friction which would keep the ball there for longer periods of time than in regions of low friction.] This mechanism alone, however, would generate only very broad fixation histograms, as was shown by calculations using the local open loop optomotor response values of *Musca* and the *Musca* "torque noise" (Reichardt and Poggio, 1976).

For *Drosophila* at the torque compensator we distinguish two closed loop situations resulting in a stabilization of the stripe in frontal positions. One is "forced fixation" during which the fly takes the torque fluctuations it generates as "disturbances" and applies its object response. This type of fixation is associated with a deep potential trough for torque at $\psi = 0°$ (Fig. 74). Forced fixation frequently occurs at the onset of closed loop experiments in particular during oscillating and noise flight modes.

The second, more common, situation is "voluntary fixation" occurring during menotaxis (see Chap. 8). In this case the stripe is kept in frontal position by the suppression of torque spikes which otherwise would shift the stripe backward. It is characteristic for this behavior that the torque fluctuations are sufficiently small not to elicit the object response. No significant potential trough is found around $\psi = 0°$ and the stripe is not stabilized around one particular position (Fig. 74). This is the reason why with *Drosophila* individual stripe position histograms (4 min) are "spiky". Even after averaging 40 min of flight time from 10 flies, no smooth Gaussian distribution of stripe positions is obtained (Fig. 56 a).

With two stripes in the arena the fly alternates between fixating one of the stripes and performing menotaxis (see Chap. 8). If the stripes are separated by $\Delta\psi_{sep} < 180°$, positions along the short connection are more likely to be found in front of the fly than positions along the far connection. At separation angles $\Delta\psi_{sep} \leqslant 30°$ one-peaked stripe position histograms are obtained with the maximum in the middle between the two stripes (Fig. 75). The striking difference between this result and the one obtained with *Musca* (Fig. 61) is the fact that with *Drosophila* in two-peaked histograms the sepa-

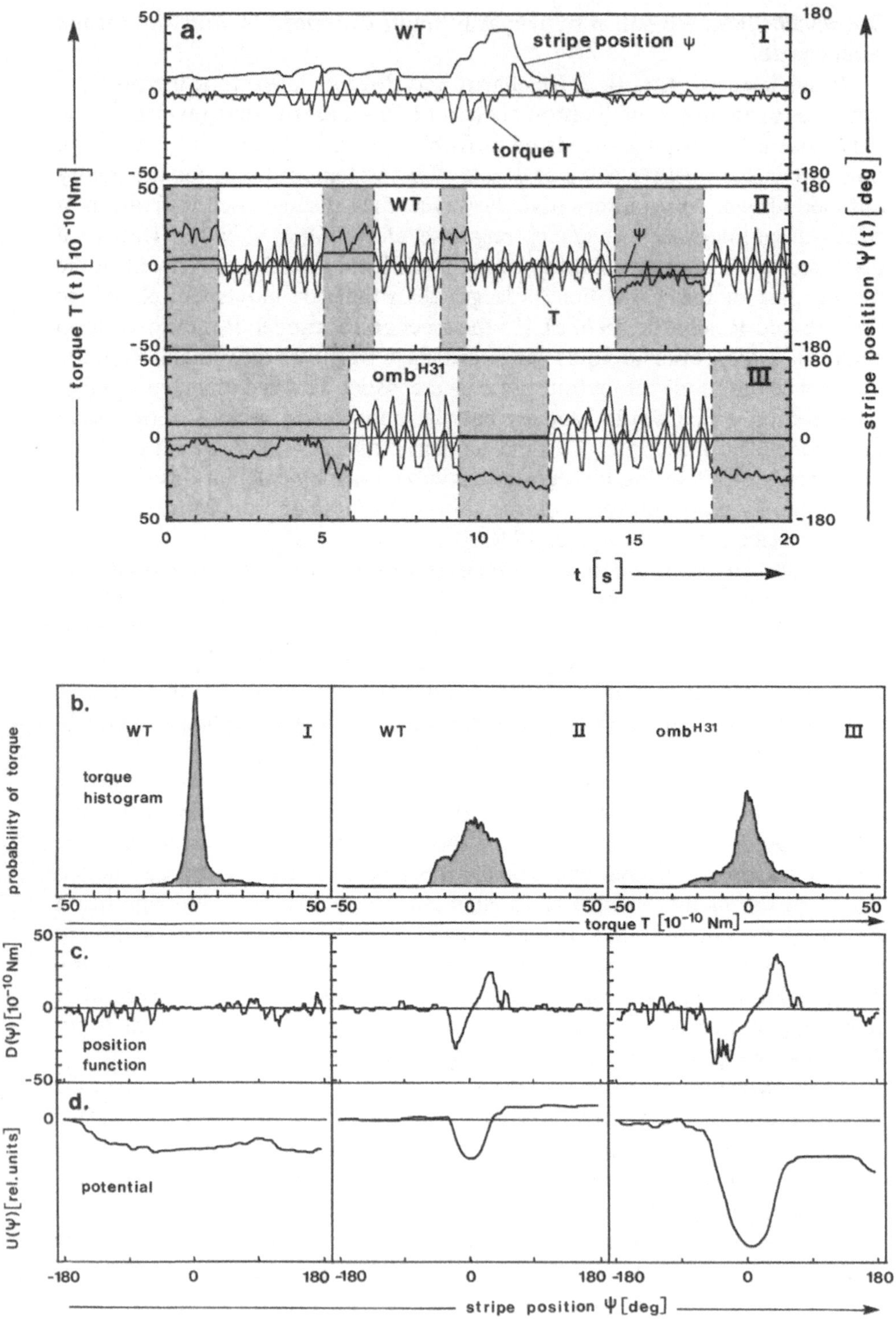
a.
WT
stripe position ψ
torque T
I
WT
II
ψ
T
omb^{H31}
III
torque T (t) [10^{-10}Nm]
stripe position Ψ(t) [deg]
t [s]
0
5
10
15
20
b.
WT
I
WT
II
omb^{H31}
III
torque histogram
probability of torque
torque T [10^{-10} Nm]
c.
D(ψ)[10^{-10}Nm]
position function
d.
U(ψ)[rel. units]
potential
stripe position Ψ[deg]

Fig. 74. "Voluntary" and "forced" fixation at the torque compensator. (a) Three records (*I* – *III*) of yaw torque and stripe position of WT (Berlin) and *omb* during closed loop with only one vertical dark stripe ($\delta = 5°$, $H = \pm 40°$). In (*aII*) and (*aIII*) the visual feedback loop is opened (*shaded areas*) at arbitrary times for short periods to show that torque oscillations depend upon the loop being closed (see also Fig. 64b). (b) Torque histograms. (c) Position functions. (d) Potentials (see Fig. 59) for corresponding (continuous closed loop) flight episodes. Position functions are obtained by separating the torque values according to whether they occur during clockwise or counterclockwise motion of the stripe and adding the two functions for each value of ψ. In *I* the fly mostly keeps torque close to zero and the stripe in the frontal part of the visual field. Under these conditions (voluntary fixation) the position function (*cI*) is flat around $\psi = 0°$. In *II* and *III* WT and *omb* show the oscillating flight mode. They exhibit a position function (c) with a positive slope at $\psi = 0°$ and, correspondingly, (d) a pronounced potential trough (forced fixation). Since *omb* (*III*) shows normal oscillating mode and concomitant position function, we assume that this behavior is not dependent upon large field course control

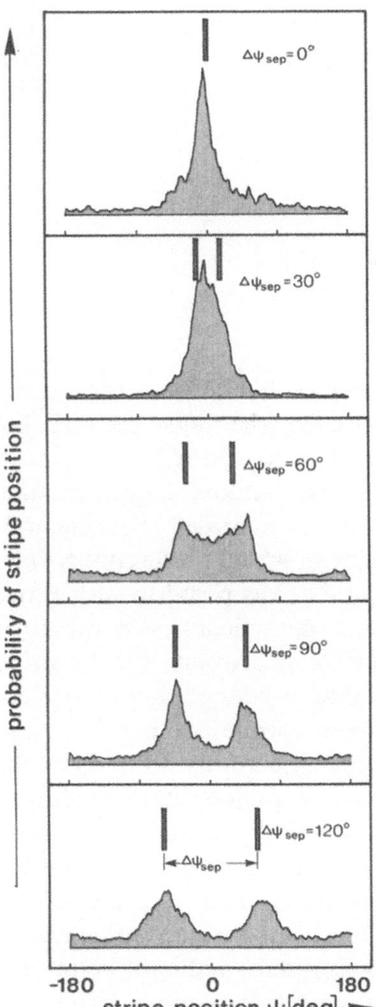

Fig. 75. Orientation histograms in stationary flight with one and two vertical dark stripes (width $\delta = 5°$; height $H = \pm 40°$) at various angular distances $\Delta\psi_{sep}$. Symmetry line of the patterns is taken as $\psi = 0°$. Angular separation of peaks is about the same as that of the stripes. At $\Delta\psi_{sep} = 30°$ flies fixate the middle between the two stripes. Histograms are averages of ten 2-min flight periods from at least seven different flies (WT Berlin) each

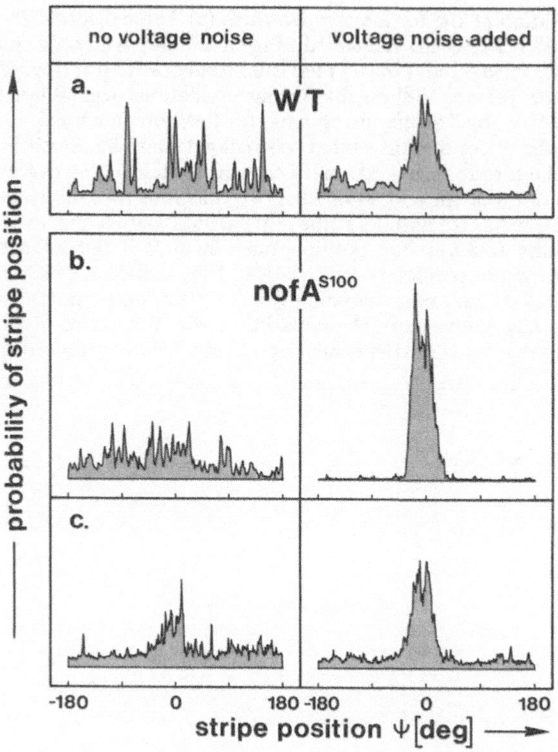

Fig. 76. Orientation histograms from 2-min flight periods of (a) 1 WT (Berlin) and (b) *nof* A^{S100} fly. *Left* undisturbed closed loop; *Right* voltage noise ($A_{max\ p\text{-}p} = 28°$; upper cut-off frequency f = 1.5 Hz) is added into the loop. In (c) ten 2-min periods each with and without additional noise are averaged for *nof* A^{S100}

ration angle between the peaks is always the same as that between the stripes, whereas for *Musca*, as we have mentioned already, the angle between the peaks is smaller than that between the stripes (for $\Delta\psi_{sep} = 60°$).

Let us return to single stripe experiments. The relative length of time during which the stripe is at a frontal position can be modified experimentally. As we have seen, mutant strains are available in which preference for the frontal position is reduced, i.e., the stripe is found at any position with about equal probability (Fig. 70 b). Optomotor balance, nevertheless, is maintained normally (Fig. 70 a). Secondly in wild-type flies the probability for the stripe in the frontal position can be increased by adding a voltage randomly fluctuating around zero to the closed loop between torque compensator and motor control unit (Fig. 76 a). This disturbance apparently suppresses the "quiet" flight mode (Fig. 53 c). Even the mutant *nof A* under these conditions shows a preference for frontal stripe positions (Fig. 76 b, c). This observation indicates that increasing the coupling constant in our closed loop might shift the behavior of the flies toward forced fixation due to increased instability.

Thirdly, also spectral composition of the light can have an influence. In the eye color mutant *white*, brief irradiation with intense red light before the experiment suppresses fixation. A similar treatment with white light restores it. We have mentioned this effect of red light already before (p. 107). Interestingly, the double mutant *white sevenless* does not show this suppression (Fig. 77). This is the only evidence for an influence of retinula cells R7/R8 on a behavioral response under stationary conditions so far recorded. What all these observations show is that in *Drosophila* a vertical dark stripe has a certain attractiveness which depends upon the state of the animal and the properties of the closed loop conditions.

The attractiveness of an object refers to the third level of orientation behavior, the valuation stage which will be discussed in Chapter 13.

Fixation in Walking Flies

When a fly with cut wings is released in the center of a circular arena carrying nothing but a vertical stripe at its periphery, the fly will in most cases head toward the stripe (Fig. 78). Occasionally, a fly turns away from the stripe and heads into the opposite direction which leads to a small second peak in the orientation histogram. This behavior is critically influenced neither by age, sex or nutritional state of the animal, nor by pattern contrast and light intensity in a wide range of stimulus conditions (Wehner, 1972; Horn and Wehner, 1975; Horn, 1978). What matters, however, is the size and shape of the pattern. For instance, if the wall of the arena (height H $= 75°$) is half white–half black ($\delta = 180°$) some wild-type flies will head toward the edges. On the other hand, if a horizontal stripe of the same width ($\delta = 180°$) is reduced in height to less than $H = 20°$, the same flies approach the center of the stripe (Wehner, 1972, 1981; see p. 209). This again leads into the field of pattern preferences which is not to be discussed here. Horn and Wehner (1975) also tested the orientation behavior in a panorama with two and three stripes. The results with two stripes (Fig. 79) show interesting parallels to the flight experiments at the torque compensator. In both cases two stripes separated by $\Delta\psi_{sep} \leqslant 30°$ are taken as one pattern, in that the line of symmetry between them is fixated. For angular separations of the stripes $60° \leqslant \Delta\psi_{sep} < 180°$ scatter is too large and the fixation peaks are too broad to decide whether their angular distance is smaller than that of the stripes.

The reliable fixation histograms obtained with freely walking flies is in contrast to the difficulty of obtaining similar histograms on the locomotion recorder (Bülthoff, 1981). An explanation is sought in the time course of the experiments. In the arena a fly is released in the center of the arena where most flies immediately start running toward the periphery. After a few seconds the fly is trapped and placed back into the center (or into its cage). For experiments with the locomotion recorder, a fly has to be carefully glued to a hook and manipulated into the right position in the apparatus. Each fly then has to perform for many minutes, if not hours or days. Glued to the

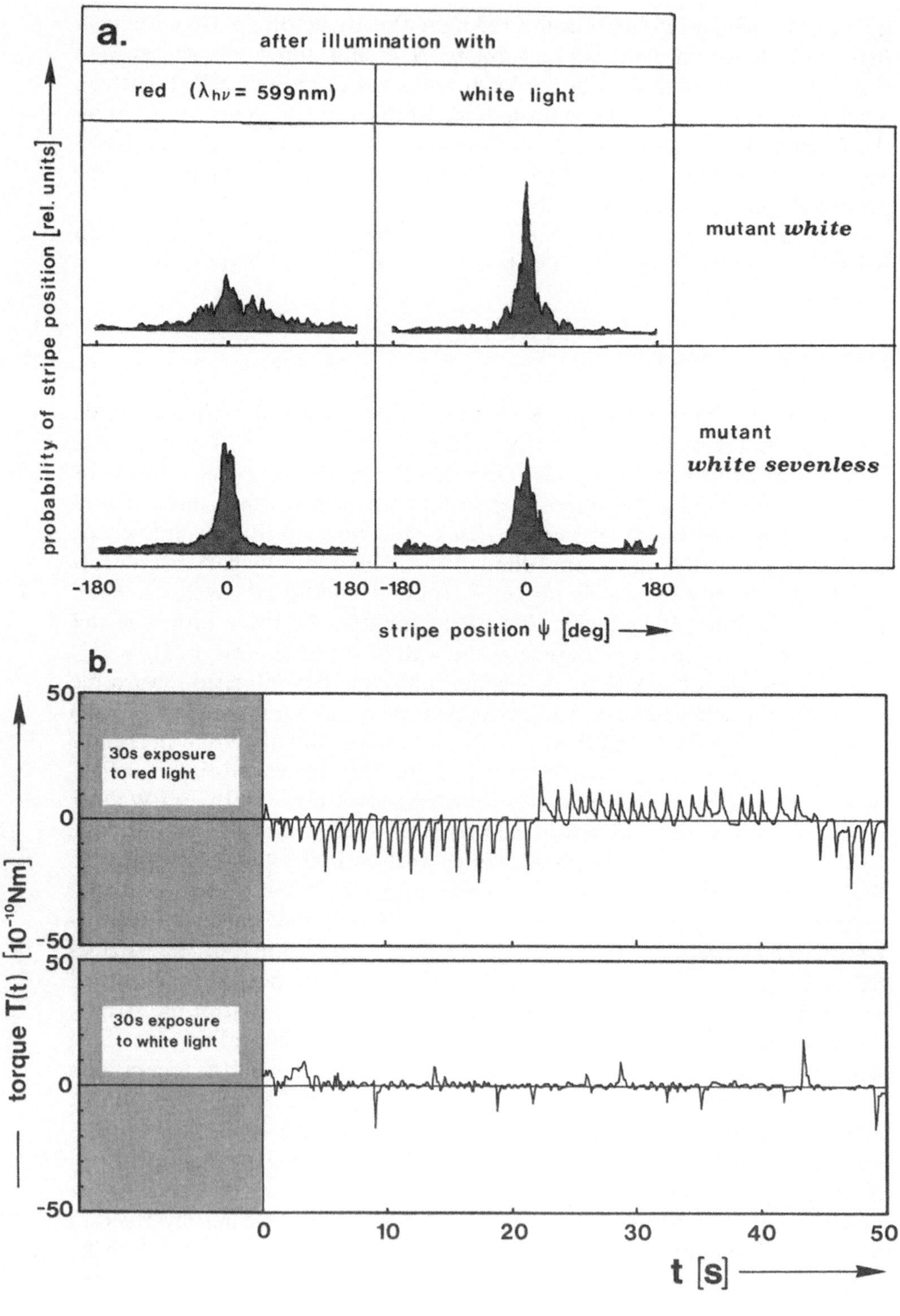
a.
after illumination with
red (λhν = 599nm)
white light
mutant white
mutant
white sevenless
probability of stripe position [rel. units]
-180
0
180
-180
0
180
stripe position ψ [deg]
b.
50
0
-50
30s exposure
to red light
30s exposure
to white light
torque T(t) [10⁻¹⁰Nm]
0
10
20
30
40
50
t [s]

Fig. 77. (a) Orientation histograms of the mutants *white* (*w*) and *white sevenless* (*w sev*) after preadaptation with red ($\lambda_{h\nu} = 599$ nm) and white (tungsten iodine, 12 V, 100 W) light. Preadaptation with red light suppresses fixation in favor of menotaxis and increases torque spike frequency. Preadaptation with white light restores normal state (quiet flight mode). Fixation behavior is tested in dim blue ($\lambda_{h\nu} = 476$ nm) light. In the double mutant (*w sev*) this switch is not observed. Histograms are obtained from ten 4-min flight periods each. The same seven *w* and ten *w sev* flies respectively are used with both kinds of preadaptation. (b) Torque traces of short flight periods of the mutant *white* after 30 s illumination with red and white light

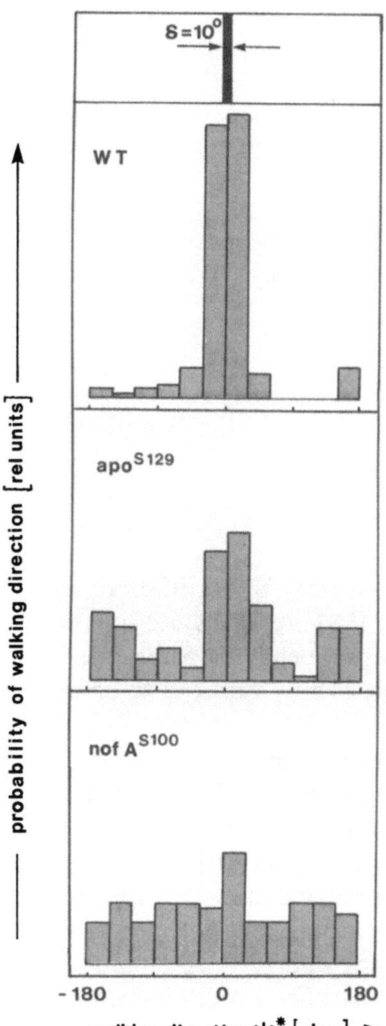

Fig. 78. Orientation histograms of freely walking flies facing a vertical dark stripe ($\delta = 10°$, $H = +60°$) in an otherwise homogeneously illuminated arena (radius $r = 280$ mm). Single flies with shortened wings are released in the center. The angular position (relative to the stripe) where the trajectory of the fly crosses an arbitrarily chosen "measuring circle" ($r = 50$ mm) is recorded. The angular position of the stripe in the arena is changed after each second run (200 runs, 20 flies). (a) WT (Berlin); (b) mutant *apo*^{S129}; (c) mutant *nof A*^{S100}

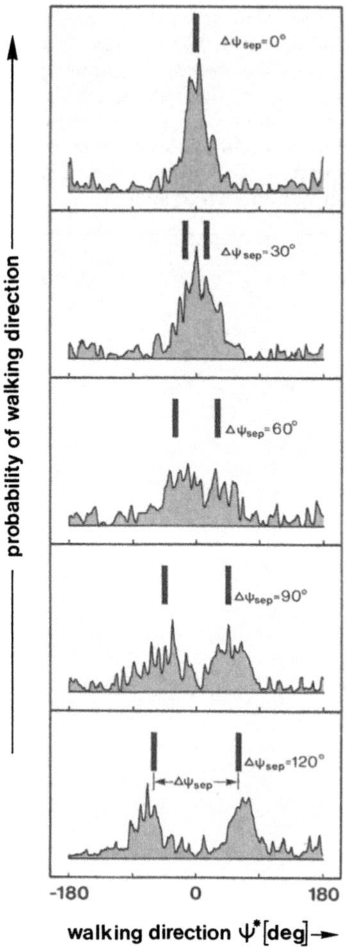

Fig. 79. Orientation histograms of freely walking WT (Sevelen) flies. The arena carries 1 or 2 vertical black stripes ($\delta = 6°$; $H = +68°$) on its periphery. Radius of arena $r = 96$ mm; measuring circle $r = 30$ mm. (After Horn and Wehner, 1975). Compare two-stripe experiments with flying *Drosophila* (Fig. 75) and *Musca* (Fig. 61). Note, however, that in flight each fly is tested (continuously) for several minutes and generates several hundred values for the histogram. In the walking mode each experiment lasts a few seconds only and provides one value for the histogram

hook and clamped to the machine the fly presumably is in a different mental state than when in the arena. Bülthoff (1980; 1981) reported a 2-h period of a stationarily walking fly mainly showing fixation behavior. Subtleties in the stimulus situation and the closed loop properties may turn out to be important.

Bülthoff et al. (1982) continuously recorded the trajectories of freely walking flies on an arena using an x-y-tracking system. The arena was surrounded by a moat or a heat barrier confining the fly to a circular area of about 10 cm in diameter. Confronted with two vertical stripes separated by $\Delta\psi_{sep} = 180°$ at the circumference of the arena the flies run toward one landmark until they reach the moat, then turn around to approach the other one, and continue in this manner running back and forth for many hours.

Wehner and Horn (1975) have shown that parallactic pattern motion due to the forward velocity of the walking fly has a pronounced effect on orientation. This has not been simulated with stationarily walking flies in closed loop so far. Detailed analysis of motion pictures of trajectories in the arena (Wehner, 1979) reveals that most flies do not keep the stripe or the edge of the pattern precisely in the most frontal part of their visual field but rather at about $\psi^* = 30°$ (or $\psi^* = -30°$) thus either taking a curved approach or walking slightly sideways in respect to their longitudinal body axis. This observation is backed up by an experiment of Wehner and Fukushi (mentioned by Wehner, 1981) in which the eyes of *Drosophila* flies are occluded by light-tight paint while only a vertical sector of 30° to 40° is left uncovered. Flies in which this sector extends from about $\psi^* = 30°$ to 70° show the most pronounced fixation histogram, whereas flies with the unobscured part of their visual field in the front or back are more severely impaired in their orientation toward the stripe. Conceivably it is in the fronto-lateral part of the visual field that the flies are most sensitive for motion parallax.

In summary, vertical dark stripes have only a limited attractiveness for stationarily flying or walking flies under artificial closed loop conditions. For freely walking flies this attractiveness is very high. In flight, at the torque compensator occasionally "forced fixation" is observed, which in most instances is due to large torque fluctuations generated by the fly. In this case the theory on pattern-induced flight orientation developed for fixation behavior of large flies seems to apply to *Drosophila*. Without large torque fluctuations *Drosophila* may show "voluntary fixation".

8 Menotaxis[6]

We have mentioned already on several occasions that *Drosophila* can maintain an arbitrary angular position relative to a visual object (e.g., a vertical stripe). If one tries to translate menotaxis at the torque compensator into the free flight situation, one is reminded of hoverflies hovering at a fixed position and orientation in mid-air. Hover flight of *Drosophila* is less perfect, but it too is characterized by stable orientations changed only by means of body saccades. Might the "quiet" flight mode (Fig. 53c) at the torque compensator correspond to hovering flight? The absence of visual cues relating to forward motion, the lack of motion parallax and the low streaming velocity of the air (Fig. 26) most probably inform the fly about its "slow" forward speed. In fact, the discovery of Götz (1983) that differently oriented sets of EMD's are mediating lift/thrust control, depending on whether the fly is flying in an air current or in still air, argues in favor of this conjecture. However, the fly's horizontal body posture does not conform to hovering flight and gravity as well as the horizontal rotatory motion of the panorama keep the fly informed about its real orientation. Thus the fly at the torque compensator may be in a conflict as to its appropriate flight behavior (see similar argument on p. 46f and 107f).

Whatever the behavioral basis of menotaxis in stationary flight may be, we should try to fit it into what we know already about visual behavior at the torque compensator. Superficially, one may see an inconsistency between the open loop position function (see Fig. 62), which should drive the stripe to the frontal position, and the occurrence of menotaxis. The object response is, however, normally elicited only by motion which is not induced by the fly (as an exception see Fig. 74II). During menotaxis the fly shifts the stripe by means of torque spikes, which by way of an efference copy suppress the responses to the concomitant displacements of the stripe (see Chap. 6). Thus, no pattern movement which might stimulate the object response occurs during menotaxis. How, then, is the stripe stabilized? We are back at the unsolved problem of how optomotor balance is maintained.

For a while we considered the possibility that during menotaxis the zero torque periods between torque spikes might be due to apathy rather than to

[6] The term "menotaxis" will be used here for a behavioral mode in which the fly keeps the stripe at any angular position including the front (voluntary fixation; see Chap. 7)

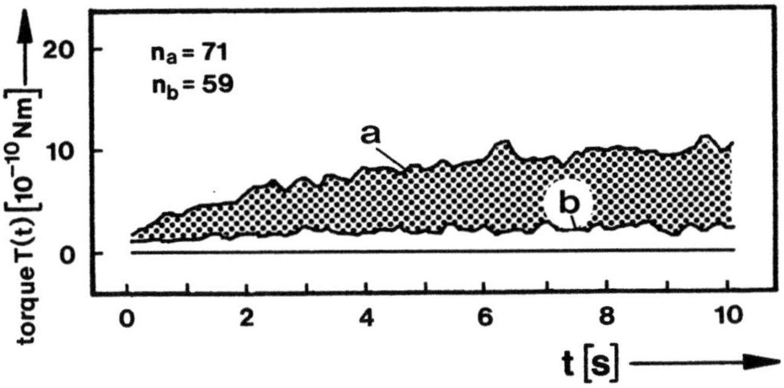

Fig. 80. Experiment showing that *Drosophila* stabilizes its course during menotaxis. At arbitrary moments during menotaxis torque is recorded for 10 s. In (a) the visual feedback loop is opened at $t = 0$, in (b) the feedback loop is left undisturbed. For averaging negative torque values are multiplied by -1. n_a and n_b give the number of experiments averaged for traces *a* and *b* respectively. *Shaded area* indicates the average effect of opening the feedback loop. The first data point was taken at $t = 0.1$ s (from Heisenberg and Wolf, 1979; closed loop coupling: 10^{-9} Nm $\triangleq 55° s^{-1}$)

a precise flight control mechanism. A simple experiment ruling out this explanation is the following: we open the loop at an arbitrary moment and observe for how long afterward the torque stays at zero. Occasionally flies take no notice of the lack of visual feedback, in particular after a very long flight time at the torque compensator. In most cases, however, it takes them less than 100 ms to notice the difference (Fig. 80). Thus the flies continuously monitor the small angle oscillations of the pattern.

Small Angle Oscillations

Next we looked at the torque and position traces at high resolution (Fig. 81). Even during the most quiet flight periods, the fly modulates its torque sufficiently to generate small angle fluctuations of the pattern in the amplitude range of $A_{p\text{-}p} = 0.5°$ to $3°$ with significant frequency components up to $f = 10-20$ Hz.This is the upper frequency limit of our torque compensator. The apparatus itself adds voltage fluctuations in this frequency range with about the same amplitude as those generated by the torque signal of the fly. Fortunately, the high frequencies contribute little to the motion of the pattern. Thus most of the small angle fluctuations are still due to the yaw torque of the fly, as control experiments show.

The divergence angle between vertical rows of visual elements in the lateral part of the visual field is about $\Delta\varphi^* = 5°$ (Götz, 1964; Buchner, 1971; 1976). Thus, the small angle oscillations of a pattern will cause intensity modulations in neighboring visual elements, which are mostly in phase or in antiphase [i.e., phase-shifted by $\Phi = 180°$ (Fig. 82)] or have twice the frequen-

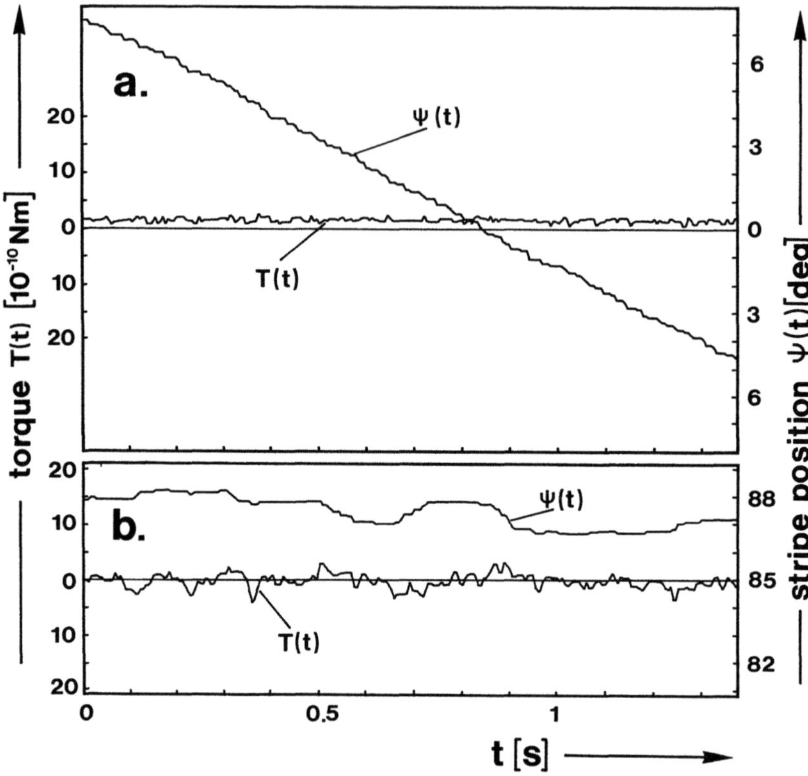

Fig. 81. Yaw torque and stripe position traces at high resolution (a) without fly, (b) with flying fly during "quiet" flight mode (Fig. 53c). In (a) a slight offset is added to the loop. Note in (b) the distinct eigen-behavior of the fly even during quiet periods of menotaxis. The "noisier" appearance of this trace compared to the one in Fig. 53c is explained by the fact that Fig. 53 was recorded with a conventional chart recorder (mechanical low pass filter), whereas Fig. 81 was drawn by a digital plotter from a computer record

cy. Thus even asymmetric presentation of patterns cannot be expected to generate net torque toward one side by means of direction-sensitive movement detection combined with a different gain for front-to-back and back-to-front movement (Pick, 1976). We have seen already that small angle oscillations of a stripe in lateral position under open loop conditions cause little torque toward that side (see Fig. 63b). Such considerations had prompted us to ignore the fly's small amplitude torque fluctuations during menotaxis.

In the following experiment the fly keeps the stripe menotactically while we add to the closed loop a voltage randomly fluctuating around zero. This causes a slight, continuous disturbance in the angular position of the stripe. The A.C. voltage consists of a repetitive 5.8 s period and is a sythesis of 15 frequency components of equal amplitude between $f = 0.17$ Hz and

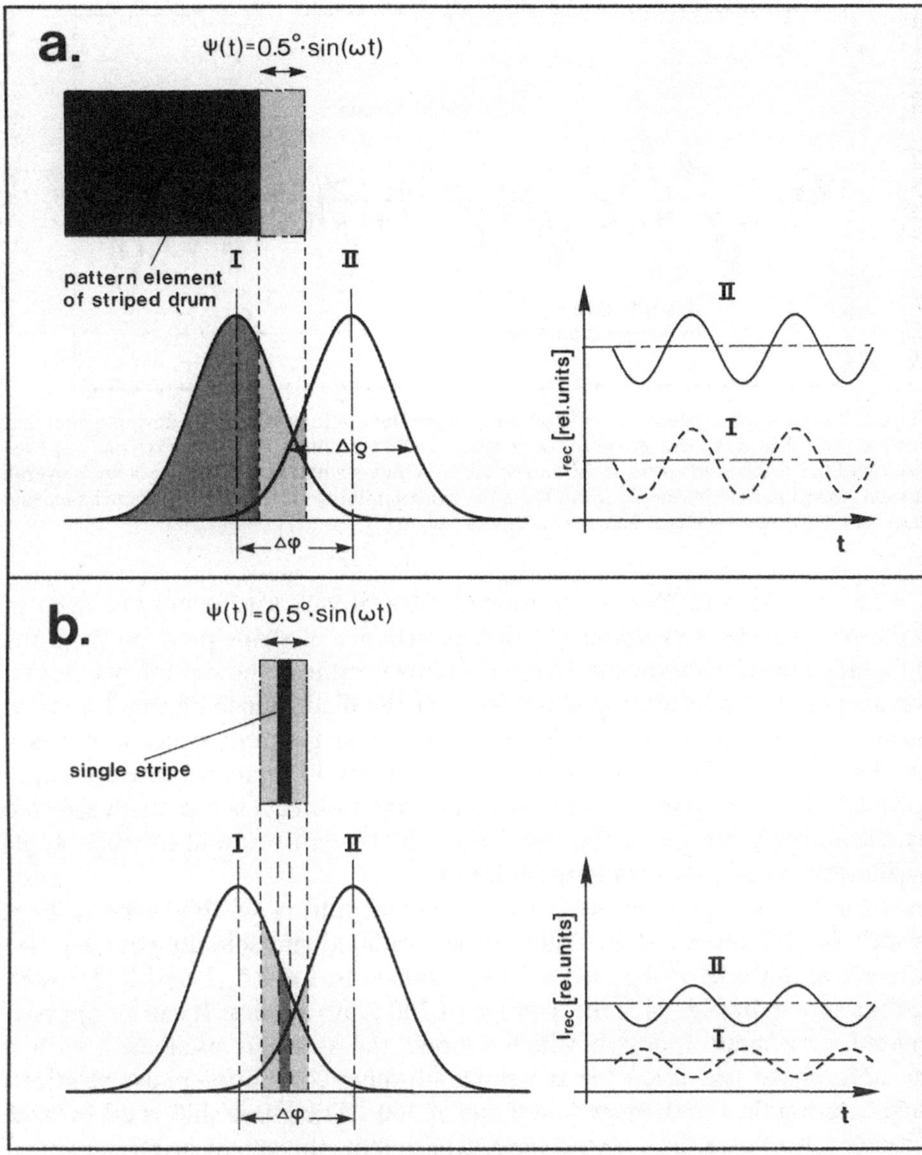

Fig. 82. (a) Modulation of adjacent visual elements due to small angle oscillations of a black–white boundary. The response-time functions of the two receptors (*right*) are in phase but mean response levels ($\bar{I}_{rec}$) are different. (b) With small pattern element ($\delta \ll \Delta\varrho$) modulation is in antiphase. If pattern element moves across peak of the angular sensitivity distribution ($\Delta\varrho$), modulation of visual elements has twice the frequency of the oscillation and the modulation in the second visual element

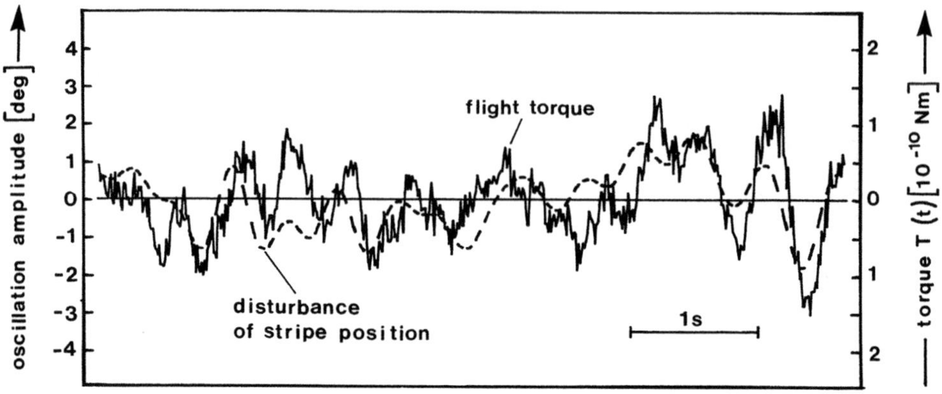

Fig. 83. Yaw torque response to artificial small angle fluctuations of a stripe during menotaxis. Torque trace shown is the average of the response to 20 stimulus cycles. The externally applied motion of the stripe (*dashed curve* shows one cycle) is only a small fraction of the stripe's overall motion caused mainly by the fly itself. Note the close similarity of the external disturbance and the average torque trace (see text for further details)

f = 2.6 Hz (Fig. 83). Twenty periods of 5.8 s of yaw torque are averaged in order to separate the response to the disturbance of stripe position from the fly's endogenous yaw torque activity. Figure 83 shows the similarity between the averaged torque trace and the trace of the disturbance of angular stripe position. The correlation coefficient is r = 0.38 for zero delay and has a maximum value of r = 0.51 for a delay of 120 ms. (The actual stripe position to which the fly responds has not been averaged, since it is too much affected by the active behavior of the fly). Thus, the fly can respond to small angle oscillations with yaw torque modulations.

In a subsequent open loop experiment a grating of thin vertical lines (width $\delta = 1°$) spaced at 20° intervals is oscillated sinusoidally with a peak-to-peak amplitude of $A_{p\text{-}p} = 2°$. The response to f = 0.5, 1 and 2 Hz oscillations shown in Fig. 84 is the average of 500 (250) sweeps. It can be approximated by a sinus function which follows the stimulus oscillation with a variable phase lag. Even for the same stimulus conditions phase relations differ among flies over the whole range of 360°. The phase shift is often large enough to increase, in a closed loop experiment, the actual displacement of the stripe instead of reducing it. Whatever the interpretation of this result, it again shows that *Drosophila* is able to perceive the self-induced small displacements of the pattern occurring during the most quiet phases of menotaxis.

We should briefly divert from our discussion of menotaxis and consider this small angle displacement response. Without having calculated the detailed properties of the correlation model of motion detection (Hassenstein and Reichardt, 1956, Poggio and Reichardt, 1973 b) for this stimulus situation we assume that it does not account for the response to small angle

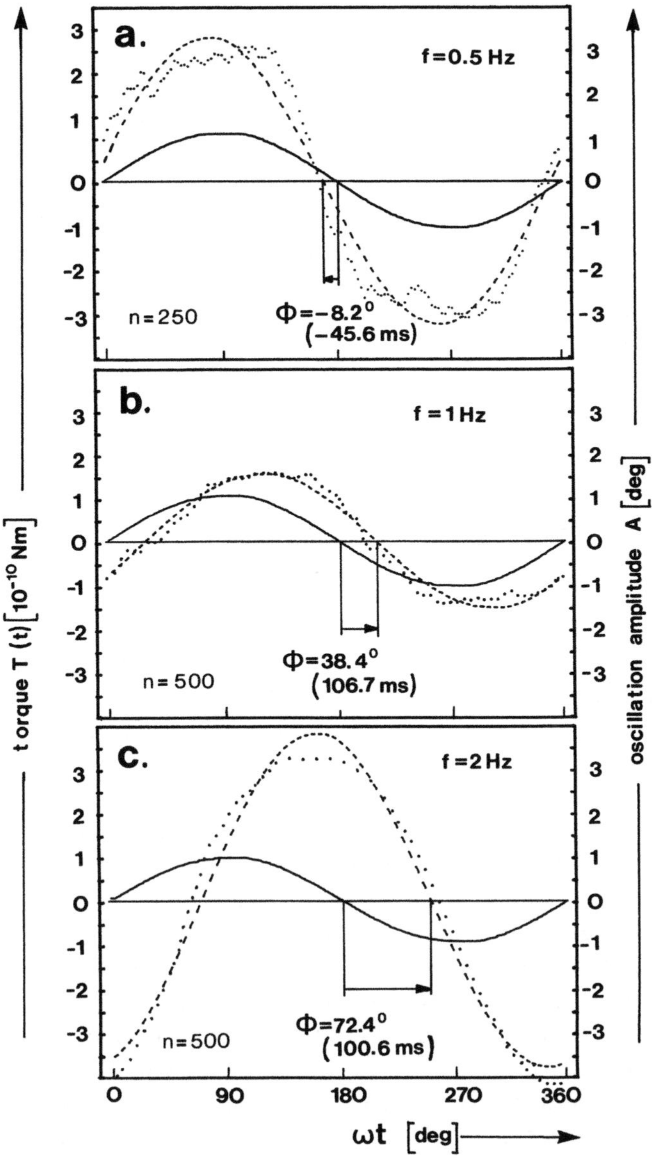

Fig. 84. Open loop yaw torque responses (dotted curves)to small angle oscillations (drawn curves; $A_{p\text{-}p} = 2°$) of a periodic grating ($\delta = 1°$ stripes spaced at 20° intervals). Stimulus frequencies and no. of cycles (*n*) are indicated in the figure. Each trace is from a different fly. For evaluation of phase and amplitude of responses regression sinus functions are calculated (*dashed curves*). *Φ* indicates phase difference between the two sinus functions

displacements of narrow lines (width $\delta < 5°$). More than 15 years ago Thorson (1964, 1966 a, b) studied isometric neck torque responses to small angle displacements of visual patterns in the locust. With large, strongly contrasted objects the locust perceives displacements as little as 20 s of arc, although the divergence angle and angular sensitivity of visual elements are in the range of $\Delta\varphi = 3°$ and $\Delta\varrho = 3°$. Thorson (1966 a) pointed out that displacements as small as that would modulate the individual photoreceptors by a factor of 0.003. Since the locust has to evaluate coherent motion of the whole visual field, the signal-to-noise ratio is improved by the modulation of many photoreceptors. Originally, Thorson proposed (1964) that the model of Hassenstein and Reichardt (1956) was sufficient but not necessary to account for the displacement response. A detailed analysis (Thorson 1966 a, b) however, showed that gain and phase properties do not conform with this model. More recently Poggio and Reichardt (1976 b), in the context of foreground–background experiments, confirmed that an array of elementary movement detectors should respond to small angle oscillation with an equidirectional oscillation as output signal (torque), provided the horizontal angular diameter of the moving pattern elements is larger than $1.5\,\Delta\varphi$. This is not the case for the experiment of Fig. 84.

An array of visual elements like the compound eye seems to be well suited to mediate responses to small angular displacements. For this type of stimulus temporal comparison of the intensity distribution in the visual field can provide a directionally selective torque signal without even the necessity for interactions of the cross-correlation type required for the motion detection model described in Chapter 3. It seems most important to find out whether it is the elementary movement detectors (EMD's) involved in large field course control which mediate the response to small angle oscillations, since this may help specify the neural mechanism underlying the EMD in more detail than the present correlation model does. Also the context in which this mode of motion detection is important remains to be studied.

Let us return now to menotaxis. The main significance of small angle fluctuations may lie in the self-control the fly can obtain perceiving the visual effect of its own torque fluctuations. We were led to investigate this effect by an apparent discrepancy between two results obtained under closely similar experimental conditions. In a series of open loop experiments we noticed that moving a stripe back-to-front often had either no or very little effect until the stripe reached the frontal region of the visual field (see p. 120 ff). On the other hand, adding to the closed loop a back-to-front displacement without opening the loop causes the fly to react immediately and to reestablish optomotor balance within 250 ms (Fig. 43). We thus compared open and closed loop conditions under as similar conditions as possible. Figure 85 shows the following three experimental situations. In Fig. 85a a vertical dark stripe (width $\delta = 5°$, height $H = \pm 40°$) is positioned at $\psi = +$ or $-120°$ and is kept

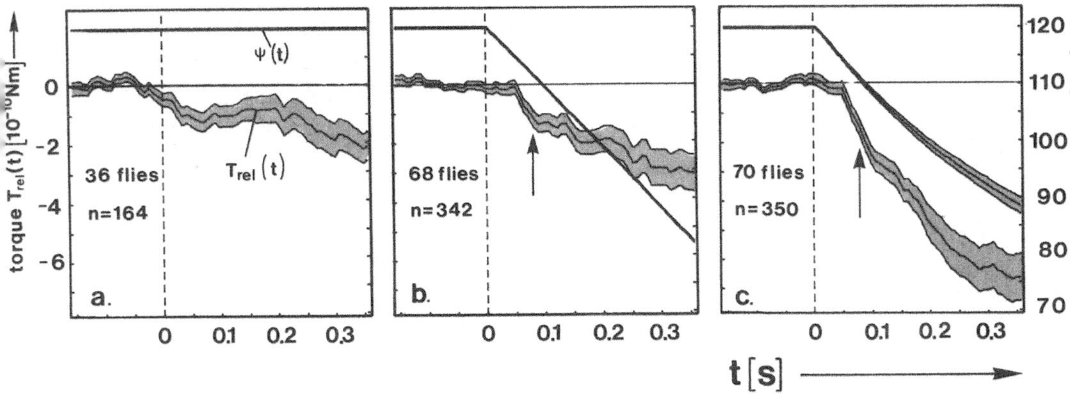

Fig. 85. Dependence of closed loop behavior upon reafferent visual stimuli. Back-to-front motion of a single vertical dark stripe ($\delta = 5°$; $H = \pm 40°$) is presented during stationary flight under open or closed loop conditions. (a) As a control the stripe remains fixed at $\psi = 120°$. Record starts when torque happens to be around $T = 10^{-9}$ Nm which in closed loop would move the stripe towards the front at a speed of $w = 110° s^{-1}$. Due to averaging a slight drift toward the mean of the torque distribution (see Fig. 57b) is observed. [Note that in (a) only about half as many sweeps have been averaged as compared to (b) and (c). Thus trace is slightly "noisier".] (b) The stripe is rotated in open loop from $\psi = 120°$ toward $\psi = 0°$ starting at $t = 0$ with angular velocity $w = 110° s^{-1}$. For $t < 0$ conditions are as in (a). (c) At $t = 0$ the feedback loop is closed. Zero torque is set to provide an initial angular velocity of the stripe of $w = 110° s^{-1}$ as in (b). Again for $t < 0$ conditions are as in (a). Although stripe movement is very similar in (b) and (c) the flies respond differently about 30 ms after onset of response (*arrows*). This time apparently is sufficient for the fly to realize the difference in the feedback situation

there without being moved. The computer registers 0.58 s of yaw torque if the yaw torque reaches a certain value, which in closed loop would move the stripe with an initial velocity of $w = 110° s^{-1}$ toward the front. In Fig. 85b the recording is started at the same criterion, but then the stripe moves in open loop with $w = 110° s^{-1}$ toward the front. In Fig 85c the computer switches to closed loop at $t = 0$ when the criterion is reached. Then the torque of the fly drives the stripe toward the front with the initial angular velocity $w_b = 110° s^{-1}$ until the fly readjusts its optomotor balance.

Significance of Reafferent Stimuli

The experiments reveal the dependence of the fly's behavior upon reafferent visual stimuli. When the stripe starts moving, the fly responds with a delay of about 50 ms. In Fig. 85b it takes another 30 ms until the fly "realizes" that its response has no effect and stops responding while in Fig. 85c it continues to respond in an effort to slow down the movement of the stripe. What the fly extracts from the reafferent signal is not yet clear, but the fly must rely on small angle fluctuations since during the latencies involved the stripe does not move very far. Responding with equidirectional yaw torque to back-to-front motion of a landmark is reasonable if the perceived motion is due to a rotation of the animal itself. If the yaw torque, however, has no effect on the perceived motion, the fly might as well abandon this strategy. *Drosophila* follows the scheme of trial and error elimination (see p. 2).

Several problems remain for the interpretation of these results. In our experiments the head of the fly was glued to the thorax. Would the yaw torque fluctuations in free flight cause the head to move in register with the thorax, or would the head oscillate alone? What about a resting fly? Does the head show long-lasting small amplitude oscillations? Finally, our torque compensator does not record frequencies much above f = 20 Hz, whereas the flicker fusion frequency for the visual system may lie considerably above that value. We also have to keep in mind that the coupling coefficient in the artificial closed loop situation is still about ten times below the "natural" value. Are we still overlooking the most significant aspects of the visual feedback loop? We have demonstrated that small angle fluctuations are of some significance during menotaxis; but even if they are eliminated by low pass filtering the fly can maintain menotaxis. Thus the role of the small angle fluctuations needs to be further characterized.

The Object Response During Menotaxis

During undisturbed menotaxis the object response is not actuated. The fly suppresses its response to the "expected" displacements of the panorama. Compensator noise may have some effect at the beginning of the closed loop experiment but soon is ignored entirely. A sudden "unexpected" movement of the stripe, however, is immediately counteracted. We have seen already (Fig. 43 on p. 89) that the addition of a D.C. voltage into the closed loop is swiftly and smoothly counteracted by the fly in the case of a single stripe whereas with a large field texture or grating the readjustment takes in the average about 5 s during which the fly often generates fast, huge torque fluctuations as if it had difficulty in mastering this situation. The fly certainly reacts very differently to the off-set, depending on whether it is confronted with a single stripe or with a large field pattern. But can the fast restitution of optomotor balance with the single stripe be attributed to the object response as it is known from open loop experiments? The answer is no, at least for the reason that the object response is very small to movement from back to front in the lateral part of the visual field. The fly finds its new optomotor balance in a fraction of a second, irrespective of the direction of motion externally applied. We have seen in the previous section that the restitution of optomotor balance is a closed loop response which does not occur in the same way under open loop conditions.

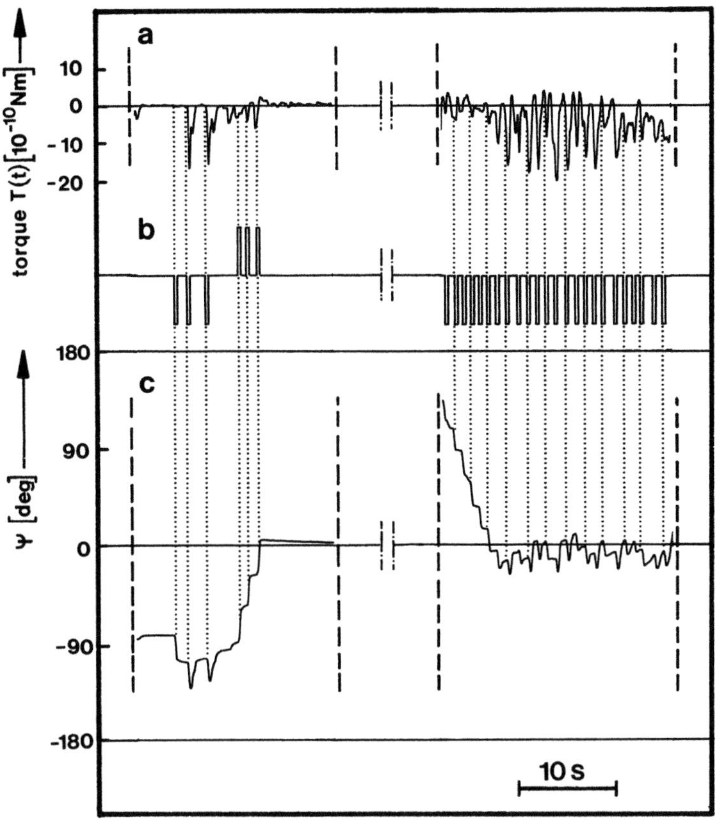

Fig. 86. Artificial displacements of a stripe during menotaxis are often corrected when the stripe is shifted front-to-back and only rarely when displacement is in the opposite direction. (a) Torque trace; (b) D.C. voltage pulses; (c) angular position of stripe. The first pulse is "ignored" although the stripe moves front-to-back. (Heisenberg and Wolf, 1979). Note that correction responses have about the right size to shift the stripe back to its previous position; closed loop coupling: 10^{-9} Nm $\triangleq$ $55° s^{-1}$

Closer inspection of the individual torque responses to back-to-front and front-to-back movement added to the movement of the panorama under closed loop conditions does after all reveal an asymmetry, but an unexpected one. The responses fall into two classes, those which counteract only the motion of the stripe and those which counteract not only the motion, but also the change in position of the stripe, shifting it back to about the position it had before. These strong responses may be the closed loop equivalent to what had been observed as object response under open loop conditions.

We have already noticed in the reafference experiment described in Fig. 51 that during stable orientation in periods of menotaxis the fly may respond to sudden displacements of the stripe with a strong phasic burst of torque. In this experiment the loop is opened for 200 ms and the stripe shifted

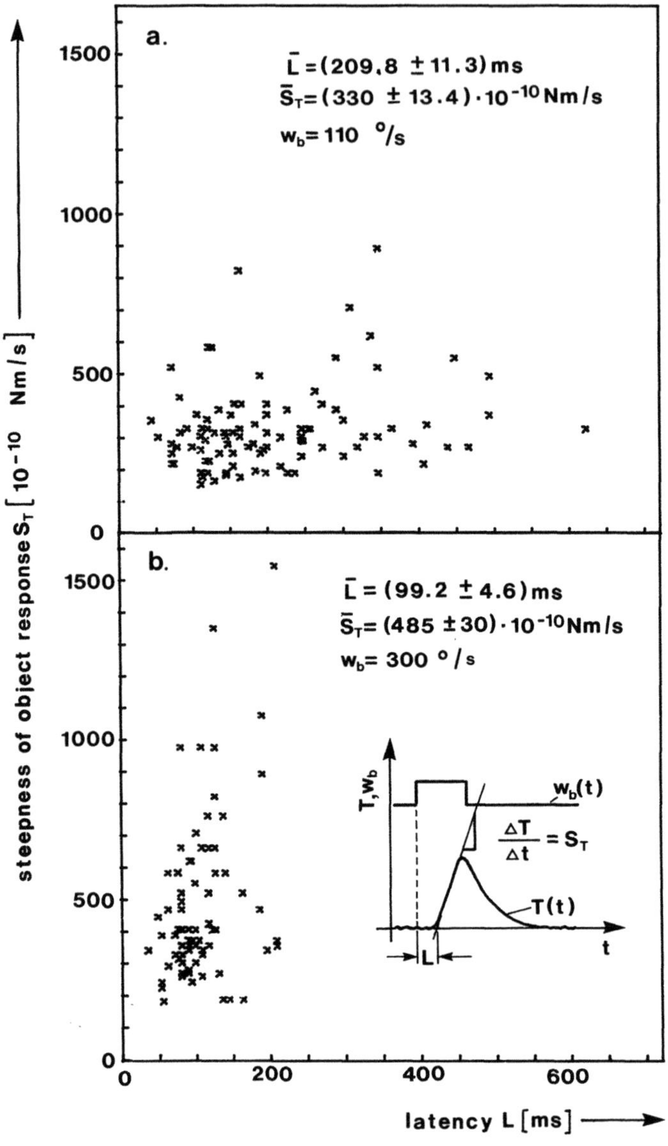

Fig. 87. Scatter diagrams for steepness S_T and latency L of object response during menotaxis. A stripe is displaced from front to back with an angular velocity (rotatory bias) of a) $w_b = 110° s^{-1}$ and b) $w_b = 300° s^{-1}$ for 200 ms. At low angular velocity (a) the response varies mostly in latency whereas at a higher velocity (b) variability in the steepness of the rising phase is more pronounced

during this period externally by $\Delta\psi = 30°$ with constant angular velocity, whereupon the loop is closed again. The torque pulses of the fly have the polarity and about the right size to counteract the effects of our voltage pulses (Fig. 86). As the example in the figure shows, the flies counteract the displacements only with a probability $p<1$. This probability depends upon the direction of displacement and to a lesser extent upon the position of the stripe in the visual field. Individual flies do both, they may respond or they may not respond. The average probability for responses to front-to-back displacements is about 80 to 90%, whereas for responses to back-to-front displacements it is 10 to 15% (for response probabilities at defined positions see Chap. 11).

In addition to their probabilistic occurrence, the responses have a most variable latency and steepness. At an angular starting velocity of $110^\circ\,s^{-1}$ (as set by the amplitude of the D.C. pulses) their latencies range from 35 to 600 ms, but the responses follow a fairly uniform time course. With an angular velocity of $300^\circ\,s^{-1}$, long delays in the responses are rare, the latencies ranging from 35 to 200 ms, whereas the steepness of their rising phase is most variable (Fig. 87). In addition we shall see in the next chapter that their time course depends upon the duration of the stimulus. We will tentatively call these responses "object responses", assuming that irrespective of whether the visual feedback loop is opened or closed the fly has a strong tendency to generate yaw torque in response to front-to-back movement of visual objects.

A series of displacements like the ones shown in Fig. 86 should cause the fly to perform fixation, since most of the disturbances shifting the stripe forward should go uncorrected, whereas most disturbances in the opposite direction would be compensated. This is in agreement with the earlier finding that the injection of large random voltage fluctuations into the closed loop lead to (forced) fixation on account of menotaxis (Fig. 76). Periodic oscillations, incidentally, lose their effectiveness in inducing forced fixation much faster than random ones. It seems that it is the degree of predictability which matters.

In summary, we have seen that during undisturbed menotaxis the object response is not elicited. If disturbances occur, the fly counteracts them either by simply reestablishing optomotor balance or by shifting the stripe back to about its previous position. The latter behavior seems to be related to the object response of open loop experiments, since its frequency of occurrence is high for front-to-back displacements of the stripe and low for displacements in the opposite direction. Menotaxis is under reafference control.

9 Foreground–Background Experiments

In Chapter 5 we discussed the degree of orientedness which a fly can acquire at the torque compensator. Closing the visual feedback loop provides a considerable increase of orientedness since it enables the fly, at least marginally, to tailor its actions according to the goals it pursues and – equally important – to keep reference to its own actions by means of reafferent stimulation. However, with a single stripe as the only moving pattern in the arena many situations are ambiguous, in particular if we "inject" additional motion into the loop. A disturbance of the stripe position leading to front-to-back motion, for instance, could stem from an involuntary angular displacement of the fly, from the fly's forward motion in the vicinity of a stationary object, or from movement of the object itself. The fly cannot distinguish between these possibilities. It has to take its chances. We now turn to more complex visual stimuli which lessen the ambiguity for the fly.

Figure–Ground Discrimination in *Musca*

Relative movement distinguishes objects. The cricket in the bush is well hidden as long as it does not move. While our eyes wander around, jumping from one feature to the next, the whole scenery is in permanent motion relative to the eye. An object can be perfectly obscured if it is textured similarly to its surround, but the motion of the object relative to the background makes it suddenly visible. This phenomenon has long been known to psychophysicists and Gestalt psychologists but it is only recently that its computational base has been elaborated in experiments with the large fly *Musca* at the torque compensator (Virsik and Reichardt, 1974, 1976; Heimburger et al., 1976; Reichardt and Poggio, 1979; Reichardt et al., 1983). The apparatus used for stimulation is derived from Reichardt's original one (Fig. 40, p. 84f) by adding a second coaxial cylinder independently driven by a servomotor and equipped with a ring potentiometer. Both cylinders can be rotated by applying external voltages to the servomotors (open loop) or by feeding back to the motors the appropriately processed torque signal of the fly (closed loop) or by a combination of both (Fig. 88). *Musca* fixates a vertical stripe cut out of a random dot texture not only when this is moving in front of a white background, but equally well when the stripe moves

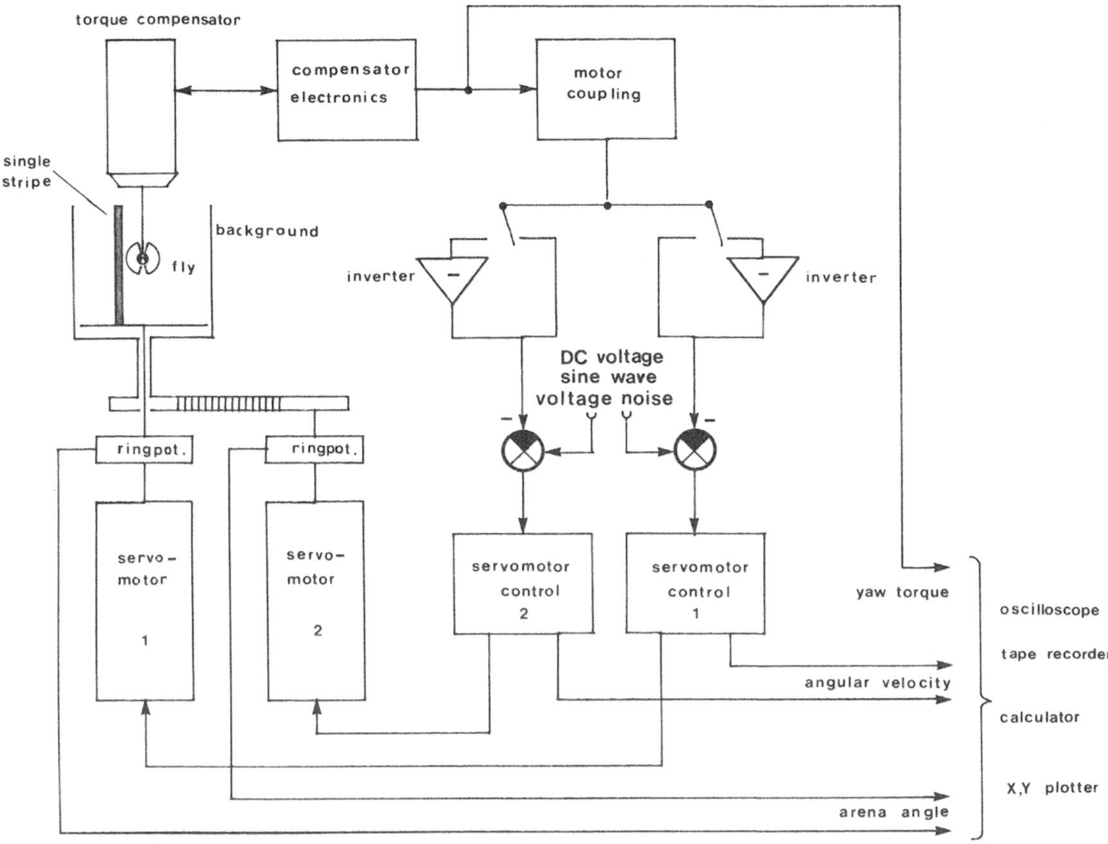

Fig. 88. Block diagram of set-up used in foreground–background experiments. (After Virsik and Reichardt, 1976). Apparatus is derived from that shown in Fig. 40 by duplicating the servomotor and motor control unit. In addition, the signals in each of the two parallel lines can be inverted if necessary

relative to a similarly textured surround. The relative motion necessary to support fixation may be very small. If both cylinders are coupled to the yaw torque the remaining imperfection in the electronic synchronization is enough to make the stripe detectable. The fly ignores the stripe, however, if it is mechanically attached to its background (Virsik and Reichardt, 1976). It also ignores a stationary stripe when only the background texture is moved. Open loop experiments show effects similar to closed loop experiments in this set-up. For instance, in open loop experiments the stripe and the texture can be independently oscillated sinusoidally with small amplitude ($A_{p\text{-}p} = 1°$ to $10°$). In most cases *Musca* generates yaw torque toward (or away from) the position of the stripe as soon as the oscillations of stripe and texture have a phase or amplitude difference (Heimburger et al., 1976).

In a similar experiment, the fly is facing two displays showing electronically generated random dot textures which move in a given direction. In order to generate the figure–ground effect, a certain area in the texture can be made to move relative to its surround (Reichardt and Poggio, 1976). Both procedures of stimulation generate basically similar responses.

Models describing figure–ground discrimination go beyond those for flicker and movement detection. The new step in the data acquisition procedure of the visual system revealed by this behavior requires a homogeneous array of nonlinear interactions between movement detectors.In previous chapters we have discussed the idea that the weighted output of elementary movement detectors is summed over large parts of the visual field. Clearly such a model does not take into account the special "attractiveness" of incoherence in the motion field. What are then the minimal requirements for an adequate model?

One Output–Multiple Input Systems

It may be useful at this point to refer briefly to the classification proposed by Poggio and Reichardt (1976 a, b) for the possible interactions in systems with many inputs and one output like the one discussed here (consisting of an array of visual elements as inputs and yaw torque as the only output). The basic idea is that in many cases such systems can be formally decomposed into the sum of simpler systems which are amenable to a mathematical treatment and can perform certain types of information processing. These subsystems which are represented by the graphs shown in Fig. 89 (modified from Reichardt and Poggio, 1979) are listed according to the number of inputs, their symmetry and the computational order of the operation performed. On the right side of the table the visual stimuli are listed which would require interactions of this type (or of higher order) in order to elicit yaw torque. For instance, an array of first-order graphs depicting "noninteractions" would be sufficient to explain the steady response to a stationary light source or pattern. Second order one-input systems ("self-interactions") would be a minimum requirement for responses to flickering light or toward a stripe oscillating with small amplitude ($A_{p\text{-}p} < \Delta\varphi$). Second order two-input systems (cross-interactions) are the lowest class needed to generate directionally selective movement responses. Here, two types can be distinguished,

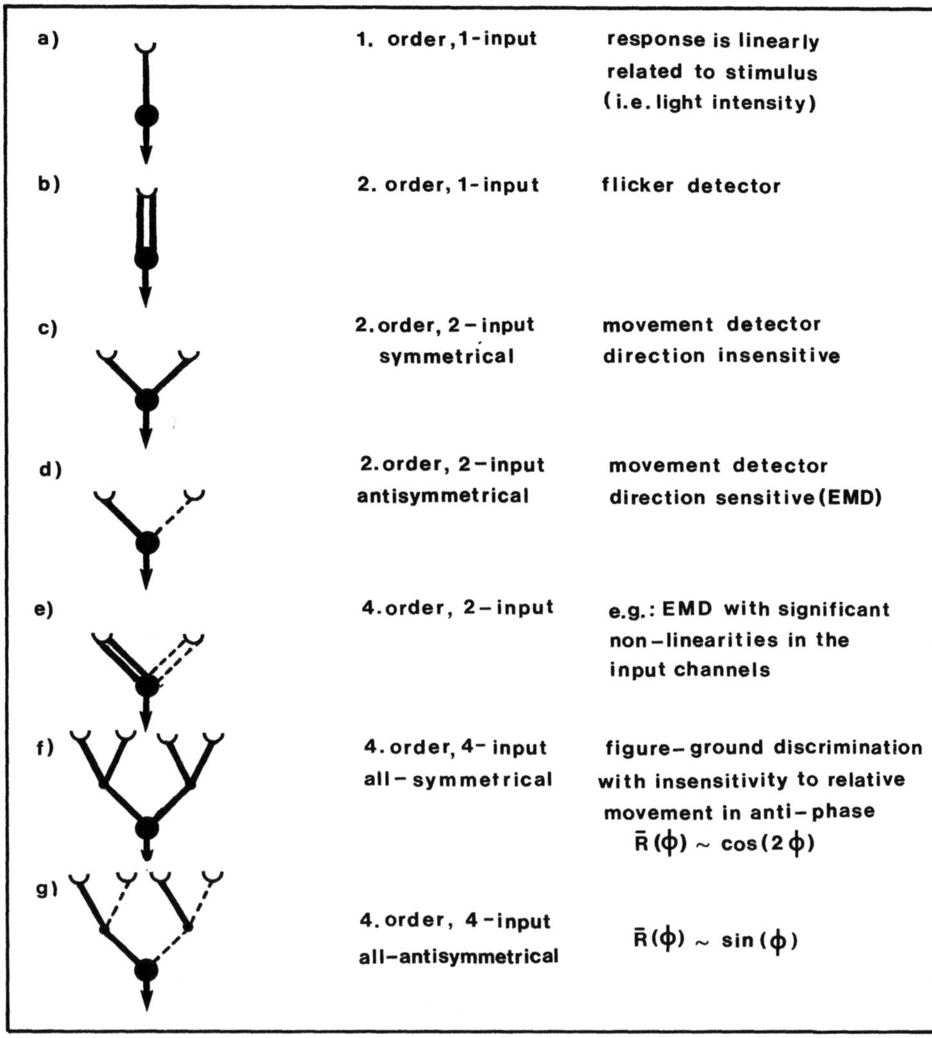

Fig. 89. Graphs depicting formal components of one output-multiple input systems in order of increasing computational complexity and number of input channels. On the right stimulus–response relations are listed for which the corresponding interactions are a minimum requirement. Φ indicates the phase difference between two oscillating patterns (After Reichardt and Poggio, 1979)

a symmetrical (Fig. 89c) and an antisymmetrical one (Fig. 89d), depending on the properties of the two input channels. As we have seen before, antisymmetry is required for movement detection. The symmetrical cross-interaction merely provides direction insensitive responses to motion.

Nonlinear interactions between movement detectors as postulated for the figure–ground effect in the present scheme are at least of the fourth order. Reichardt and Poggio (1979) tried to specify experimentally which of the many fourth-order interactions might underlie this behavior in *Musca*. Oscillation of a stripe and background in open loop showed that the average yaw torque response in this situation depends upon the phase difference Φ between the two oscillations. At high oscillation frequency (e.g., f = 2.5 Hz) the response is proportional to cos 2Φ, implying that when stripe and background texture oscillate in antiphase ($\Phi = 180°$) the torque toward the stripe is as small as for $\Phi = 0°$ (Fig. 91c; Heimburger et al., 1976). At low oscillation frequency ($f < 1.5$ Hz) this response pattern changes to a dependence upon sin Φ with negative torque at large phase differences. Reichardt and Poggio (1979) proposed an interaction of order >4, consisting of a succession of direction-sensitive movement detectors which at high oscillation frequencies become direction-insensitive (type f of Fig. 89). In fact, a simple low-pass filter in one of the input channels to each movement detector would account for this property.

A Simple Model Generating Figure–Ground Discrimination

On the basis of this analysis, Poggio et al. (1981) proposed a neural circuit implementing the figure–ground algorithm (Fig. 90). An essential part of this model are large field neurons which pool excitation of the ipsi- and contralateral EMD's and inhibit (by the shunting type of inhibition) each individual EMD. Two additional findings led to this model: the torque modulation in response to an oscillating pattern is dependent upon the oscillation amplitude and frequency, but is rather invariant with respect to the number of movement detectors stimulated. Thus a powerful gain control mechanism must be involved. Secondly, the oscillation experiments have the same outcome when the stripe segment is presented to one eye and the background texture to the other. The circuitry is well suited for gain control in respect to spatial integration and it allows the modelling of several of the properties of the open loop behavior recorded in the oscillation experiments just cited.

Hausen (1981) has shown in *Calliphora* that the input to the HSE-neuron in the lobula plate serving spatial integration of local motion sensitivity is in fact gain-controlled. But as will be shown in the remaining section of this chapter, at least for *Drosophila* the output of the HS-neurons alone does not

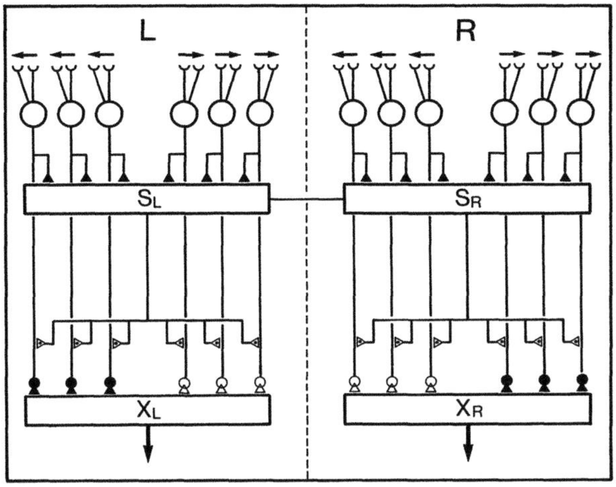

Fig. 90. Network providing gain control for spatial integration and concomitantly some degree of figure-ground discrimination. *L* and *R* denote the two eyes. Large field neurons X_L and X_R receive excitatory ▲ and inhibitory ♁ inputs from two antiparallel sets of unidirectional movement detectors. These in turn are inhibited (△) by a pair of pool neurons (S_L, S_R) which receive their excitatory input ▲ from all movement detectors. (Poggio et al., 1981, see there for further details)

account for figure–ground discrimination during tracking (although it may be an essential part of it).

Figure–Ground Discrimination in *Drosophila*

In *Drosophila* little is known about the algorithms underlying figure–ground discrimination. The main results are as follows: in stationary flight (unpublished) as well as in the walking mode (Bülthoff, 1981), *Drosophila* is able to detect a small object moving relative to a moving surround. If fore- and background are oscillating (open loop) with the same frequency and amplitude, the phase difference Φ between the two oscillations determines the strength of the torque or turning response and, as Bülthoff (1981) has shown, also the forward speed of the walking fly. In flight, torque is zero at $\Phi = 0°$ and positive for $\Phi = 180°$ (i.e., the fly tries to turn toward the side of the figure). Fourier analysis of the phase dependence shows a strong contribution of the fundamental and a weaker one of the 2nd harmonic indicating in the formalism of Poggio and Reichardt (1976 b) a predominant influence of direction-sensitive but an additional contribution of direction-insensitive interactions (Fig. 91 a, b). The result is largely independent of the oscillation frequency (between f = 0.5 to 5 Hz; Bülthoff, 1980; 1981; and own observation). The similarity to the *Musca* results is strengthened, however, by the observation that the weight of the two components (1st and 2nd harmonic) varies considerably during the experiment and from fly to fly (Bülthoff, 1980; 1981). In flight experiments with the mutant *sol* the 1st harmonic seems to be missing since at $\Phi = 0°$ and 180° the yaw torque is zero (Fig. 91 b). (Responses at phase differences $\Phi = 270°$ are similar to those at $\Phi = 90°$ but have not been included in Fig. 91 b since they are from a different sample of flies). In *sol* flies about half the cells in the medulla and lobula complex are

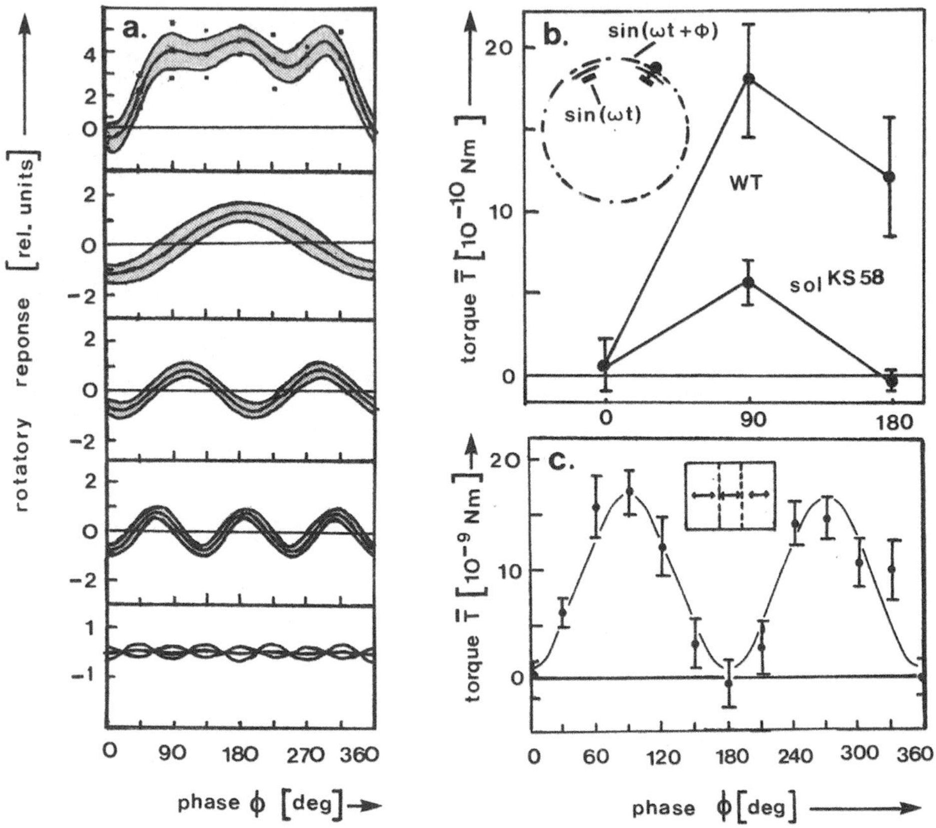

Fig. 91. Torque and turning responses to oscillating pattern and background texture under open loop conditions as functions of phase angle Φ between fore- and background oscillations. (a–c) show data obtained under slightly different stimulus conditions in different laboratories. (a) *Drosophila* walking on locomotion recorder (see Fig. 16). Double arena as shown in Fig. 88 is used for stimulation. Background texture is similar to that displayed in Fig. 45d except that squares are 15° × 15° and foreground figure consists of a vertical stripe of the same texture, two squares wide. Figure and ground both oscillate at fixed angular position $\psi = \pm 45°$ with amplitude $A_{p\text{-}p} = 10°$ and frequency f = 1.5 Hz at the relative phase angle indicated. Data points in the uppermost row are fitted by the four Fourier components below. (Bülthoff, 1981). (b) Yaw torque in flight. Random squares pattern (3° × 3°; with a 60° periodicity) is used as background and for stripes which are two pattern elements wide. Stripes are positioned at $\psi = 30°$ and $\psi = -30°$, one oscillating in phase with the background, one with the relative phase inidicated. Frequency f = 2.5 Hz; amplitude $A_{p\text{-}p} = 5°$. Two stationary white screens ($\delta = 15°$) are placed between stripes and background. (c) Yaw torque of *Musca* in stationary flight. Figure (stripe, $\delta = 10°$) and ground both consisting of a random dot texture are displayed on two oscilloscope screens in front of the fly centered at $\psi = +30°$ and $\psi = -30°$. Oscillation amplitude for figure and ground $A_{p\text{-}p} = 10°$; frequency f = 2.5 Hz (Reichardt and Poggio, 1979)

missing. It is tempting to speculate that an extra set of neurons might be required to make the relevant movement detectors direction-sensitive; but in our opinion this would be an overinterpretation of the data, since too little is known about the behavior underlying these effects.

The behavioral properties of *sol* lend no support to the idea that gain control in the optomotor pathway and figure–ground discrimination are provided by the same neural mechanism. Large field course control and object response appear to function normally in *sol*. If an as yet hypothetical gain control mechanism for spatial integration were affected (see argument on p. 162), such a defect would have shown in the contrast frequency and intensity functions with stimuli subtending large or small parts of the visual field (Fischbach and Heisenberg, 1981 and unpublished).

In closed loop experiments we encounter again the main difference between *Drosophila* and *Musca* at the torque compensator. Also in figure–ground experiments *Musca* always fixates a small pattern element, whereas *Drosophila* part of the time performs menotaxis. This holds for a variety of combinations of a figure and a textured background: (a) a stripe moving coherently with the background which carries less contrasted pattern elements; (b) a stripe segment or textured stripe moving in front of a stationary background carrying the same pattern elements; (c) similarly textured stripe and background both in closed loop with the yaw torque of the fly but moving in slight asynchrony (for *Musca*: Virsik and Reichardt, 1976; for *Drosophila*: Bülthoff, 1980; 1981).

Tracking

An interesting situation arises when the figure is rotating with a fixed angular velocity relative to the ground. Such situations have been called "tracking" experiments (Virsik and Reichardt, 1976). Let us for a moment recall the respective experiments without a background texture. Here the differences between the responses of *Drosophila* and *Musca* show already in the nomenclature. The ambiguous situation in which the fly is in closed loop with nothing but a single stripe and a D.C. voltage added into the loop has been treated in *Musca* as a tracking situation. At moderate angular velocities of the stripe the large fly still fixates the stripe with the fixation histogram being shifted to the side to which the stripe would rotate if it were not "tracked" (see Fig. 60). The shift of the fixation peak is interpreted as error angle which is proportional to the open loop angular velocity. To account for tracking against a textured background one merely has to change the proportionality constant between the tracking angle ψ_{tr} and the angular velocity of the relative motion.

In *Drosophila*, D.C. voltages producing angular velocity and yaw torque in a range corresponding to the normal position function (see Fig. 62) have little effect on fixation and menotaxis, since they are entirely compensated by optomotor balance. Thus in *Drosophila* tracking is confined to situations with relative motion between a pattern and its surround. In Fig. 92a a single stripe without background is kept menotactically. We inject into the loop a sinusoidally modulated voltage and record the yaw torque as function of this disturbance. As expected according to optomotor balance the fly follows with its torque base line in order to maintain the panorama at rest except for its own intended turns. We repeat the experiment with a stripe and a background texture both moving coherently. The result is the same (not shown). But if we now add the voltage to the stripe alone, leaving the closed loop with the background texture undisturbed, the behavior of the fly is entirely different. The large field pattern determines the torque base line, whereas in the episode shown in Fig. 92b torque spikes counteract the front-to-back motion of the stripe. The stepwise motion of the background thus takes up the disturbance which the experimenter had imparted to the stripe. But this motion does not offset optomotor balance, since torque spikes, as we have seen, are under reafference control. The fly transforms the disturbance into an actively controlled movement. If a small D.C. voltage is used as a disturbance instead of a sinusoidally modulated one, the same behavior is found.

Tracking in *Drosophila* resembles saccadic tracking reported to occur for instance in free flight of *Fannia cannicularis* (Land and Collett, 1974) and hoverflies (Collett and Land, 1975a), as well as in head movements of mantids (Lea and Mueller, 1977; Rossel, 1980). Especially in the latter case the similarities to the situation in *Drosophila* are striking. Mantids track small targets by smooth head movements if these are moving in front of a homogeneous background, but they use head saccades when the target moves against a textured background which seems to "clamp" the head (Rossel, 1980). A similar effect of background texture on eye movements has been observed with crabs (Sandemann et al., 1975).

Interplay Between Object Response and Optomotor Balance

In the following we investigate the interaction between responses toward objects and large field course control under closed loop conditions. We use the experimental procedure which previously served to demonstrate optomotor balance (Fig. 41). A slowly increasing voltage is injected into one of the loops which control the movements of the stripe and the background texture. In Fig. 93c the loop with the background is undisturbed, whereas to the movement of the stripe the increasing angular velocity is added. The figure shows the angular velocity of the stripe which is only slightly reduced by the torque of the fly. The details of the experiment reveal that the torque base line is not measurably influenced by the stripe. This is surprising since the stripe

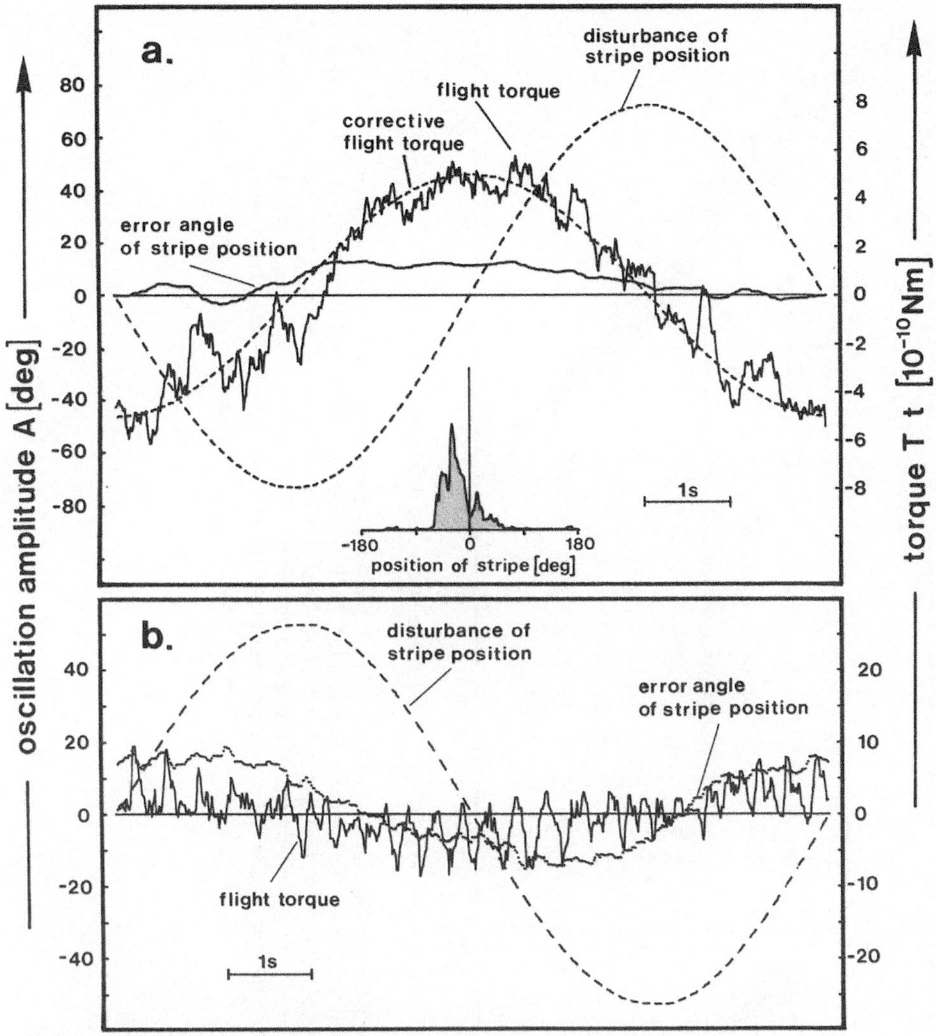

Fig. 92. Saccadic tracking occurs only if an object moves relative to a background. In (a) a fly performs menotaxis with a single stripe in a homogeneously illuminated arena. Into the loop a sinusoidally modulated voltage is added (disturbance of stripe position) which would cause a slow oscillation of the stripe if the fly would not interfere. However, the fly maintains optomotor balance generating just the appropriate torque to compensate the disturbance. The torque trace and the continuous function marked "error angle of stripe position" indicating the deviation between actual and undisturbed stripe position are averages from 20 oscillation cycles. Actual stripe position during the recording period is shown as orientation histogram in the inset. In a similar experiment (b) with a textured background in addition (random dot texture, Fig. 45c), the fly is in closed loop with stripe and background and only stripe position is disturbed externally, causing relative motion between stripe and background. The fly stabilizes the stripe by means of torque spikes but at the same time keeps optomotor balance in respect to the background. Selected single trace from WT (Berlin)

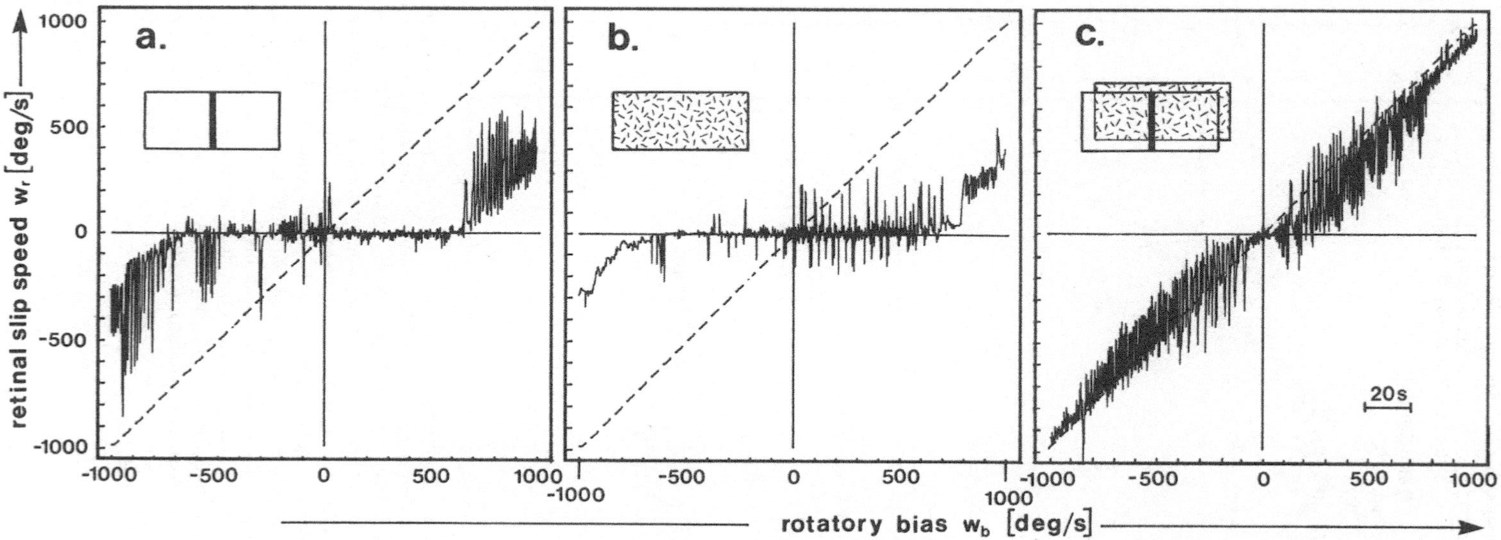

Fig. 93. Optomotor balance in figure–ground experiment. Optomotor balance (see Fig. 41) with (a) a single stripe ($\delta = 5°$; $H = \pm 40°$) and (b) little stripes pattern (see Fig. 45e). In (c) both the stripe and the texture are presented together; the slowly increasing D.C. voltage (rotatory bias, w_b) is added to the circuit driving the stripe; trace shows the retinal slip speed w_r for the stripe. Optomotor balance is maintained in respect to background texture. Stripe as foreground elicits object responses or alternatively the fly manipulates it by means of torque spikes. (d–f) Yaw torque (*indented curves*, scales on the *left*) and stripe position (*smooth curves*, scales on the *right*) traces from the experiments (a–c) on an extended time scale. *Dashed vertical line*: start of acceleration

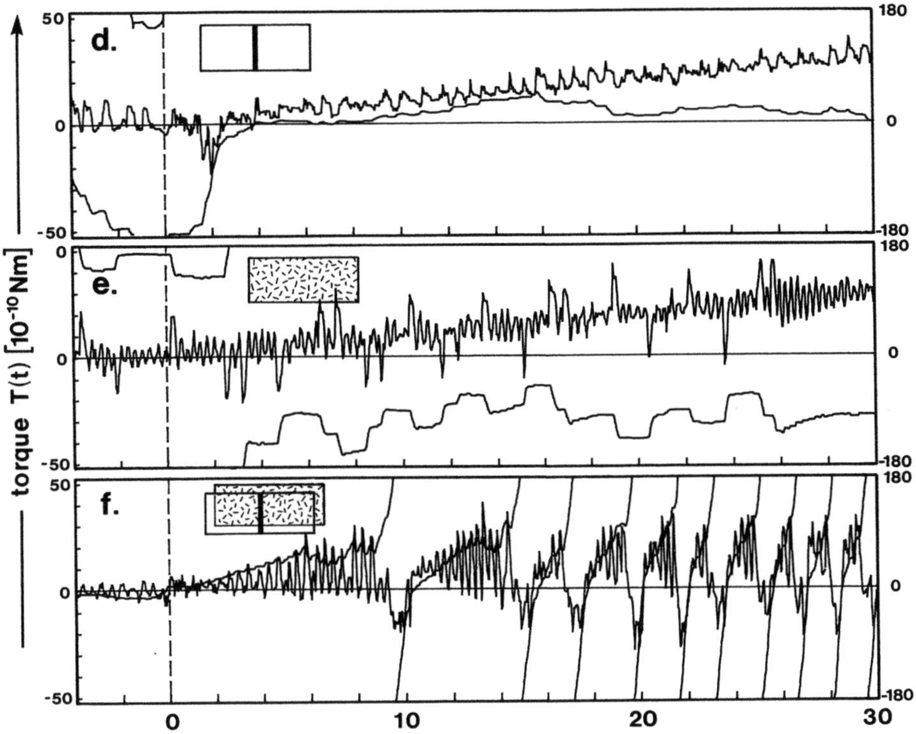

Fig. 93 d–f

alone can be stabilized at least as well as the background texture (Fig. 93 a, b). In the figure–ground experiment at slow angular velocity the fly generates torque spikes to stabilize the stripe (saccadic tracking), but soon the torque spike frequency obviously reaches saturation. Now the fly responds with one big phasic burst of torque each time the stripe crosses the direction of flight and starts moving front-to-back. The response remains phasic even if the movement of the stripe is not fully stopped or reversed. This again may be somewhat unexpected, since in open loop experiments the torque response to a single stripe moving front-to-back is a sustained response (object response, Chap. 7). However, we assume that the phasic response in the present case has the same neural origin as the object response. Presumably it is the background texture which shifts the torque back to baseline as soon as the stripe moves out of the frontal half of the visual field. The difference between open and closed loop experiments might, in addition, be due to the changes in angular velocity in the closed loop case. This idea will be pursued later.

The role of the background in the "tracking" response is illustrated by repeating the above experiment with the mutant *omb*. We see in Fig. 94 that the mutant can maintain optomotor balance with a single stripe over a

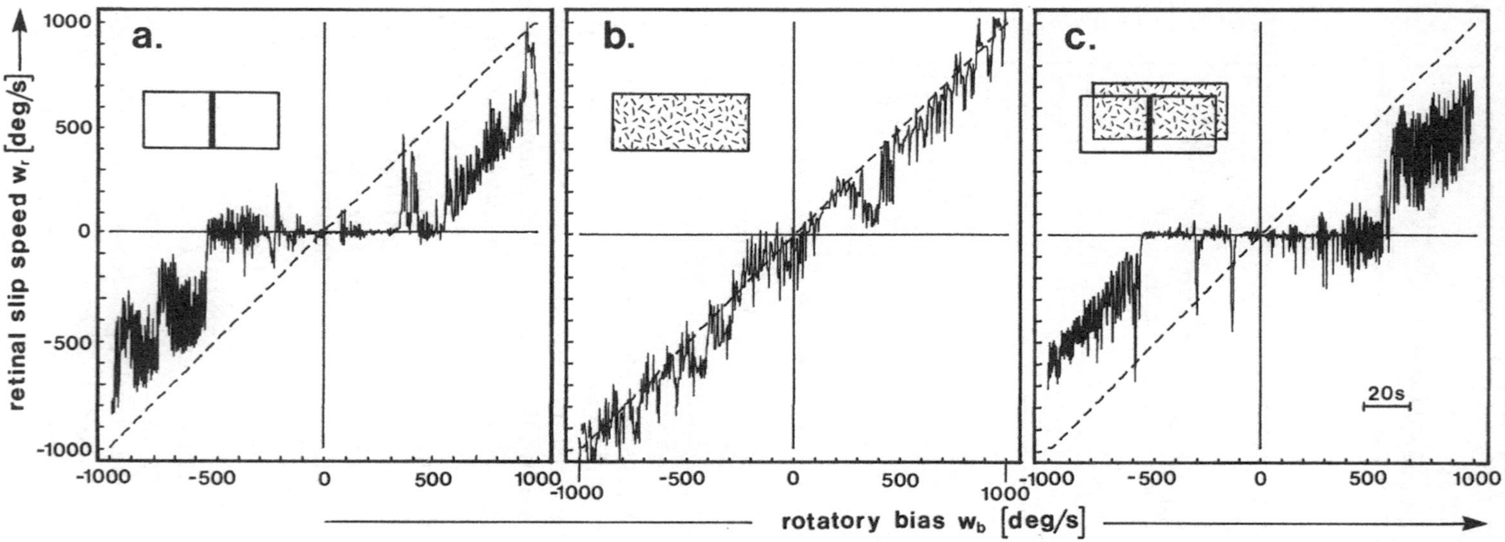

Fig. 94. (a) *omb* flies are able to maintain optomotor balance with a single stripe ($\delta = 5°$; $H = \pm 40°$) in a considerable range of angular velocities . (b) They are unable to stabilize a large field texture without landmarks (here: little stripes pattern, LT 131, see Fig. 45e). (c) The background texture has no effect on the stabilization of the stripe. (d–f) Yaw torque (indented curves, scales on the left) and stripe position (*smooth curves*, scales on the *right*) are shown on an extended time scale. *Dashed vertical line*: start of acceleration

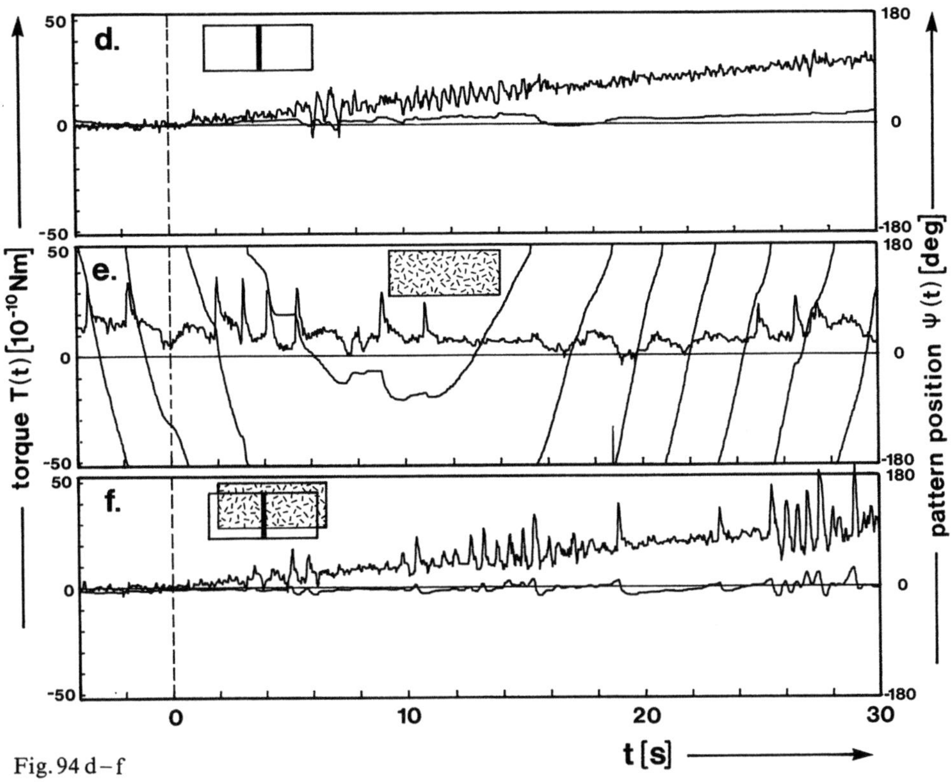

Fig. 94 d–f

considerable range of angular velocities, but is unable to use the "little stripes pattern" (Fig. 45e) in this task. In the figure–ground experiment just presented, the background texture has no influence. The *omb* flies stabilize the stripe as if it were presented alone.

It is interesting to look into the details of this process in *omb* (Fig. 95). The stripe can be stabilized only in the vicinity of $\psi = 0°$. For motion at positions $\psi_{abs} > 90°$ and also for back-to-front motion in all other positions, the fly appears to be insensitve (see p. 126ff). The yaw torque fluctuates spontaneously without reference to visual stimuli (Fig. 95). Only front-to-back motion in the anterior visual half-field leads to torque responses which arrest or revert the motion of the stripe. Thus, when the stripe crosses the frontal midline it often elicits a sufficient "object response" to be kept in frontal position. This leads to narrow fixation and torque histograms, to the establishment of optomotor balance, and to open loop optomotor yaw torque responses despite the virtually complete absence of the lobula plate giant neurons. This and the experiments reported in Figs. 65, 66, and 67 show that optomotor responses and optomotor balance with large field gratings in *omb* are brought about by the object responses to the pattern elements moving front-to-back in the anterior part of the visual field.

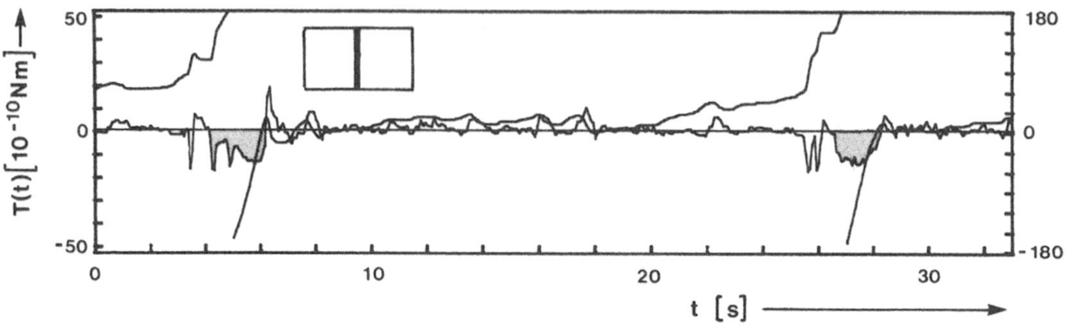

Fig. 95. Typical closed loop episode of *omb*^{H31} with a single vertical dark stripe ($\delta = 5°$; $H = \pm 40°$, no D.C. voltage added). Torque drifts away from baseline ($T = 0$) as soon as $\psi_{abs} > 90°$. While the stripe moves back-to-front the *omb* fly does not respond to the movement

We now consider the interaction between large field course control and the object response. From the experiment shown in Fig. 93 we can see that during the object response the movement of the background does not change optomotor balance, even if we take into account the long time *Drosophila* needs to readjust optomotor balance with a large field texture or grating. The same suppression of responses to movement in the "expected" direction which was already observed during torque spikes (Fig. 51) may be operating during the object response.

The interplay between object response and large field course control is nicely documented in the following most artificial situation: In Fig. 96a the "foreground" is a single gray stripe and as background we use the random dot texture of Fig. 45c. The fly is in closed loop with both. Into the loop with the stripe we add the slowly increasing angular velocity as before, but the random dot pattern is now linked to the yaw torque in a positive feedback loop. Thus torque to the right leads to clockwise rotation of the panorama. The fly is not able to stabilize the background alone. In the attempt to reduce its angular velocity it accelerates it. However, with the foreground stripe being coupled correctly (negative feedback loop), one might expect to find that the fly now should be able to ignore the background and to fly straight with the stripe. This is not the case. If the zero is carefully adjusted in the loop, the fly is caught in a cycle which keeps the stripe oscillating around $\psi = 0°$ with large amplitude and low frequency. This oscillation can be completely understood on the basis of the closed loop behavior discussed

→

Fig. 96. Optomotor balance with stripe and random dot texture (see Fig. 45c). Experimental conditions are the same as in Fig. 93c except that fly is in a positive feedback loop with the background texture (for details see text). (a) Retinal slip speed w_r of stripe as function of its rotatory bias w_b. (b) 60 s of yaw torque (*indented curve*, scale on the *left*) and stripe position (*smooth curve*, scale on the *right*) on extended time scale. *Dashed vertical line*: start of acceleration

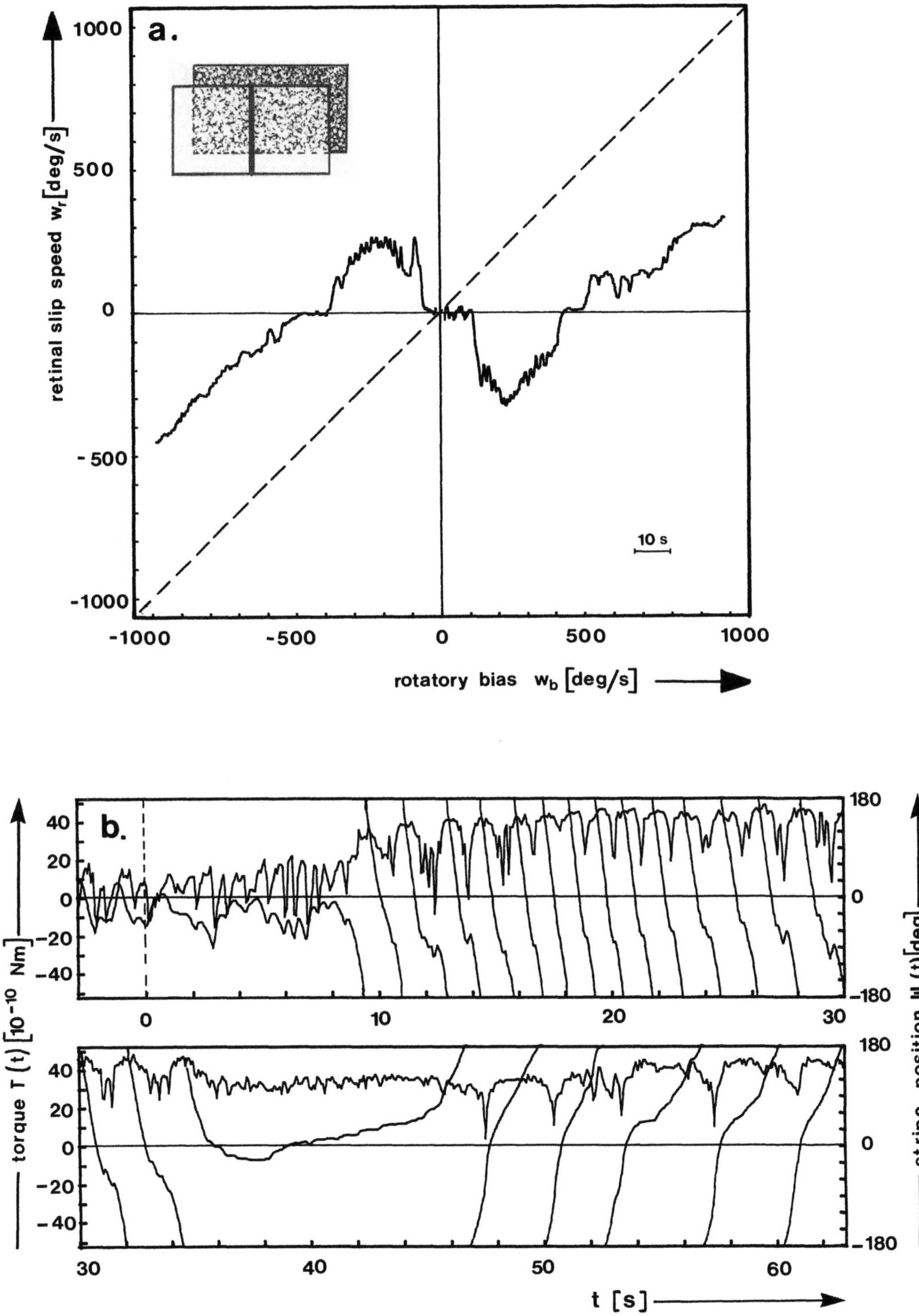
a.
1000
500
0
- 500
-1000
retinal slip speed w_r [deg/s]
10 s
-1000
-500
0
500
1000
rotatory bias w_b [deg/s]
b.
40
20
0
-20
-40
0
10
20
30
180
0
-180
30
40
50
60
torque T (t) [10^{-10} Nm]
stripe position ψ (t) [deg]
t [s]

so far. The course control response is accelerating the background until the stripe which rotates in the opposite direction starts moving front-to-back. This triggers the object response aimed at setting the stripe back to a frontal position. When the stripe changes its direction of motion so does the background, which as soon as the figure does not move front-to-back, is determining the torque output. The positive feedback accelerates the background, and this again drives the stripe to move front-to-back on the opposite side.

The same cycle can be observed with a single arena if a periodic grating with spatial wavelength $\lambda_{sp} = 6°$ is used in which one space between stripes is closed. A conventional negative feedback loop between yaw torque and arena leads to inverted coupling conditions for the apparent motion of the periodic grating [due to geometrical interference (Götz, 1964)] and to normal coupling with the single 9° stripe. Again the stripe oscillates with large amplitude around $\psi = 0°$ and the fly is unable to switch to menotaxis.

Returning to the double arena experiment, we now add an increasing rotatory bias to the stripe. The torque the fly has to generate in order to stabilize the stripe now differs from that necessary to stop the background. While stabilizing the stripe the fly has to cope with an increasing rotatory bias from the background texture, which in addition is positively coupled to flight torque. For a few seconds the object response is sufficiently strong to keep the stripe in the frontal visual field but then both patterns start rotating. As a consequence of the fly's attempt to stabilize the stripe, it moves against the externally applied rotatory bias at first. With steadily increasing rotatory bias the actual movement of the stripe slows down. When it is sufficiently small the fly again manages to fixate the stripe for a little while (8–9 s) by a slight modulation of its torque, which is otherwise "clamped" by the background motion. The mechanism for this stabilization is unknown. With the further increase of the external bias the stripe starts rotating anew, but now in the same direction as the background and the bias. This experiment can be performed in a number of permutations. They all show the striking ability of the fly to use large field course control and object response independently for improving its orientedness.

Dissociation into Object and Background

In the figure–ground experiment in which stripe and texture are coupled normally to yaw torque, the background obviously provides course stabilization and the stripe is used as landmark for the actual orientation. Without the background the stripe takes over both functions. However, the tendency of the fly to distinguish between object and background is very strong. If two identical stripes are presented with a fixed angular velocity relative to one another, the fly often uses one to stabilize its course and from this base line it pursues the other by saccadic tracking (Fig. 97). Thus one stripe is used as

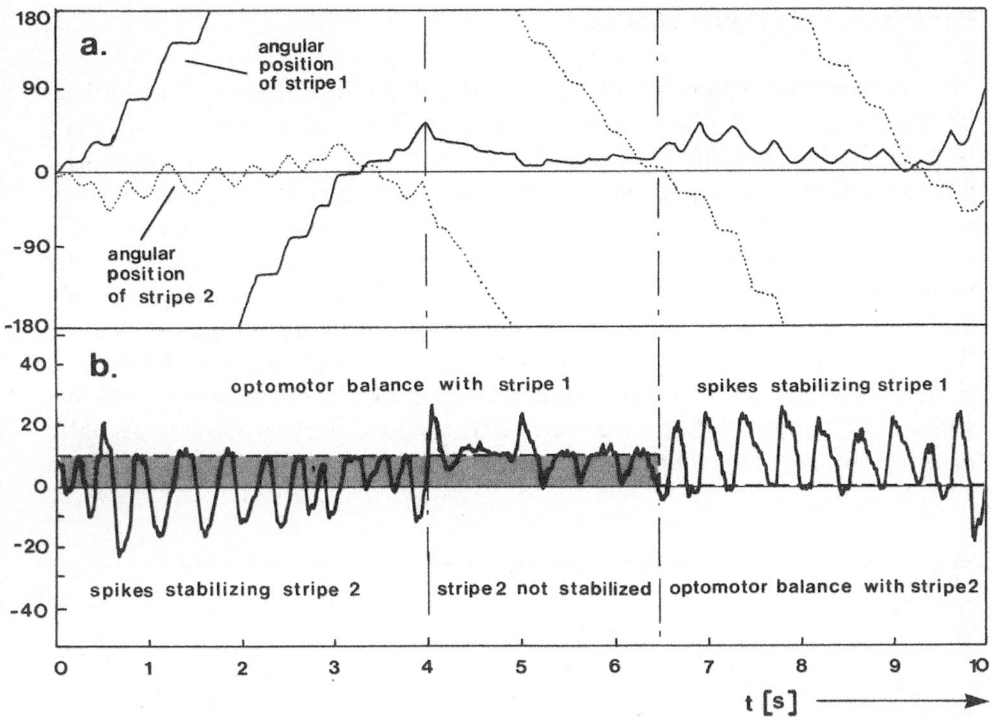

Fig. 97. "Dissociation" of two stripes into "foreground" and "background". Closed loop conditions with two identical looking stripes ($\delta = 5°$, $H = \pm 40°$) made to rotate at an angular velocity $w = 90° s^{-1}$ relative to one another. The fly often adjusts its optomotor balance to one of the stripes and (as in Fig. 92 b) stabilizes the other by means of torque spikes. From time to time the fly switches as to which stripe it attributes the background function. (a) Position traces of the two stripes; (b) torque trace of same flight period. *Shaded area* indicates the off-set of the torque base line, while stripe 1 serves as "background"

"foreground", one as "background". Which of the stripes has the one or the other function changes from time to time.[7] The switch is likely to occur when the "background stripe" moves front-to-back through the anterior part of the visual field. We propose that the distinction between figure and ground is made whenever possible according to the size, pattern contrast, coherence of motion and other properties which still have to be defined, and that the fly applies these "concepts" tentatively in order to gain as much orientedness as possible. In the experiments described in this chapter, the concepts of "figure" and "ground" are revealed to the experimenter by the large field course control and object responses.

[7] This behavior is not observed all the time. Alternatively the fly may adjust its torque to the mean of the two values. In this case the stripes rotate relative to one another with equal speed and in opposite direction (see also Chap. 11).

Interpretation of Object Response

Over several chapters we have been postponing the question of the biological meaning of the object response, since we have no definite answer. We have deliberately avoided the term "tracking". After all, vertical stripes like grass, branches, fences, edges of houses, garbage cans etc. in reality do not move by themselves, but they do move from front to back through the fly's visual field as soon as the fly has a forward velocity. We propose that at the torque compensator the fly interprets front-to-back motion as being possibly due to its own forward flight. In particular, if a background texture allows the fly to maintain a stable orientation, front-to-back motion is likely to signal a nearby object (motion parallax). Not all objects are attractive. We will see in Chapter 13 that for certain patterns the object response is partially inverted. But the fact that most patterns do elicit a positive object response

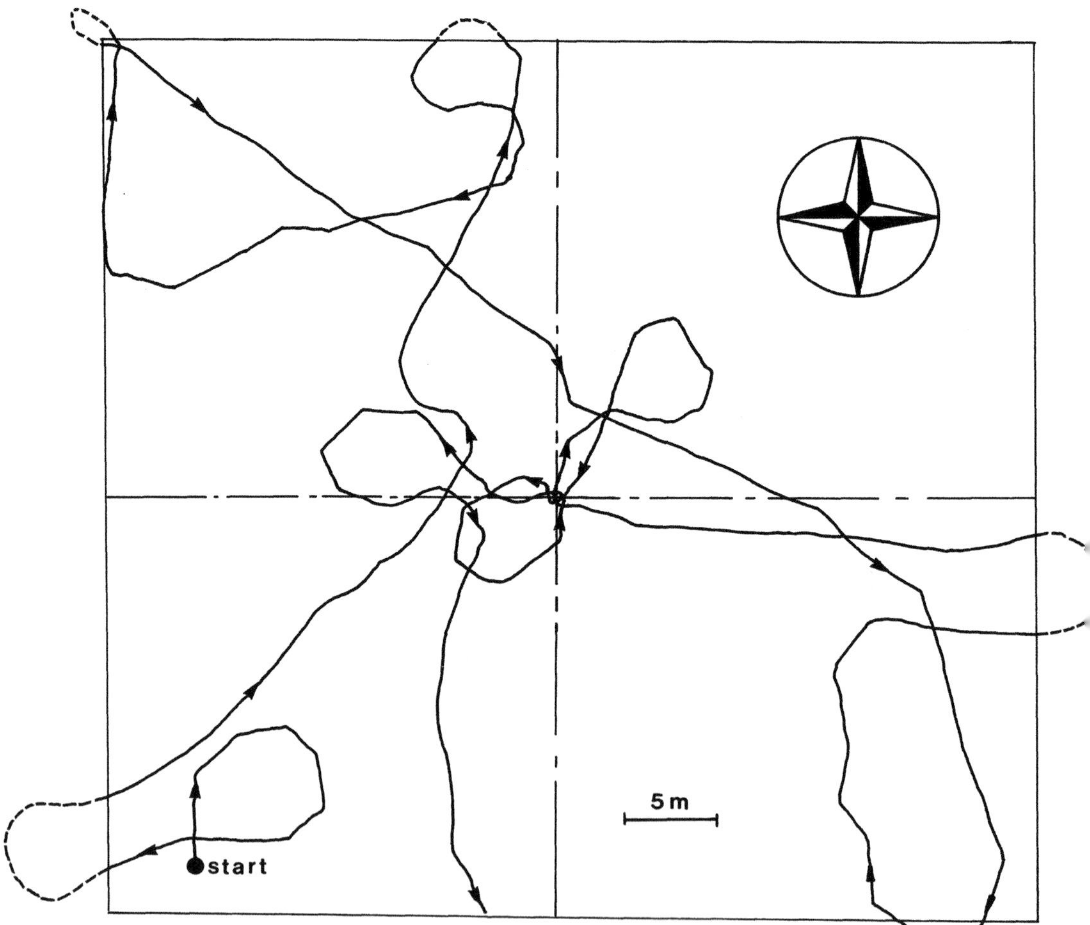

may have a simple explanation. A fly should take even the slightest chance to escape from the bizarre situation of tethered flight by trying to approach a nearby object.

However, this seems not to be the whole story. It may be the lack of context making front-to-back motion in the previous experiments so irresistible for the fly. Front-to-back motion due to motion parallax follows a certain pattern of acceleration. In our experiments involving artificially applied front-to-back motion, the dynamics of motion parallax so far has not been simulated. In forward flight of a given speed the angular velocity of a stationary object depends upon two variables: its angular position and its distance; but the two are not independent. Starting at a certain distance from an object, the fly determines by its angular orientation the distance changes and thus the angular velocity of the object. This situation will now be imitated in the double arena (Fig. 98). Figure and ground are in closed loop with

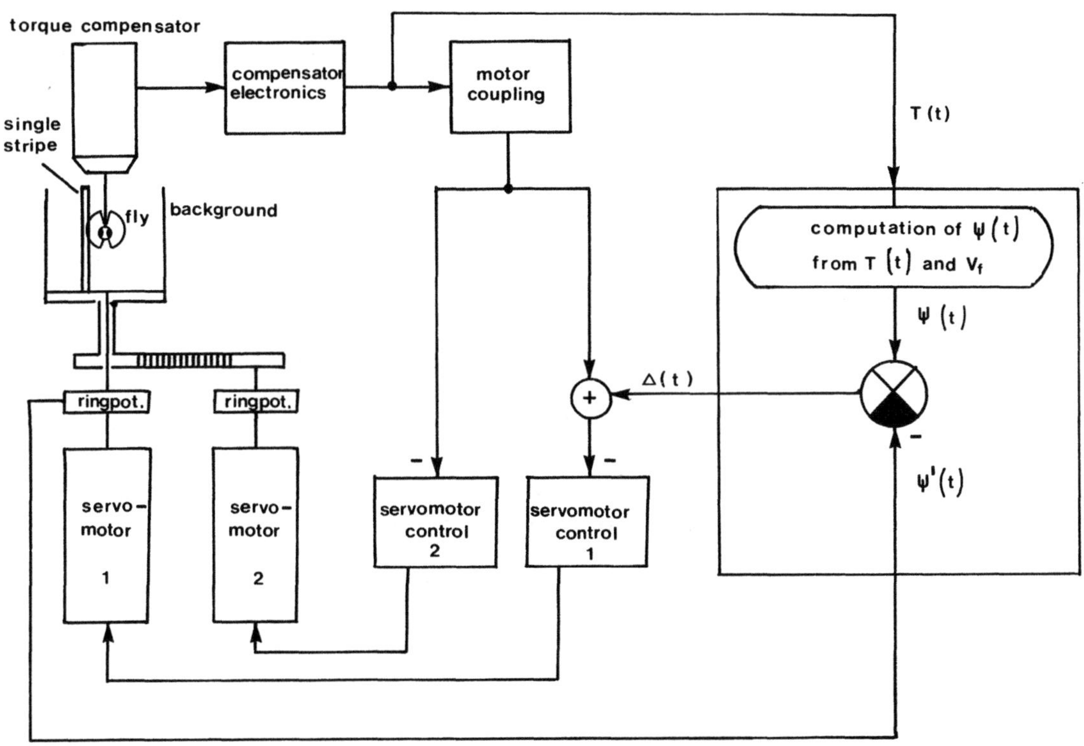

Fig. 98. Artificial flight trajectory (*left diagram*) in stationary flight at the torque meter. Stripe is in the center of the field. A constant forward velocity of $V = 0.5\ ms^{-1}$ and an arbitrary starting position are chosen. Motion parallax of the stripe relative to the background texture (random dot pattern, Fig. 45c) is simulated by the computer (*right diagram*). Required angular position of stripe $\psi(t)$ is calculated from relative direction of flight [$\psi^*(t)$], flight velocity V_f and position coordinates of the fly. Calculated angular position is compared with real one [$\psi'(t)$], the difference $\Delta(t)$ is added into the closed loop of yaw torque and stripe. Note that simulation is poor in the vicinity of stripe since retinal size of visual object does not expand

yaw torque but the figure is, in addition, moved according to the relative motion it would undergo at its position in an imaginary (two-dimensional) space. What the double arena cannot provide is the concomitant expansion or diminution of the object and the landing site where the flight should end. So to speak, the fly always misses the object when getting close. Exclusively front-to-back motion of the stripe (calculated from the estimated constant forward velocity of the fly and the momentary position of the stripe) is added into the loop. A compensatory response of the fly (object response) should show up in the flight trajectory of Fig. 98 as a spiral around the "object" in the center of the graph. In none of the experiments has this been found. It seems that the fly's behavior in respect to objects is less stereotyped when the visual effect of forward flight is simulated.

In summary, large field course control and object response appear to operate independently, being elicited by background and foreground respectively. The background determines the torque base line from which the object response starts. This holds true also when optomotor balance is not reached. The movement of the object does not offset optomotor balance. Movement of the background caused by flight maneuvres directed toward objects (torque spikes, object response) does not elicit course control responses. Thus the object response has an active component.

Can the *Musca* Theory on Pattern-Induced Flight Orientation Be Applied to *Drosophila*?

We have described the striking similarity between *Drosophila* and *Musca* in their basic performance at the torque compensator in open loop experiments. Both fly species show optomotor responses to large field stimuli and they possess an independent yaw torque response to front-to-back motion of objects in the lower part of the visual field. Relative motion between a figure and its surround leads to torque toward (or away from) that figure. Finally, without stimuli yaw torque consists of spontaneous fluctuations and of occasional torque spikes. If one goes into the details one does, after all, observe some differences even in open loop behavior; but these do not seem to be crucial. We have, for instance, mentioned differences in the spontaneous torque fluctuations. They are distinctly nonrandom in *Drosophila* and were interpreted as searching behavior. But one might determine the "spectral composition" of these fluctuations and define them as "pseudo-noise" in Eqs. (3) or (4) on pp. 115, 119.

The problems arise if one turns to closed loop behavior. An important prerequisite of the theory on pattern-induced flight orientation is that in *Musca* closed loop behavior can be quantitatively predicted from open loop experiments. With *Drosophila* at present this is not the case. Closed loop behavior at the torque compensator is still too rich and the stimulus-response analysis in open loop experiments is too coarse to fit together. But more than that, the torque fluctuations, which themselves are influenced by sensory stimuli, reveal initiation, fixed action patterns, and internal states (moods) of the fly (see *Synopsis*). The different states of the fly cannot so far be adequately controlled experimentally. According to its object response and the "spectral composition" of torque fluctuations in open loop *Drosophila* should fixate a single stripe in a frontal position as *Musca* does. This happens only part of the time, and as we have discussed before, most of fixation behavior is not even due to the object response. Fixation of the type observed in *Musca* is an infrequent event occurring mainly at the onset of closed loop experiments or with additional noise added to the closed loop. As mentioned on p. 111, fixation prevails, however, in experiments in which wing beat amplitude is used to control the panorama (Götz, 1983). It is likely that certain phases of *Drosophila* closed loop behavior like "forced fixation" and "forced" oscillation (Fig. 96) in the figure–ground experiment in which the coupling for the ground is inverted will be amenable to the type of analysis chosen for visual flight orientation of *Musca*.

The main difference between open and closed loop behavior in *Drosophila* is due to the fact that the momentary yaw torque output modifies the processing of the visual input. We have seen this in the suppression of responses to reafferent stimulation during torque spikes and in the suppression of the input to the course control system during the object response in closed loop. In addition, the perception of the reafference modifies the responses to exafferent stimuli. This property has been most apparent during torque spikes and has been inferred also for optomotor balance. Responses to back-to-front motion in many flies depend critically upon self-induced motion of the pattern.

In summary, the behavior for which the *Musca* theory has been designed is most probably part of the behavioral repertoire of *Drosophila* at the torque compensator, but it clearly marks a state of the animal more related to the time averages of open loop experiments than to the multi-faceted maneuvres of free flight.

10 Visual Control in Free Flight

Experiments at the torque compensator inevitably lead to the question of how the behavior observed relates to free flight. Detailed accounts of visually guided free flight behavior have been provided in a series of most elegant papers by Land and Collett (1974; Collett and Land, 1975 a, b; 1978; Collett 1980 a, b) for the muscid fly *Fannia cannicularis* and in particular for hoverflies. Flies display an elaborate repertoire of visually controlled flight maneuvres of which very little has been detected as yet with *Drosophila* and *Musca* in stationary flight. Since in free flight studies data are obtained through frame by frame analysis of motion pictures, normally only short episodes are recorded, showing the dependence of the flight trajectory upon certain visual stimuli such as, for instance, another fly, a stationary object among other objects, or a large rotating drum with vertical stripes surrounding the fly.

Many such episodes show male flies chasing other flies in the attempt to catch a conspecific female for copulation. This is a real tracking situation, in which the fly is confronted with a target moving relative to its surround, and both fore- and background are moving relative to the chasing fly. In the case of male *Syritta pipiens*, three different tracking strategies have been described. Toward targets outside the fovea, a *Syritta* male mostly turns by means of accurately aimed, ballistic body saccades. Objects within the fovea he pursues by a smooth tracking mechanism in which the error angle of the target image on the retina is converted into angular velocity ("position servo"). In addition, he can track an object by modulating his sideways velocity instead of making a turn. Again he uses the position of the target image on the retina to derive an average sideways velocity. Interestingly, *Syritta* occasionally shows saccadic tracking while keeping a moving object at a lateral position in the visual field.

For approaching stationary objects *Syritta* uses saccadic turns and the adjustment of the ratio between forward and sideways velocity, but apparently not the smooth tracking mechanism. Approaches again involve relative motion between the target and the background due to motion parallax.

In a rotating drum *Syritta* can stabilize its direction of gaze by turning smoothly, by body saccades or by translatory sideways motion. It often applies a combination of these (Collett and Land, 1975 a; Collett, 1980 a). Similar to the approaches toward stationary objects, these maneuvres are highly variable. Hoverflies can estimate their distance from a conspecific fly.

Males use this to keep a fixed tracking distance and to circle around resting flies by adjusting their sideways velocity to their angular velocity.

In houseflies chasing behavior seems to be also male-specific. With the leading fly in the frontal part of the chasing fly's visual field, the angular velocity of the chasing fly is determined by the angular velocity of the leading fly, in more lateral parts by its angular position (*Fannia*: Land and Collett, 1974). In chasing behavior of *Musca* males angular velocity is also linearly related to the angular position of the target. This dependence holds for the whole visual field. Like *Fannia*, *Musca* chases from below, keeping the target in the upper part of the visual field. Chasing behavior is not found in stationary flight at the torque compensator. In contrast, free flight tracking in female *Musca* is confined to the lower part of the visual field and the linear relationship between the excentricity of the target and the angular velocity holds only for the range $-20° < \psi < 20°$. Free flight tracking of females can be well accounted for by the properties of tracking behavior measured in stationary flight (Wehrhahn et al., 1982).

Large hoverflies (*Eristalis* and *Volucella*) have adopted a special chasing strategy which takes account of their larger momentum and their consequently slightly poorer maneuverability. They intercept conspecific flies by "extrapolating" their flight trajectories from their angular position, direction of motion and angular velocity and by "knowing" the species-specific flight speed, the distance at which a conspecific can be detected, and their own acceleration. Interception is an open loop "strike", which normally is followed by a continuously guided chase unless the fly abandons the project (Collett and Land, 1978).

In *Musca* and *Calliphora* the different tracking performances of males and females have found a structural correlate in a set of male-specific large field tangential neurons in a part of the lobula subserving the anterior frontal visual field (Hausen and Strausfeld, 1980). It is in this region that *Musca* males try to keep their quarry during their pursuits, and it is just this part of the eye in which the rhabdomeres of retinula cells R7 of *Musca* were found to be specialized for maximal sensitivity, thus enabling them to support retinula cells R1–6 in the detection of low pattern contrast (Franceschini et al., 1981; Hardie et al., 1981). This sexual dimorphism in neuronal structure and behavior, reminiscent of the dorsal eyes of Bibionidae (Zeil, 1979), may provide the opportunity to investigate whether the male specific neurons are a separate neuronal pathway or rather a modification of the existing network. Furthermore, the male-specific giant fibers in the lobula may be good candidates for the study of the influences of behavioral states on the performance of single neurons.

No such neurons have been detected in *Drosophila*. The prelude to sexual intercourse, which in large flies may occur in flight and appears more like a rape, in *Drosophila* is a cordial pursuit on the ground accompanied by a love song (Bennet-Clark and Ewing, 1970) and the exchange of chemical and tactile signs of affection (for visual components of courtship tracking in *Drosophila*: see Chap. 13). Each species has its own ways; but what has been learned for *Drosophila* from the study of free-flight behavior in larger flies is the abundance of behavioral "subroutines" for course maintenance, tracking a moving target, and approaching a stationary object. This gives real behavior a high degree of flexibility and unpredictability. The behavior of *Drosophila* females at the torque compensator, where it is restricted to one degree of freedom (i.e., rotation around the vertical body axis) nevertheless is surprisingly similar in several respects to that of male *Syritta*, although male-specific chasing, of course, is not found. *Drosophila* at the torque meter, like *Syritta* in free flight, can and normally does fly straight in a stationary surround. What may not be obvious from the motion picture analysis is that flies stabilize their orientation in an active process in which they react to self-induced displacements of the image on the retina. The closed loop gain of $G = 0.8$ found for *Syritta* in a rotating drum (Collett, 1980 b) may possibly be due to stationary landmarks outside the drum and may well be $G = 1$ if they are eliminated as in the case of *Drosophila* (see discussion on p. 92 ff). Voluntary changes of orientation in both species are often performed by body saccades. The additional ability of *Syritta* to change the ratio between forward and sideways velocity cannot be detected with *Drosophila* at the torque compensator. But C. T. David (personal communication) observed in the wind tunnel that even *Drosophila* does not always keep its longitudinal body axis aligned with its direction of flight.

If saccadic tracking at the torque compensator is directed toward objects (as we have proposed), it may have its parallel in the saccades by which *Syritta* changes its course when approaching a flower. The stimulus parameters by which saccadic tracking is controlled at the torque compensator have still to be investigated and may, in fact, be as variable as the behavior itself in free flight. Finally, as expected, the object response does not seem to occur as regularly in free flight as at the torque compensator.

Collett and Land (1975 a) emphasize that the flexibility of behavior, well known to the ethologist, is most apparent also in the microstructure of visual guidance of flight. They deliberately use terms like "intentionality", "voluntary behavior", "selective attention", "behavioral states" etc. One theme of this book is to argue that these terms designate, in fact fundamental properties of behavior based upon initiation. However, as mentioned in the introduction, the problem with observations of freely moving animals in a complex environment is to prove beyond doubt that meaningful behavior may indeed be independent of releasing stimuli. In this respect experiments on stationarily flying and walking animals may offer a distinct advantage, because of their potential precision in the determination of input–output relations, even if the price, the perturbation of the animal's motivational state, is high.

11 Selective Attention

Considering the huge amount of sensory information reaching the nervous system at any moment, one of the major tasks of the brain is to separate the important from the irrelevant. If all information were processed continuously with equal scrutiny, the depth of analysis would be seriously limited by the size of the brain. Among the most powerful means of a brain to cope with this situation is the ability to limit some of its operations to a momentarily selected fraction of the sensory input. Such a mechanism may be particularly appropriate in the case of compound eyes, which unceasingly sample almost all of visual space.

Free insects certainly show many signs of attention. When you approach with your hand a housefly which is standing on the table grooming, it may at a sudden movement of your finger stop grooming and lower its body slightly in order to be ready for a quick start. Collett and Land (1975 a) describe how *Syritta* in a box full of other flies and flowers actively selects a target for pursuit or approach, and that during this maneuvre it ignores other flies or landing sites in the vicinity. Already Helmholtz (1866; cited edition 1910) noted that a human subject can shift its attention to any part of the visual field without shifting its fixation point to this position:

"Es ist in dieser Beziehung die Aufmerksamkeit ganz unabhängig von der Stellung und Akkommodation des Auges, überhaupt von irgendeiner der bekannten Veränderungen in und an diesem Organe, und demgemäß kann sie mit einer selbstbewußten und willkürlichen Anstrengung auf eine bestimmte Stelle in dem absolut dunklen und unterschiedslosen Gesichtsfelde hingerichtet werden. Es ist dies einer der auffallendsten Versuche für eine künftige Theorie der Aufmerksamkeit." [8]

Recently this phenomenon has been investigated in more detail (Posner et al., 1980; Del Pezzo and Hoffman, 1980), and was compared to "a spot light that enhances the detection of an event within its beam" (Del Pezzo and Hoffman, 1980).

Drosophila displays selective attention at the torque compensator in a manner similar to that which Helmholtz has described for humans. Under

---

[8] "In this respect attention is completely independent of the position and accommodation of the eye, generally of any of the known changes in and at this organ. Thus it can be directed by a self-conscious, voluntary effort to a certain position in the absolutely dark and contour-less visual field. This is one of the most conspicuous experiments for a future theory of attention."

certain conditions *Drosophila* restricts its yaw torque responses to parts of the visual field, be it spontaneously or guided by other sensory stimuli (Wolf and Heisenberg, 1980). We will briefly review the experiments demonstrating this phenomenon.

The available data do not go far beyond the original findings. The reason for this is that instead of working out the details, we have been preoccupied with the interpretation of these results. True selective attention requires the existence of intentions, of centrally represented goals pursued by the animal, and thus of a deep structure of behavior, which so far has been out of reach for a more rigorous system analysis. This will be discussed later.

Displacement Experiments

Let us consider the following situation: The fly flies under open loop conditions at the torque compensator. It is surrounded by a transparent cylinder brightly illuminated from behind. At an angular position $\psi = +45°$ a dark vertical stripe ($\delta = 5°$, $H = \pm 40°$) is positioned. From time to time it is suddenly shifted to $\psi = +75°$ with a constant angular velocity of $w = 150° s^{-1}$. In most cases the fly responds with a short burst of torque, which in closed loop would shift the stripe back to about the position where it had been before (object response). In open loop, the experimenter has to move the stripe back between trials. The response amplitude is variable, but in most cases responses can be clearly distinguished from cases where the fly does not respond. Now we place a second stripe into the arena, which remains at the fixed angular position $\psi = -45°$ throughout the experiment. This stationary second stripe reduces the frequency of responses to the displacement of the first stripe by about 50%. This result is remarkable, since it shows that in sationary flight, patterns which are neither moving nor flickering are by no means ignored, although they may not induce a strong yaw torque response toward them (see also p. 82). We consider here the "inhibitory" effect of the stationary stripe upon the object response.

As a next step we perform a similar experiment under closed loop conditions. Again one stripe is displaced by $\Delta\psi_{disp} = 30°$ in 200 ms from front to back (while the loop is opened) and the response frequency is measured in the presence or absence of a second stripe. The result (Fig. 99) is similar to that in the open loop case. The fly responds to the displacement of one stripe alone about twice as often as when it is accompanied by a second stripe which is not displaced. The angular separation ($\Delta\psi_{sep}$) and position (ψ_1, ψ_2) of the stripes is variable in this experiment. The only condition is that each of the two frontal quadrants contains one stripe at the onset of the displacement. In an additional experiment (Fig. 99c) the angle between the two stripes is

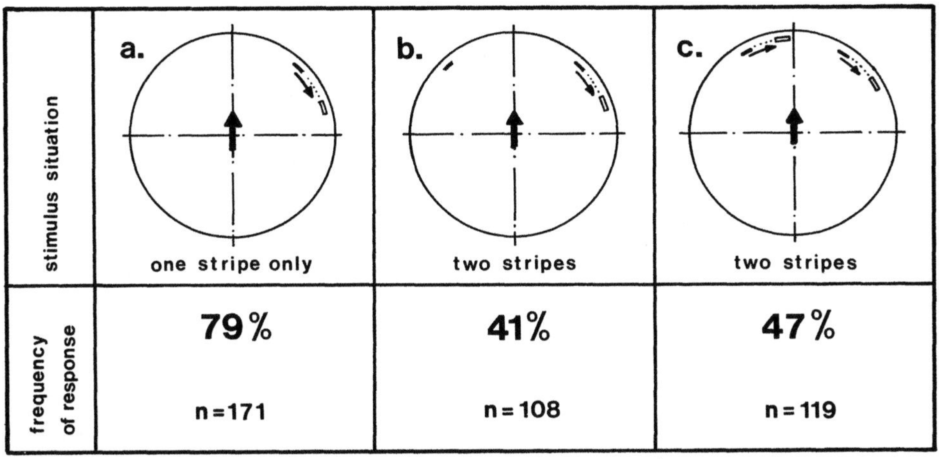

Fig. 99. Displacement of a vertical dark stripe ($\delta = 5°$; $H = \pm 40°$) from front to back during closed loop. Angular velocity $w = 150°\ s^{-1}$; displacement angle $\Delta\psi_{disp} = 30°$; position of stripe at onset of displacement $-60° < \psi < 60°$ for (a) and (b), $-30° < \psi < 30°$ for (c). In (a) only one stripe is presented (cf. Fig. 86). In (b) a second stripe is presented simultaneously at a variable angular distance (due to the displacement of the first stripe). The fly is under closed loop conditions with both stripes but the position of the second stripe remains undisturbed. In (c) a fixed angular distance of $\Delta\psi_{sep} = 60°$ is maintained during the displacement. For a displacement to be generated each stripe has to be in one of the frontal quadrants of the visual field. Presence of a second stripe in the frontal visual field reduces the response frequency

kept constant ($\Delta\psi_{sep} = 60°$), and both stripes are displaced together, one from front to back and the other from back to front, i.e., both stripes stay in their quadrant during the displacement. This condition restricts the position of the two stripes. The one to be displaced from front to back has to be at positions $-30° < \psi < +30°$. This is important, since stripe position which is a critical variable is not explicitly recorded. Thus even with the displaced stripe in the frontal part of the visual field, the response frequency in the presence of a second stripe is significantly lower than in the one-stripe experiment. The readiness to respond to the displacement seems to be divided between the two stripes.

In the following experiment the stimulus is slightly different. The angular velocity of the stripe during the 200 ms displacement now is $w = 300°\ s^{-1}$ (and thus $\Delta\psi_{disp} = 60°$). When two stripes are presented they are kept at a fixed angular distance $\Delta\psi_{sep} = 90°$. Only due to the displacement is the distance briefly increased to $\Delta\psi_{sep} = 150°$ or decreased to $\Delta\psi_{sep} = 30°$, whereupon it is slowly readjusted to its original value with a constant angular velocity of $w = 30°\ s^{-1}$. It is always the same stripe which receives first the fast and then the slow antagonistic part of the disturbance. The flies are tested in closed loop and except for stripe positions in the rear of the fly a stripe may be displaced at any position relative to the fly and to the other stripe. Stripe positions are recorded, together with the response of the fly. Thus a large variety of stimulus constellations can be distinguished. We have analyzed 1485 displacements with 28 flies. We will briefly summarize the main results. Some of them, drawn from a different (smaller) set of data have been published already in a somewhat preliminary form (Wolf and Heisenberg, 1980).

The responses clearly fall into six types with very few ambiguous cases (Fig. 100). The fly either generates a large phasic response (object response) during the fast displacement (D–F) or it does not (A–C). During the slow readjustment of the angular distance between the stripes the fly has three options as to its optomotor balance. It can ignore the angular motion of the stripe which is being displaced and keep the other stripe stabilized (A, D), it can do the opposite, i.e., stabilize the stripe which receives the (slow) disturbance by readjusting its optomotor balance and thus cause the other stripe to rotate (B, E). Finally, it can adjust its torque to the mean of the values in the two preceding cases, causing both stripes to rotate with half the angular velocity in opposite directions (C, F). We find here the same response strategies the fly applies when flying in closed loop with two stripes which have a constant angular velocity relative to one another over longer periods of time (see Fig. 97). In the present context, each of the three responses may or may not be preceded by the object response to the fast displacement. (In the control experiment with a single stripe type C and F responses are not found. The fly has no reason to generate only half the torque required to stabilize the stripe. The fact that we do not find such cases is a reassurance for the validity of our classification).

Unlike in the previous experiment, the object response to the fast and large displacement is suppressed by a second stripe only under certain conditions. The displacement has to occur in lateral positions or from back to front. The second phase of the experiment, the adjustment of optomotor balance during the slow reverse displacement, however, is influenced in most cases. With a second stripe in the arena the slow displacement is ignored in nearly 50% of the cases. Furthermore, if the slow displacement is compensated, this event very often is preceded by an object response to the fast displacement; if the slow displacement is ignored often also the fast displacement had not elicited a response.

Interestingly, of the 251 cases in which the fly adjusts its optomotor balance to a mean torque value which causes both stripes to rotate with equal velocity and opposite direction (Fig. 100C+F) 81% are "converging" displacements. The two stripes are separated by $\Delta\psi_{sep} = 30°$ at the onset of the slow phase of the disturbance. Obviously, the small distance between them favors this response type.

We take these results to suggest that *Drosophila* can restrict some of its responses to varying parts of the visual field, and that this selective attention

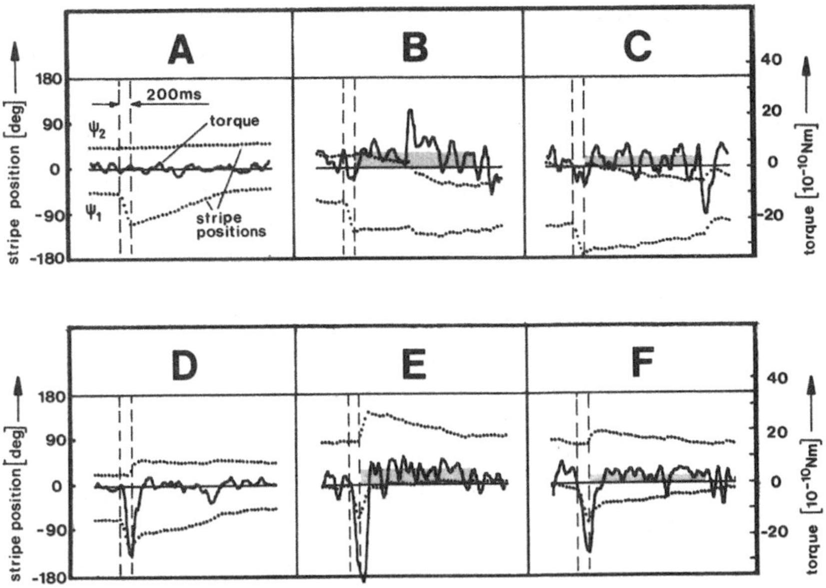

Fig. 100. Fast displacement of a stripe ($\delta = 5°$; H = $\pm 40°$) in one direction, immediately followed by a slow displacement in the opposite direction during closed loop with two stripes elicits six types of responses (A–F). Displacement angle $\Delta\psi_{disp} = 60°$; angular velocity w = $300° \, s^{-1}$ and $30° \, s^{-1}$. In between trials angular distance of stripes is $\Delta\psi_{sep} = 90°$. *Dotted curves* show position of stripes (scale on the *left*); *dashed vertical lines* mark the fast stimulation period. See further explanations in the text. (After Wolf and Heisenberg, 1980)

is guided visually. However, before we evaluate the data further we will review still another set of experiments.

Open Loop Oscillations

As described in Chapter 7, slow open loop oscillations of a stripe in the fronto-lateral visual field cause a characteristic torque pattern consisting of the object response during front-to-back motion and torque spikes toward the position of the stripe during back-to-front motion (Fig. 68). Superimposed on this fine structure one finds a shift of the torque base line toward the position of the stripe (Fig. 101, after Wolf and Heisenberg, 1980). If one shifts the stripe to the other side, one obtains a mirror image of the previous torque pattern. (The scores in Figs. 101 and 102 are particularly difficult to read. The relevant behavior is superimposed by the small angle oscillations of the fly and the compensator noise. The polarity of torque spikes is not obvious, but the reader can determine it by looking for their steep rising phase. The traces shown are drawn from a large sample of similar data.)

We now present two stripes at once, one at the position $\psi_1 = 45°$ the other at $\psi_2 = -45°$, oscillating coherently. The fly is now confronted with back-to-front and front-to-back movement simultaneously. As is to be expected assuming selective attention, the torque pattern is very similar to that in the single-stripe oscillation experiment, except that now the fly spontaneously switches from one side to the other, inverting its torque fine structure and shifting its base line of torque to the other side.

A similar but somewhat less regular torque trace is recorded with the pattern of random squares (Fig. 45d) as panorama. In this case the fly is confronted with several accumulations of dark or bright squares to which to attend. We can still distinguish whether the attended patterns are on the right or on the left side of the fly (Fig. 102a). We do not know yet what the torque trace looks like when the fly follows oscillations in the front or back part of the visual field.

Visual attention is not only directed voluntarily, it can also be guided by stimuli. In fact, in single-stripe experiments we observe that the fly's attention stays with the stripe for most of the time, whereas with two stripes the time is divided about evenly between them. With only one stripe the fly appears still to be free not to attend to it. We noticed in the displacement experiment (see Fig. 86) that the fly occasionally does not respond, without any obvious reason. Also in the oscillation experiment with one stripe, short episodes are found during which the fly shifts its torque base line to the other side and suppresses the characteristic fine structure. We designed two pilot experiments to show the influence of external stimuli directly. In the first one the random squares pattern again is slowly oscillating and we record yaw torque. There are two stationary white screens (width $\delta = 15°$) at the angular positions $\psi_{1,2} = +30°$ and $-30°$ in front of the pattern. In front of one of them a vertical stripe is oscillating with small amplitude ($A_{p\text{-}p} = 3°$) and high frequency (f = 2.5 Hz). This small amplitude oscillation elicits a small torque modulation which can be extracted from the torque trace by averaging. However, if the stripe oscillates in front of the right screen, the fly preferentially follows the slow oscillation of the background pattern elements on the right side, and when the stripe oscillates in front of the left screen the fly inverts its torque pattern and shifts its base line of torque as if flying a

→

Fig. 101. (a) Yaw torque under open loop conditions in view of an oscillating stripe as described in Fig. 68c. *Shaded areas* indicate times of clockwise rotation. Sine functions show stripe position (frequency f = 0.3 Hz; amplitude $A_{p\text{-}p} = 30°$, *left hand scale*). Stripe is oscillating either around $\psi = 45°$ or $\psi = -45°$. In (b) two stripes are presented simultaneously and oscillate around the positions $\psi_1 = 45°$ and $\psi_2 = -45°$ in phase and with the same amplitude. The fly maintains the same typical torque pattern it shows with one stripe alone but alternates as to which stripe it selects. Remember that torque spikes are best identified by their steep rising phase. Also note the sudden decay (at *) of torque just prior to torque spike. (Wolf and Heisenberg, 1980)

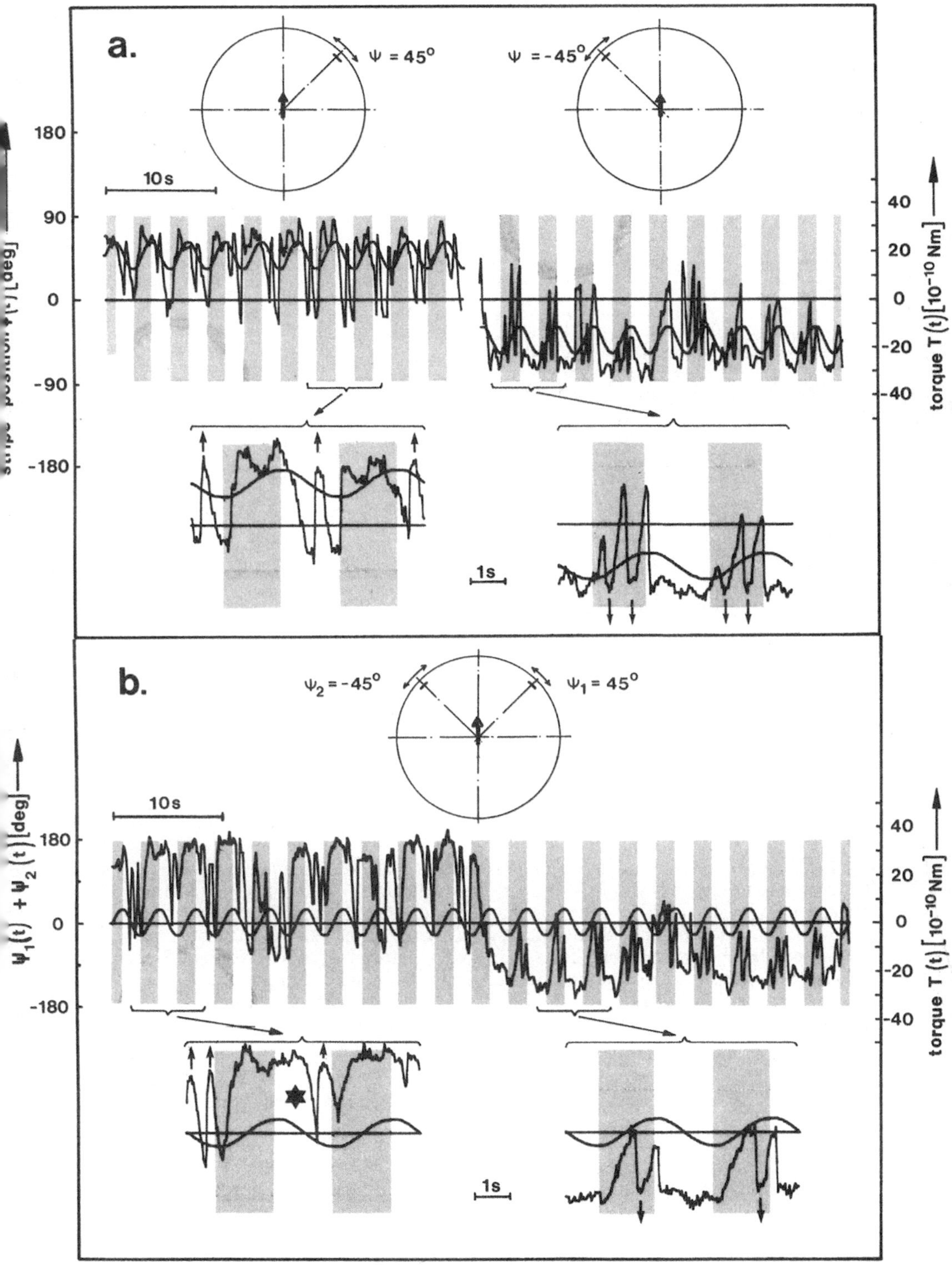
a.
Ψ = 45°
Ψ = -45°
10s
180
90
0
-90
-180
40
20
0
-20
-40
torque T(t) [10^-10 Nm]
1s
b.
Ψ2 = -45°
Ψ1 = 45°
10s
Ψ1(t) + Ψ2(t) [deg]
180
0
-180
40
20
0
-20
-40
torque T(t) [10^-10 Nm]
1s

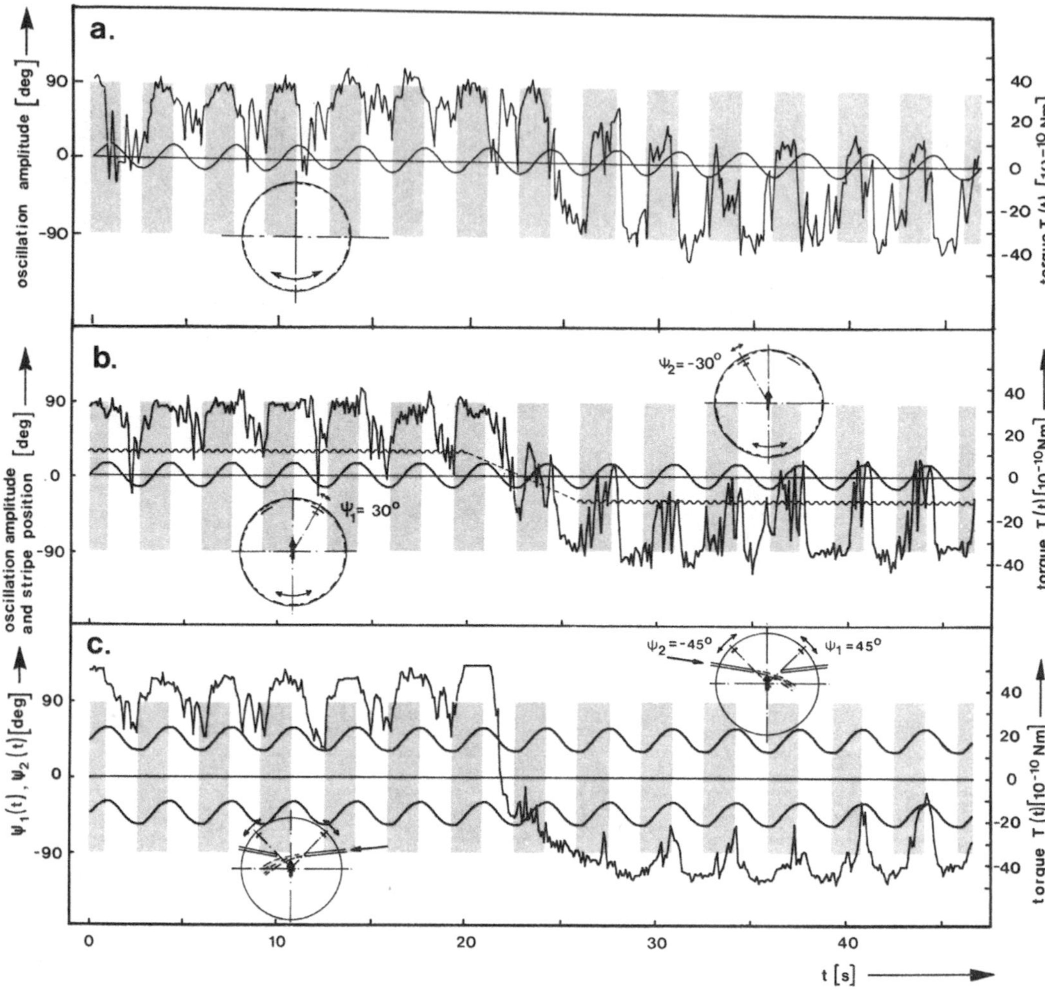

Fig. 102. Guidance of visual attention. (a) Surrounded by the random squares pattern (Fig. 45d) the fly is able to respond with its torque to the oscillation ($f = 0.3$Hz; $A_{p\text{-}p} = 30°$) of pattern elements in the right or in the left visual half-field. (b) The fly can be prompted to attend to features on one side if on this side a vertical stripe in front of the texture (and a stationary white sreen) is oscillating with frequency $f = 2.5$ Hz and amplitude $A_{p\text{-}p} = 3°$. (Wolf and Heisenberg, 1980). (c) Stimulus situation as in Fig. 101b except that either from the right or from the left odor of fermenting banana is blown onto the fly. The fly responds to the oscillations of the stripe on that side from which odor is blown. (Heisenberg et al., 1983)

continuous left turn (Fig. 102b). It seems that the fly directs its attention to the side of the stripe while responding to the background pattern.

We were curious whether visual attention could be guided by other sensory stimuli besides visual ones. This would suggest that sensory as well as voluntary control of attention is a central process. We tried sound, wind, heat, and odor with little success. What finally did work was to blow a jet of air scented with the odor of fermenting banana past the fly just in front of it. The air current produced by the beating wings obviously draws enough of the odor onto the antennae to tell the fly its quality and the direction from where it comes. When the air is blown from the right side, the fly tries to turn right and if confronted at the same time with two slowly oscillating stripes, it follows the oscillation on the right side. If we switch the jet of air to blow from the left, the fly inverts its response pattern and tries to turn left (Fig. 102c). We do not know whether the odor is essential, nor have we determined the range of stimulus conditions at which this effect occurs, nevertheless this experiment is in agreement with our hypothesis that attention is centrally controlled.

We have noted in the displacement experiment that phasic responses to large, fast displacements ($\Delta\psi_{\mathrm{disp}} = 60°$) in the frontal part of the visual field are not significantly suppressed by the presence of a second stripe. Thus, certain stimuli escape the "screen", which limits the response to a particular area in the visual field. We find a similar phenomenon in the oscillation experiment. If two stripes are oscillating with different frequencies, the fly most of the time responds to the sum of the movements (Wolf and Heisenberg, 1980).

Taking the displacement and oscillation experiments together, it seems reasonable to conclude that *Drosophila* under certain conditions restricts some of its responses to selected parts of the visual field. The properties of this screening process have yet to be worked out, but some of them are already apparent from the data just presented. What we find most remarkable is that selective attention can be measured under open loop conditions. It must be a dominant trait of the visual system, since it depends neither upon a high level of orientedness which the fly cannot attain in tethered flight, nor upon a special motivational state. It concerns the responses toward objects as well as the adjustment of optomotor balance. During the whole recording period of at least 30 min the fly does not cease to suppress its responses to stimuli in parts of the visual field.

Still, attention is largely an active behavior. The fly selects the part of the visual field to which it attends. Stimuli have an influence on the length of time the fly spends attending to a certain region. Certain stimuli, like the large, fast displacements seem to attract attention instantaneously. This effect is well known from selective attention in humans. Even without significant external stimuli the fly does not distribute its attention evenly, but preferentially directs it to the front. This bias may reflect the importance of the frontal part of the visual field during forward flight.

A slightly different interpretation was suggested by Collett (1980 a), who observed a similar phenomenon for the control of free flight in *Syritta*. Hoverflies can fly in any direction; during forward flight they use the frontal part of the visual field for the control of their angular velocity, but when flying sideways they ignore movement in this part, instead controlling their angular velocity according to the motion on the side which they approach. Collett points out that during cruising flight in a stationary surround, motion at the poles of the flow field is unambiguously due to self-rotation, whereas in other parts of the visual field image motion is due to the sum of translatory and rotatory components of the self-motion (see *Optomotor Equilibrium* on p. 43 f; Götz, 1975 a). Thus it is a valuable strategy to limit angular velocity control to motion in the pole areas of the visual field. We are tempted to interpret Collett's data to indicate that *Syritta*, like *Drosophila*, preferentially shifts its attention into the direction of flight. It would be interesting to monitor forward and sideways forces of *Drosophila* in stationary flight (sideways flight in *Drosophila* may, however, require a lateral tilt of the body which a tethered fly would not be able to perform), and to compare the intended direction of flight with the direction of attention.

It is conceivable that spontaneous shifts of attention occur at a fixed frequency if no additional cues are provided. For freely moving *Drosophila*, however, attention will be mostly guided by the situation. A telling example is the experiment of Bülthoff et al. (1982) described on p. 144. The fly, which is walking back and forth between two landmarks, fixates one of them as long as it can approach it. On being stopped by a barrier, it turns around and fixates the other landmark. This switch could well be mediated by a shift of attention in response to the "problem" the fly faces arriving at the obstacle.

We like to think that attention as an active process should be centrally controlled. If the stimuli guiding it were purely visual, one might still assume this variable weighting function to be an intrinsic property of the visual ganglia. As we pointed out in the introduction, initiation on which behavioral activity is based is a very general property of central nervous systems and it would not be too astonishing to find it to be an autonomous process in the visual ganglia. On the other hand, this assumption would be difficult to prove. Also, our finding that olfactory or wind signals can guide attention also seems to indicate control of attention outside the visual system. However, much more has to be done on this issue.

As cited in the opening remarks of this chapter, selective attention has been compared with the beam of a spotlight. The experimental paradigm used here allows one to measure the size of this beam. We have some preliminary indications as to its horizontal extent. The finding that *Drosophila* adjusts its optomotor balance preferentially to the sum of the movements of the two stripes when these are separated by $\Delta\psi_{sep} = 30°$ at the onset of the slow readjustment phase indicates that the width of the beam presumably is larger than $\Delta\psi = 30°$. This response type (C + F) is found for 42 % of all converging displacements, whereas after a diverging displacement ($\Delta\psi_{sep} = 150°$ at the onset of the slow phase) it occurs in only 11 % of the cases. These data suggest that the width of the beam is variable and can be larger than $\Delta\psi = 150°$, but direct tests are needed to settle this question.

Conditions which prevent attention from building up as well as stimuli which escape suppression and attract attention wherever they occur, are most important for the characterization of this aspect of vision. For instance, in the experiment of Collett (1980 a) with *Syritta*, translatory movement is under continuous control from most of the visual field. Some stimulus conditions in the oscillation experiment which do not give rise to noticeable attention effects may find a related explanation. If several responses are found for which attention applies, it will be most important to find out whether the areas of attention coincide. The finding that after a phasic response to the fast displacement of a stripe, this stripe is stabilized for optomotor balance much more often than when no phasic response has occurred, strongly points in this direction. Knowing some of these properties will help to understand the importance and purpose of selective attention for the fly.

12 Plasticity of Visuo-Motor Coordination

Visuo-motor coordination in humans and higher vertebrates is known to be highly modifiable by practice. For instance, wearing glasses requires a readjustment of the closed loop gain for the guidance of eye movements. Even most severe distortions of visuo-motor coordination such as are imposed by goggles which invert the whole visual scene upside down or left to right can be compensated by experience (Stratton, 1897; Kohler, 1956).

Insect behavior is thought to be rigid and stereotyped, consisting largely of fixed action patterns and reflex-like control functions. Yet plasticity of motor coordination is well documented. Many instances are known where the loss of body appendages leads to a gradual reorganization of behavior. Grooming of the antenna can be performed by the ipsilateral mesothoracic leg if the prothoracic leg is missing (de Ridder, 1974). Cutting the wings of *Drosophila* leads, after a number of fruitless attempts, to a suppression of the quick start behavior (unpublished). R. Wehner (personal communication) observed with *Cataglyphis bicolor* that, after occluding a certain dorsal eye region specialized for the detection of E-vector orientation in the sky, the ants during their foraging trips turned their heads upward as if they were looking out for polarized light from the sky to guide their course even though other eye regions apparently are not suited for this task and the ants do not seem to profit from this posture.

Compensatory modifications of walking behavior after CNS surgeries (Huber, 1955; personal communication) are also well documented. A most telling example is the male grasshopper, *Gomphocerus rufus*, which after having one cervical connective cut, displays a severe rotatory bias to the intact side during spontaneous and subject-oriented walking. In the course of several days, during which no grasshopper-like movable objects are present, the male recovers in such a way that it walks nearly straight, with only slight tendencies to rotate visible. However, when the male is confronted with a conspecific, mainly with a female, and tries to orient toward her, the initial defect suddenly breaks through again. Due to the severe rotatory bias the male is unable to start the courtship performance (Huber, personal communication).

Plasticity of visuo-motor control in freely walking *Drosophila* can be demonstrated if one covers one of the eyes with paint, cuts the wings, keeps the flies under open loop conditions in the light until they have completely recovered from anesthesia, and subsequently records their trajectories in a well-structured surround. Upon release, most wild-type flies turn toward the occluded side while walking forward, but after a few seconds most of them manage to walk straight. After being caught and released again, they resume turning, but after further handling many of them can start with a straight trajectory even after a longer interval. In contrast to previous reports with other insects (e.g., Fraenkel and Gunn, 1961), in *Drosophila* contrast perception, and thus presumably visual motion, has an important part in the initial turning behavior of monocularly blinded flies. Covering the two eyes with paint of different transparency eliminates the "forced" turning behavior (Blondeau, unpublished).

The lesson from this experiment is obvious. For monocularly deprived flies visual guidance is out of balance and the fly needs practice to equilibrate it. As in the case of the grasshopper, arousal uncovers the asymmetry. The animal has to gain experience in different motivational states in order to fully compensate the defect. Interestingly, in the mutant *optomotor-blind*^{H31} one-sided blindness causes the same rotational bias as in wild-type flies. Thus it is not large field course control alone mediating this effect (Blondeau, unpublished).

The closed loop experiments at the torque compensator can be regarded as a visuo-motor coordination task. We have discussed the problem of the missing proprioceptive feedback during torque spikes, and we have shown that *Drosophila* can adapt its closed loop gain to the situation. In the following series of experiments we show that flies (like humans) have the ability to adjust their visuo-motor coordination to artificial situations which have not occurred previously in evolution.

We use the double arena (Fig. 88) with the random dot texture (Fig. 45c) as a background and a single vertical stripe ($\delta = 5°$) as an "object" in front of it. The fly at the torque compensator is in closed loop with stripe and texture as described before, but now we invert the sign of the feedback for the stripe. When the fly tries to turn clockwise, the stripe turns clockwise as well. This situation is similar to wearing right–left inverting goggles. One sees an object on the right and tries to turn toward it, but at the same time the object starts moving to the right getting faster and faster the more one accelerates to pursue it. Unlike for a human subject with goggles, in our experiment the background texture for the fly moves as expected and provides negative feedback for flight stability. This is essential, since the tethered fly without proprioceptive feedback has to rely entirely on vision, while the human subject can rearrange the visual percepts according to the still consistent information from all other sources. As we know from Chapter 5, the fly maintains optomotor balance with the background. Starting from this base line of "zero" torque, it generates torque spikes to readjust its orientation relative to the stripe with a moderate preference for stripe positions in the

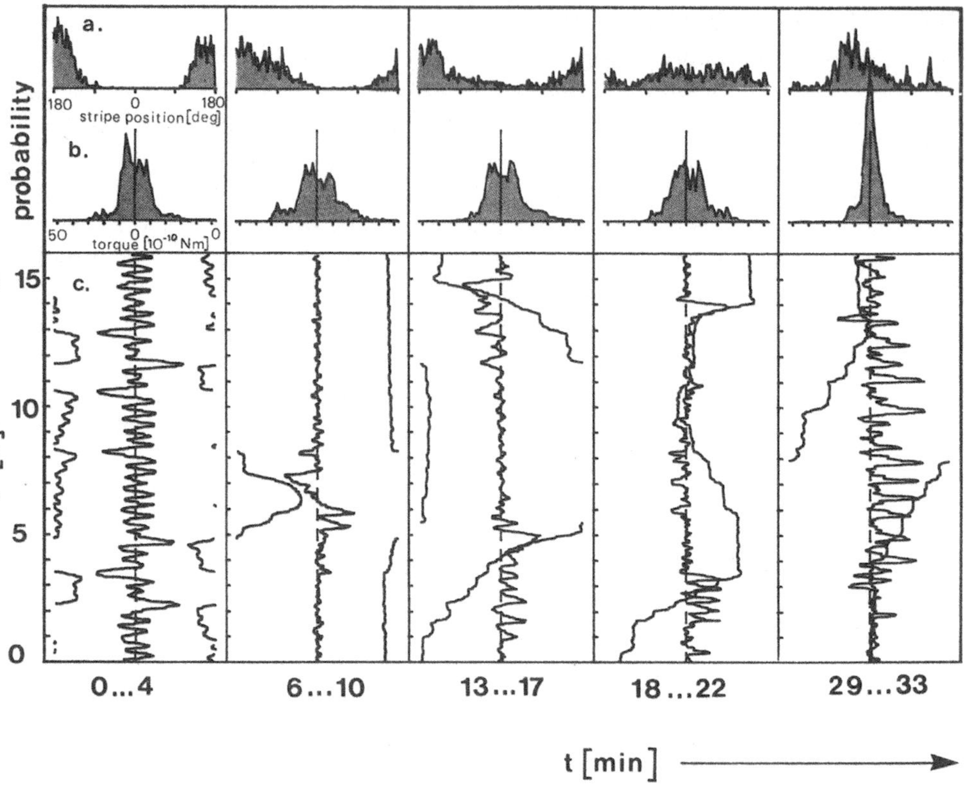

Fig. 103. Five traces at different times during a visuo-motor coordination experiment. Stimulus conditions as shown in Fig. 92b except that yaw torque is positively coupled to angular velocity of stripe and no disturbance is added. Fly learns to invert polarity of torque spikes (without inverting polarity of efference copy). (a) Orientation histograms; (b) torque histograms; (c) scores of 15 s of torque and stripe position taken from the time period shown as histograms above, illustrating typical stages of the learning process. Scores can be read like previous ones (e.g., Figs. 101 and 102) except that they are rotated by 90° in order to present the respective histograms above in the normal manner. Time runs from bottom to top. *Numbers under the scores* indicate time (min) during which histograms on top were recorded. Records in between are missing due to technical reasons (see text for details; Heisenberg et al., 1983)

frontal visual field (voluntary fixation); but now as soon as it tries to shift the stripe to a frontal position it fails. During the torque spike the stripe moves into the false direction, causing an instantaneous correction response which, of course, makes the situation worse. The stripe inevitably moves into a position in the back of the fly and is kept there for several minutes. In the rear the movement of the stripe does not evoke correction responses. As soon as the stripe moves forward the fly generates torque spikes toward its position. This, of course, shifts the stripe back or even into the other visual half-field.

In the latter case the fly again tries to turn toward the stripe with the same effect (Fig. 103, 0–4 min).

After a while the fly may cease generating torque spikes, and still some time later it manages to generate torque spikes of opposite polarity, shifting the stripe to a more lateral position. There, however, the fly cannot suppress the positive feedback. It generates a vigorous "correction response" which immediately shifts the stripe all the way to the back (Fig. 103, 6–10 min). This sequence may be repeated over and over until eventually torque spikes of normal polarity and the correction responses become less frequent and the stripe may reach the frontal part of the visual field. Typically the fly practices with the stripe in only one of the visual half-fields. As soon as the stripe crosses the frontal midline for the first time, the correction response immediately comes into play (Fig. 103, 13–17 min), but the fly tries again and again until on both sides of the visual field the behavior related to the stripe is well under control. The fly, then, seems at ease with the situation, choosing any (arbitrary) orientation relative to the stripe, again with a slight preference for frontal stripe positions (Fig. 103, 29–33 min; Fig. 104).

This sequence of events, which takes about 20–30 min, is found with many flies, although each one has its individual learning history. Some wild-type flies are not able to suppress their correction responses even after more than 1 hour of flight time. This indicates that neither local nor general "fatigue" is responsible for this suppression. The *small optic lobes* flies do not adjust to this situation during the 40 min of our experiment. They can shift the stripe to lateral positions, but correction responses keep occurring and the stripe is not stabilized in the front (Fig. 105). Some of the circuitry in the optic lobes missing in the mutant seems to be involved in this type of learning.

The long learning period of 20 min in wild-type flies is not entirely due to the positive feedback between yaw torque and angular velocity of the stripe. If after 20 min of a normal closed loop experiment with negative feedback we invert the coupling, only few of the above mentioned vigorous correction responses are observed, and the fly seems to learn much faster (Fig. 105). Subtle modifications of the visuo-motor coordination seem to take place during the first 20 min at the torque compensator even under normal coupling conditions. We have mentioned before that after 7 min of flight in normal closed loop (but not in open loop) with a single stripe, the open loop object response cannot be recorded any more (Fig. 69). Although the presence or absence of the object response does not always coincide with the presence or absence of the correction response under inverted coupling conditions, it indicates a gradual change of the fly in its attitude toward a moving stripe.

After the fly has been adapted to the normal closed loop situation, its learning phase is as interesting as in the first 20 min described above. Individ-

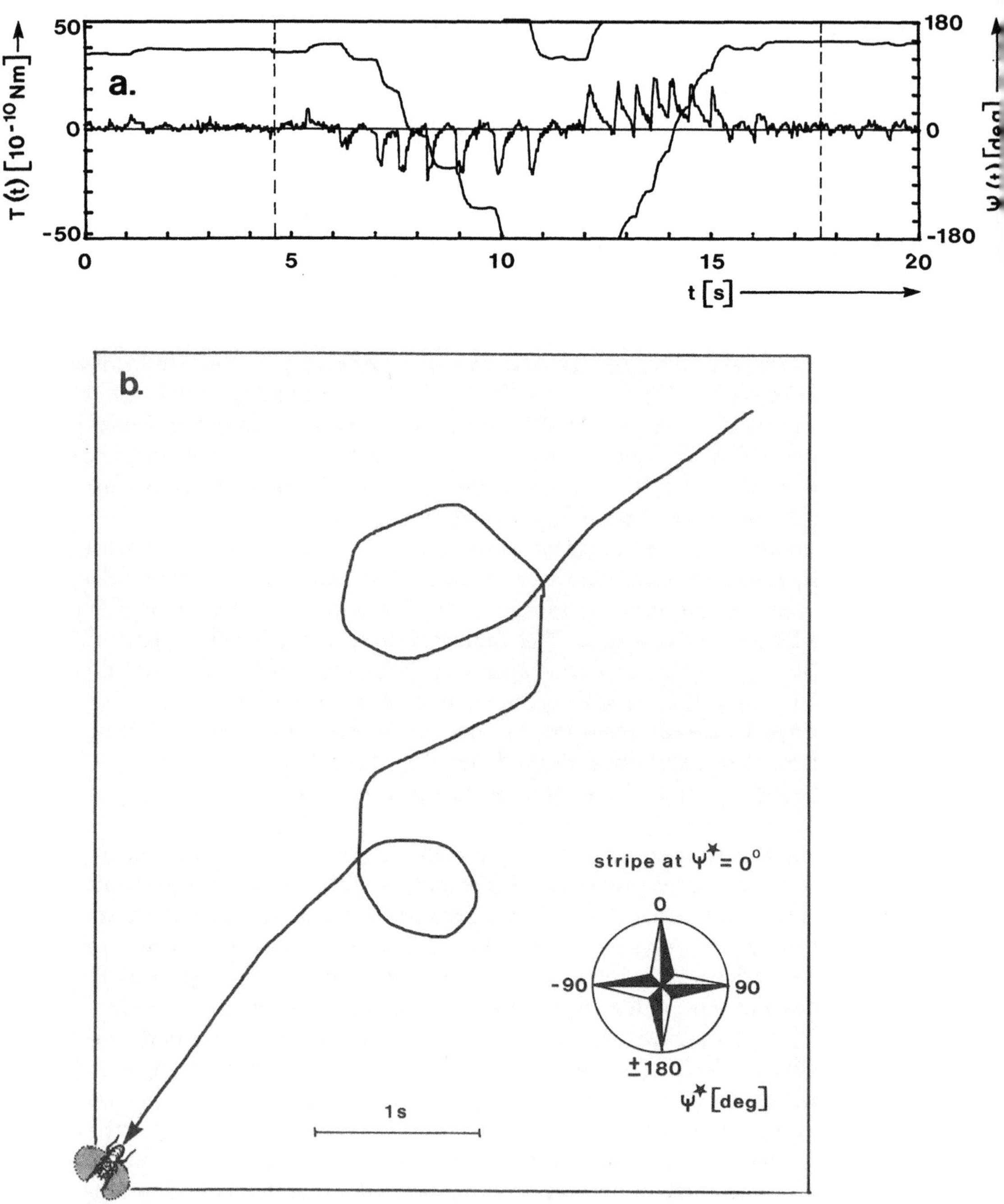

Fig. 104. Closed loop flight with random dot texture (Fig. 45c) as background and vertical black stripe ($\delta = 5°$; $H = \pm 40°$) for which coupling is inverted. The fly has "practiced" for about 30 min and is now "at ease" with this situation performing characteristic double loops as shown in Fig. 54. (a) Torque and stripe position traces; (b) reconstruction of flight trajectory of the period between *vertical dashed lines* in (a)

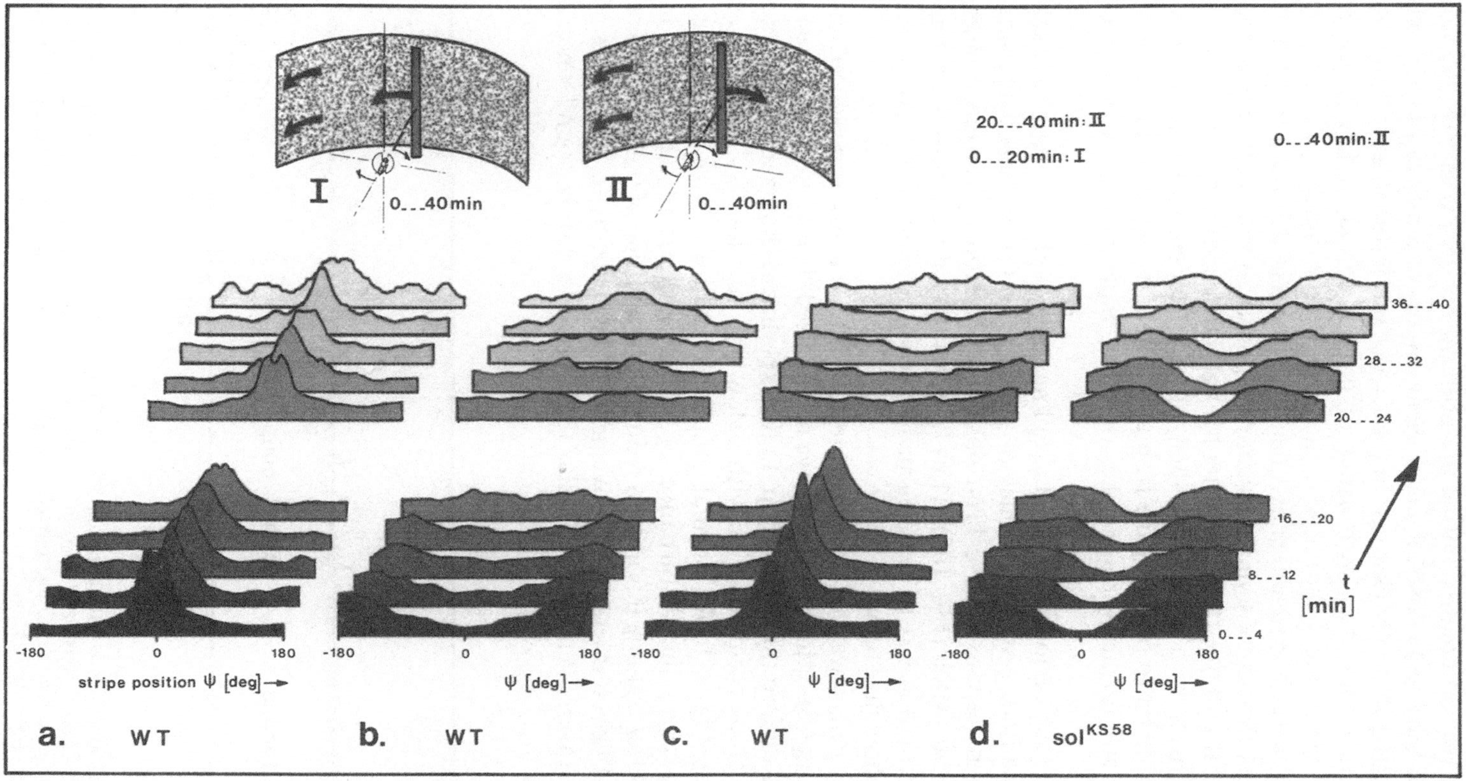

Fig. 105. Orientation histograms from ten consecutive 4-min periods of closed loop flight with a stripe and background texture (random dot texture; Fig. 45c). Histograms are averages from at least ten experiments and ten flies. (a) Normal negative feedback for stripe throughout; (b) Inverted coupling for stripe; (c) Normal coupling for 20 min, followed by 20 min of inverted coupling. (d) Mutant *small optic lobes* is unable to stabilize the stripe in the front during 40 min experiment when coupling is inverted [cf. (b)]. Above the histograms stimulus conditions are indicated. *Small arrows* underneath the flies represent torque leading to the pattern movement indicated by the *large arrows*

ual flies perform differently, but as the position histogram in Fig. 105 (an average from ten flies) shows, it takes them several minutes to prevent the stripe from being shifted backward. In some cases this is due to the correction responses; other flies start with series of normal torque spikes in either direction, yet others move the stripe to the front, keep it there for a little while and then shift it backward and subsequently perform antifixation for several minutes. Most flies show some correction responses at the onset of inverted coupling. We found one fly which after a single correction response kept the stripe in the back for a little while, then, all of a sudden, with a huge burst of torque, shifted it to precisely the frontal position, where it kept it motionless without any further torque spikes or other maneuvres for several seconds.

If, after a long period of inverted coupling, the coupling is switched back to normal one finds a slight tendency of the fly to shift the stripe backward, and correction responses have been observed occasionally during a short period after the switch (Fig. 106). Thus it seems that at least to some extent the normal situation has to be relearned after an extended exposure to the inverse coupling situation.

The most direct way to demonstrate that after having experienced inverted coupling conditions the flies have reorganized their visuo-motor coordination is to disturb the closed loop. In the experiment shown in Fig. 107 we move the inversely coupled stripe with a slow sine function ($f = 0.12$ Hz), as shown before (Fig. 92b) for normal coupling conditions. Without training, the fly aggravates the disturbance of stripe position (dotted curve in Fig. 107a). After training it is able to reduce the amplitude of the disturbance (Fig. 107b). Flies counteract the movement of the stripe by torque spikes of appropriate (inverted) polarity. Similar flight episodes can be found when a

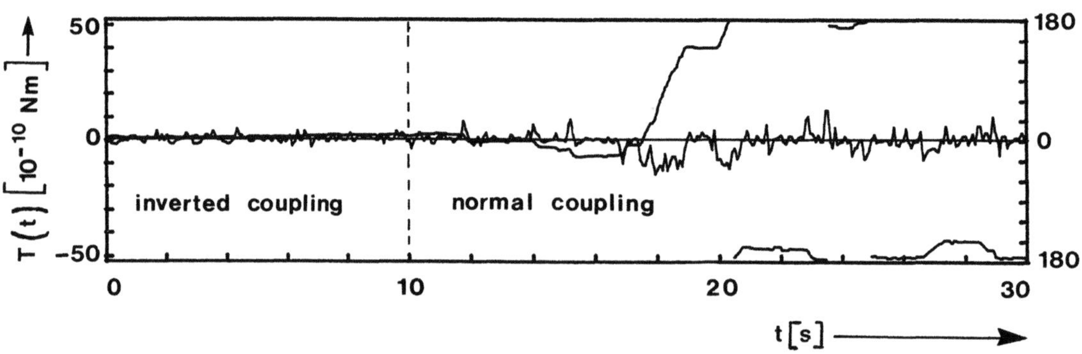

Fig. 106. Re-learning of normal coupling. After a long period of inverted coupling normal coupling is reestablished (*vertical dashed line*). After a few seconds the fly "loses" the stripe to the back of the visual field in an uncontrolled manner which is typical for the first minutes of inverted coupling (cf. Fig. 103)

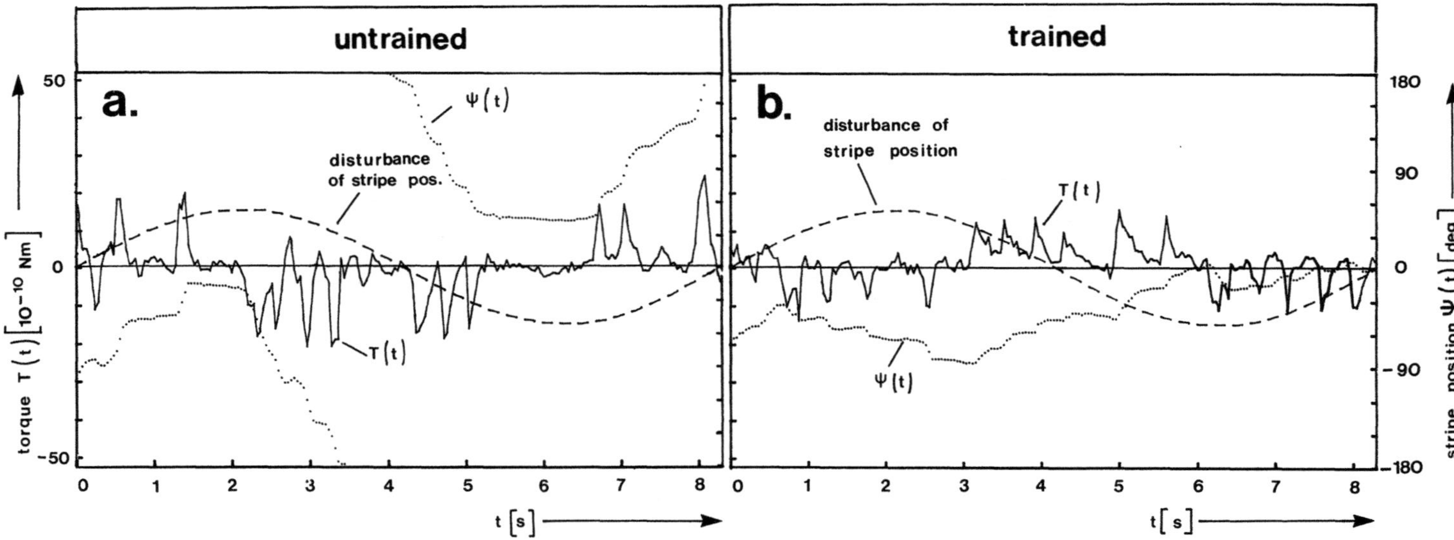

Fig. 107. Inverted coupling for the foreground stripe. Response of the fly to slow displacement of stripe is tested before (a) and after (b) training period. Diagrams are selected unaveraged traces showing that after training the fly is able to counteract the displacement (cf. Fig. 92 b). *Dashed curves* sinusoidal displacement of stripe added to its closed loop movement. *Drawn curves* torque traces showing that torque spikes are inverted after training. *Dotted curves* indicate actual position of stripe

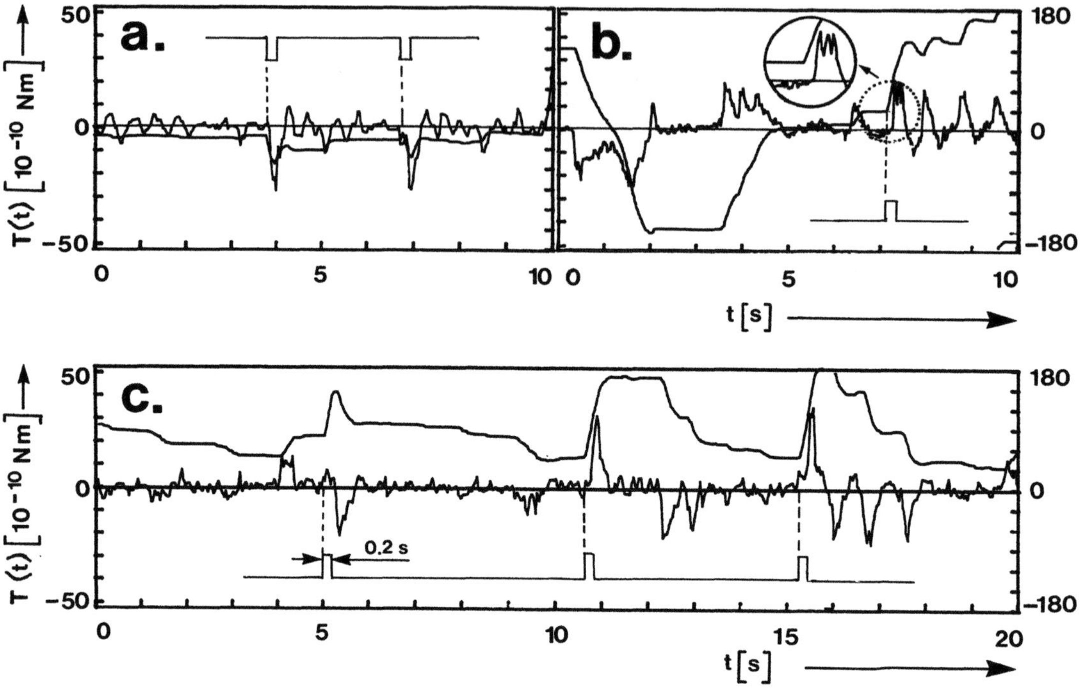

Fig. 108. (a) Fast displacements of the foreground stripe (*inset* shows voltage pulses) under normal coupling conditions lead to equidirectional phasic torque which shifts the stripe to about the position it had before. In (b) coupling for the stripe is inverted and the fly has just acquired enough experience to safely shift the stripe to any position. However, it still responds to a displacement with phasic torque of the wrong polarity and with a subsequent uncontrolled correction response. With more training (c) the same fly may either apply the phasic response with the appropriate polarity or if the phasic response still has the wrong polarity the fly may correct the stripe postion with torque spikes shortly after the disturbance

constant D.C. voltage is added to the inverted closed loop with the stripe (data not shown).

As discussed earlier, *Drosophila* may respond to displacements of a single stripe with a short, equidirectional burst of torque (object response, Fig. 108a). If we apply similar disturbances of stripe position under inverse coupling conditions, after some time of practice flies show individual response types which in most cases lead to partial restoration of stripe position but differ considerably among each other. Some flies generate an inverted phasic torque response which instantaneously shifts the stripe back (Fig. 108c). Others generate a biphasic response, the first part shifting the stripe even further, the second moving it back (not shown); still other flies answer with a normal maladjusted object response but after about a second generate one or two torque spikes in the opposite direction, moving the stripe back to its original position (Fig. 108c).

After extended training with inverted coupling conditions the responses to these displacements do not return completely to normal on switching back to negative feedback. The phasic responses often have an inverted polarity and, on the whole, responses seem more variable than those of untrained flies. Even on the next day flies having been exposed to inverted coupling conditions tend to show abnormal flight behavior.

These experiments show that *Drosophila* can reverse the polarity of its torque spikes and phasic torque responses in order to improve its visuo-motor coordination. The torque spikes and phasic responses are directed toward (simulated) objects. *Drosophila* can acquire an adaptive state or function (i.e., to stabilize stripe position) through a variety of strategies which it has at its disposition or even by constructing new ones. As in the case of selective attention, this implies a motivational organization of behavior with intentions and neurally represented goals. It may be interesting to note that the average flight time per fly is much longer during these learning sessions than during normal closed loop experiments.

This ability of *Drosophila* to use totally artificial visuo-motor coupling for orientation is not an isolated finding. One can couple the angular velocity of the panorama to the forward force of flight (instead of yaw torque) without the fly losing control (Götz, personal communication) and the same holds for coupling the angular position (instead of angular velocity) of a stripe to yaw torque. The learning performance involved in these experiments can be classified as "operant learning", if one assumes that the animal maximises its orientedness. Operant conditioning has already been demonstrated for *Drosophila* (Booker and Quinn, 1981). Flies, like cockroaches (Horridge, 1962), can be trained by electroshocks to keep a leg in a certain (unnatural) position. Headless flies perform best (Booker and Quinn, 1981). Along this line, recently Hoyle (1980) has reported remarkable achievements of locusts and grasshoppers. Partially immobilized animals learn to position the tibia of one of their hind legs so as to turn off a disturbing noise, to regulate a heat lamp or to make a motor produce a vial with fresh grass for eating. The two former tasks are again performed at least equally well by headless animals.

In the leg-lifting paradigm the insect arrives at the more agreeable state by trial and error elimination. This seems to be the procedure for visuo-motor coordination learning as well and even in the case of grooming being taken over by the mesothoracic leg a stochastic component in a goal-oriented process has been discussed (de Ridder, 1974).

The learning mutants of *Drosophila* may serve to distinguish the visuo-motor coordination learning from other types of conditioning. For instance, the mutant *dunce* which fails in an olfactory paradigm of associative conditioning (Dudai et al., 1976) and performs poorly in the leg-lifting paradigm (Booker and Quinn, 1981) appears to learn normally to adjust to reversed coupling conditions, and also the mutant *sevenless* which fails to display learning in a color or intensity discrimination task (Spatz, personal communication; Menne and Spatz, 1977; Reichert and Bicker, 1979) seems unimpaired in our experiment. Conversely, *sol*, which is deficient in learning at the torque compensator (Fig. 105) behaves normally in color and intensity discrimination learning (W. Roos, personal communication).

An important additional question concerns the behavior of visually naive flies (i.e., flies raised in the dark). Do they have to learn the sign of the feedback in normal visuo-motor coordination? In pilot experiments Reichardt and Götz (unpublished) have shown that optomotor responses are independent of visual experience prior to the test. For closed loop responses of *Drosophila* toward objects, however, this test has not yet been performed. In *Musca*, the open loop response to front-to-back movement of an object has been shown to be strongly age-dependent, the full response amplitude being reached on the 11th day after eclosion (Geiger and Poggio, 1975). Thus, some effects of maturation on the visual system may also occur in *Drosophila*.

13 Valuation

Under most circumstances a fly has certain orientations in respect to objects. This is easy to see in the behavior of freely moving animals. If *Syritta* approaches a daisy or if a *Drosophila* male zealously follows a female, orientation is not arbitrary, it serves a purpose.

With the fly at the torque meter, however, we have difficulties in distinguishing a stabilized course from a deliberately chosen orientation. At the torque meter the fly normally does not attain its goals. We are mostly ignorant about the purpose of the fly's behavior. Yet a book on visual orientation should deal with this problem: the objective in orienting, the object. This is the most important aspect of visual orientation and we imagine that the largest part of the visual system is devoted to it. We would like to assume that dispositions, drives, moods, and goals influenced by environmental and physiological parameters such as time of day, nutritional state, and temperature, set the stage for the perceptual process. The animal has to arrive at a hypothesis about the significance of a certain feature; a hypothesis is a preliminary assignment of meaning, a preliminary inference of a context from a few cues; it can be taken back and replaced by another one. Thus the fly has to keep for every pattern in the visual field a catalog of its possible significances. The catalogs have to be continuously updated with the changing sensory information and disposition of the animal. Eventually, "meaning" and "significances" will have to be operationally defined in terms of an information theory based upon the animal's value system. Value systems in brain research are not being investigated.[9] Yet vision and even visual orientation cannot be understood without them.

The value system of *Drosophila* and its neural basis is beyond the scope of this book (and is largely unknown anyway). What we can do is to bring the shortcomings of our account into better focus. In the previous chapters we have tried to study orientation, ignoring the objective of the fly. In this chapter we treat visual orientation as a whole, including the goal the fly pursues. We first reconsider the previous experiments under this aspect and then we briefly review certain pieces of visual behavior in which the objective seems to be obvious.

---

[9] This is the reason why, in this field, the concept of information has been of little use. But see, for instance *The Ecological Approach to Vision* by Gibson (1979)

Hypotheses

Already in the previous sections on course control and orientation we have seen that the fly makes hypotheses about the situation. If, for instance, it is flying in closed loop with two identical stripes which move relative to one another, it regards one as background, adjusting its optomotor balance to it, and chooses its orientation in relation to the other stripe. After some time, in particular when the "background" stripe happens to move front-to-back, the fly switches to regarding it as foreground for a while, etc. (Fig. 97).

In a similar manner a fly creates hypotheses about the significance of visual features. We can observe this in ambiguous situations in which the fly spontaneously changes between different orientations. We have described in the chapter on fixation that in additon to the fly's preference for the stripe to be in frontal positions there is also a tendency to keep it in the back. We call this antifixation. With a bright vertical stripe on a very dark background it turns out that the fly performs antifixation most of the time (Fig. 109 c). As soon as the luminance of the background is slightly increased, fixation again prevails. This gives us the opportunity to adjust the background luminance to exactly that level at which fixation and antifixation should be balanced. Now the fly tries fixation for a while, then switches to antifixation, and so forth (Fig. 109 b). The lesson to be learnt from this experiment is that at the torque compensator the same patterns can be attractive and repulsive. Some hypothesis with a negative significance is matched sufficiently well by the bright stripe on a black background to keep the fly from approaching the stripe. With a gray background the different hypotheses, positive and negative ones, are not completely satisfying and the fly tries them alternatively. The same ambiguous behavior is found with a black $10^\circ \times 10^\circ$ square in the upper part of the visual field.

We can be certain that the match between the hypothesis of the fly and the pattern presented in such a case is poor. For instance, for walking flies which are placed into the center of an otherwise bare bright arena and which have to choose their trajectory in the course of a few seconds, a black $10^\circ \times 10^\circ$ square above the horizon is strongly repellent (Fig. 110 b). When a fly runs away from the square up in the air, this may have neither the shape nor the angular velocity of a bird, dragon fly, or hornet, but the fly cannot afford to wait until it knows better what the significance of the pattern might be.

As soon as the pattern is changed into a vertical rectangle its attractiveness rises sharply (Fig. 110 a). Interestingly, the mutants *sol* (Fischbach, 1983 b) and *omb* (Bülthoff, 1980), which otherwise have little in common, both show antifixation when confronted with a vertical stripe in the arena. With either kind of defect the processing in the visual system may be too poor to warrant an approach toward the stripe in this test (for *sol* see Fig. 110). It

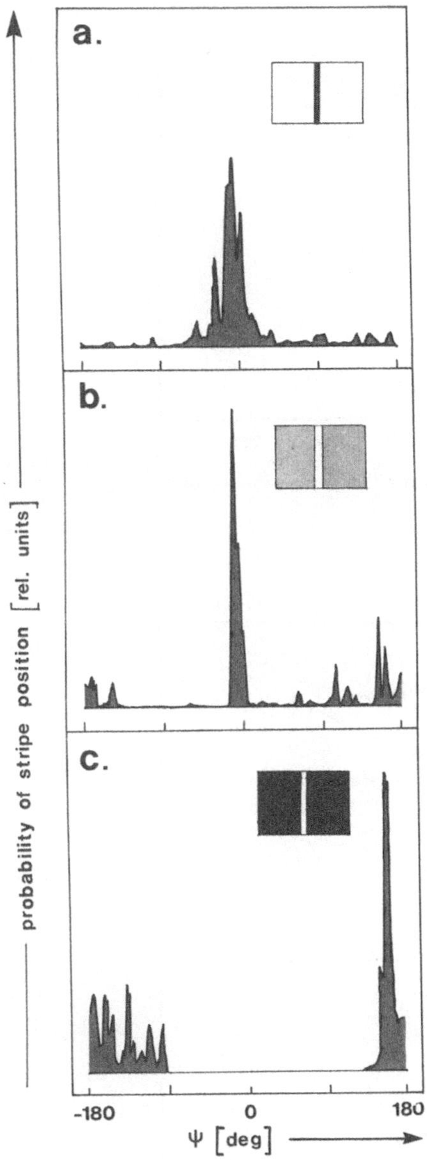

Fig. 109. Ambiguous significance of bright stripe on stationary gray background. Orientation histograms of 4-min flight periods of single flies. (a) Black stripe ($\delta = 10°$; $H = \pm 40°$) on luminous background. (b) Bright stripe ($\delta = 10°$; $H = \pm 40°$), transmittance of background $\tau = 0.7$. (c) Bright stripe on black background ($\tau = 0$). In (b) the fly switches spontaneously between fixation, antifixation and menotaxis, in (c) it shows mainly antifixation

is not a priori unlikely that mutant flies may have a specific visual deficit which renders vertical stripes repellent, but it is equally possible that it is the increase in ambiguity of the pattern which causes the fly to favor the (safer) negative hypothesis. If our conjecture is correct, many different visual mutants should have this general "topophobia". In contrast to *omb* and *sol* the mutant *minibrain* (*mnb*) is attracted by squares as well as vertical stripes. In this mutant the whole brain, including the optic lobes (with the exception of

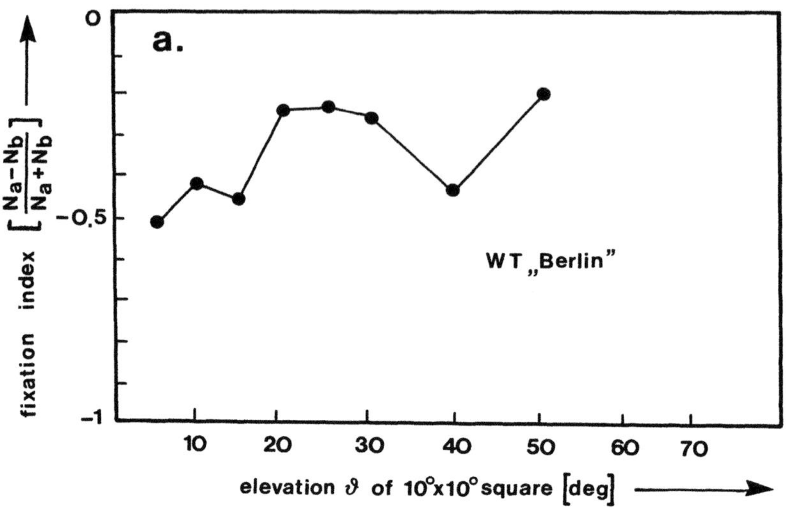
a.
0
-0.5
-1
fixation index [(Na−Nb)/(Na+Nb)]
WT „Berlin"
10
20
30
40
50
60
70
elevation ϑ of 10°x10° square [deg]

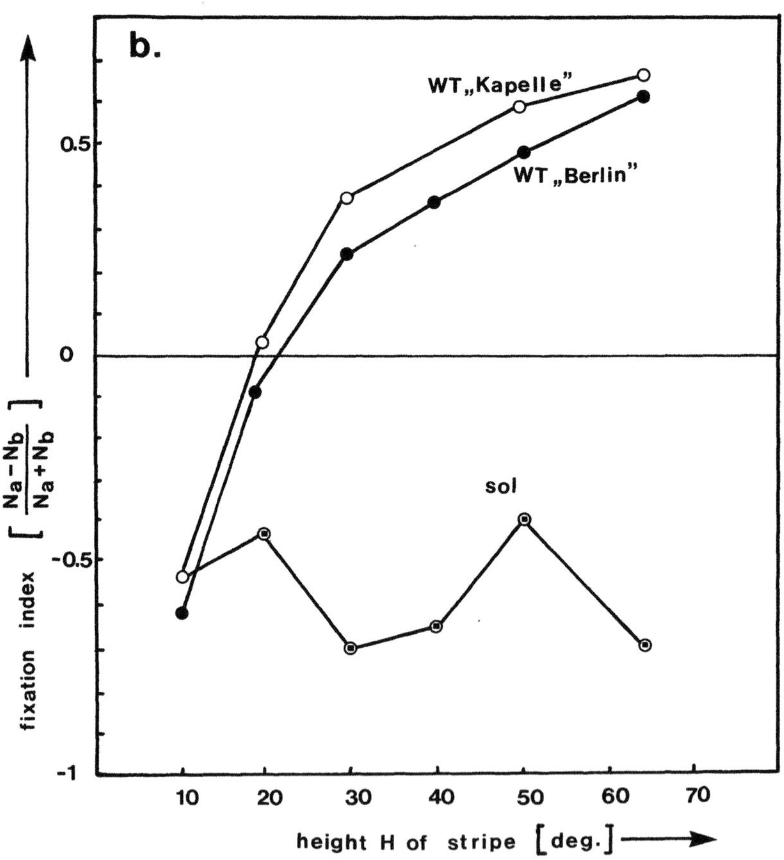
b.
0.5
0
-0.5
-1
fixation index [(Na−Nb)/(Na+Nb)]
WT „Kapelle"
WT „Berlin"
sol
10
20
30
40
50
60
70
height H of stripe [deg.]

the lamina), is reduced to about 50% of the normal volume. In the spirit of the above conjecture one could interpret the behavior of *mnb* as foolhardiness due to central brain damage.

Our proposal that flies make hypotheses as to the significance of visual patterns needs to be specified more closely. What one observes with the fly at the torque compensator are spontaneous changes in the attitude toward a pattern without any concomitant events in the sensory situation. One might be inclined to think of these changes as fluctuating motivational states. However, these changes occur on the average several times a minute and it is difficult to think of goals which would change that fast. Should we say that the sensory situation simultaneously triggers two behavioral subroutines which then alternate, maybe because of a process of mutual exclusion and habituation? We suggest a different line of reasoning: if the available information is ambiguous as to its significance (which it often is), some behavior is selected at any moment according to one of the more reasonable possibilities and the fly is ready to deal with the consequences of this "choice" which, of course, fail to present themselves at the torque meter. Whether in this case the active process takes place at the level of perception or further downstream where behavior is generated is an open question. The nature of active behavior will be discussed in the last chapter.

Significance of Simple Visual Stimuli

The significance of visual patterns is difficult to assess. As mentioned above (and in Chap. 7), the vertical extent of dark patterns on a bright background seems to be quite important for *Drosophila*. Working with the wild-type stock Sevelen, Wehner (1972) measured the trajectories of walking flies in the arena when confronted with rectangles of different heights (see p. 141), all starting at the horizon and covering half the circumference of the arena (i.e., $\lambda_{sp} = 360°$). For flat rectangles (height: $H = 2°-10°$) the flies approach the center of the dark field, whereas for $H > 10°$ they show a preference for one of the vertical edges as tethered *Musca* do (Reichardt, 1973) and *Tenebrio* (Varju, 1976). The preference for edges is less pronounced in other wild-type stocks like Canton S and Kapelle (Coombe, unpublished).

◄

Fig. 110. Orientation of walking *Drosophila* flies in relation to a black square of vertical stripe. Data are derived from orientation histograms as shown in Fig. 78, each point refers to 100 runs from ten flies. N_a (N_b) indicates no. of flies crossing the half of the "measuring circle" oriented toward (away from) the strip. (a) Flies walk away from a $10° \times 10°$ black square presented at any height above the "horizon". (b) For WT attractiveness of black pattern increases with its vertical extent, for the mutant *sol* the vertical stripe is as repellent as the square. (After Fischbach 1983 b)

In a different experiment Wehner and Wehner-von Segesser (1973) compared the attractiveness of vertical and horizontal stripes of the same spatial frequencies. The flies prefered the vertical ones. Mimura (1982) measured the preference for certain patterns composed of luminous vertical, horizontal and oblique bars on a black background. The order of preference he obtained deviates from that to be expected if the flies responded to the overall light flux or to the "verticality" in the patterns.

The only experiments with partially restrained animals come from Götz (1971, 1980), who tested preferences for patterns composed of various numbers of discs covering a constant overall surface area. Measuring turning tendency on the tread compensator (Götz and Wenking, 1973), he determined the relative attractiveness between pairs of such patterns. All these values were additively related, i.e., for patterns A, B, C... he found $\Delta R_{A,B} + \Delta R_{B,C} = \Delta R_{A,C}$ etc. This simple "unidimensional" relation suggests that the flies used a single criterion for these discrimination tasks. We would like to interpret this result as saying that each pattern has its attractiveness value irrespective of the alternative pattern presented simultaneously. In none of the experiments described in this paragraph is the biological significance of these pattern preferences understood.

When kept in the dark *Drosophila* is strongly attracted by long wavelength UV-light. In this case one can conjecture that *Drosophila* is heading for the open. Presumably the sky is the only ubiquitous source of strong UV-light in nature. Thus the positive significance of UV-light is sufficiently clear to justify the fly's approach.

Context Dependence

In our account of vision, the orientation of an animal is taken as an expression of its "hypothesis" (in behavioral terms) about the significance of the situation. Situations are not confined to one sensory modality and they comprise not only the state of the world but also that of the animal itself. Thus, when studying fixation and antifixation in stationary flight, pattern preferences in walking flies, phototaxis and spectral wavelength effects in phototaxis, one should not be surprised to find influences from other sensory modalities, age, nutritional state, overall genetic constitution, and individual life history. All of these shape the momentary goals of the fly. For instance, for flies walking inside a dark chamber the attractiveness of a light increases if the flies are exposed to a repellent odor (Fig. 111; Fischbach, 1979 b; 1983 b). The more dangerous the situation in the chamber, the more urgent is the attempt to escape, and the more does this dominate behavior.

A similar experiment has recently been reported with large flies (Burkhardt, 1983): the author adjusted the relative intensities of two spectrally

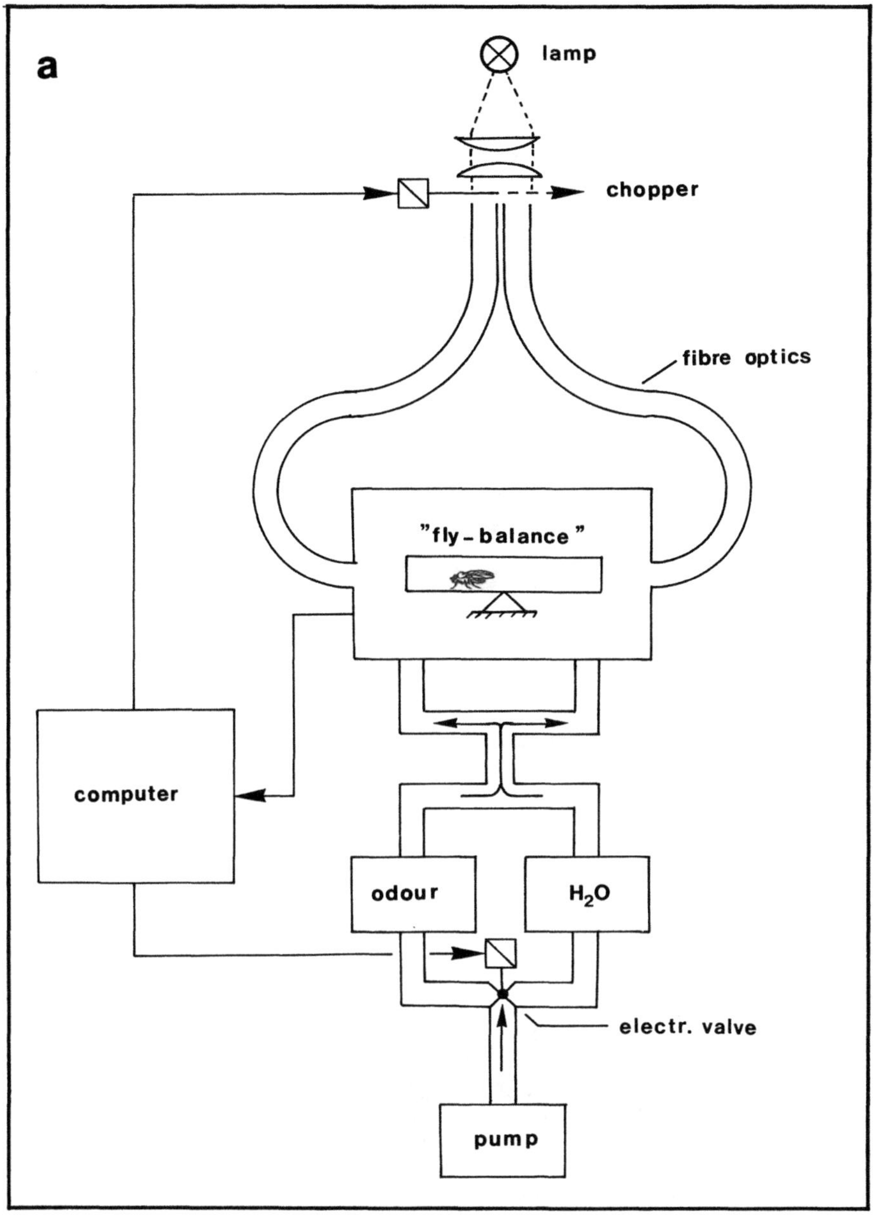

Fig. 111. Attractiveness of a test light increases when flies are exposed to a repellent odor. Five flies are free to walk inside a glass tube (D = 10 mm; length l = 106 mm) which is the beam of a balance (scheme of experimental set up shown in (a)). The center of gravity of the five flies is recorded continuously and is plotted as "position of flies" P with $-1 < P < 1$. *Insets* in (b) show phototaxis response. The test light is visible for the flies at one end of the tube and is periodically switched from one side to the other. In yellow light (top; $\lambda_{h\nu} = 575$ nm) flies are negatively phototactic (Jabob et al., 1977) whereas in UV-light (bottom; $\lambda_{h\nu} = 374$ nm) they are attracted. Traces in the insets are averages of the responses from 45 stimulus cycles (1 cycle shown). Main experiment: With the test light on one side flies eventually distribute about evenly in the tube. If now the odor of benzaldehyde is applied all flies immediately walk to the side of the light source irrespective of whether this alone elicits positive or negative phototaxis. Traces are averages of the reponses from 16 stimulus cycles (1 cycle shown). (After Fischbach 1983 b)

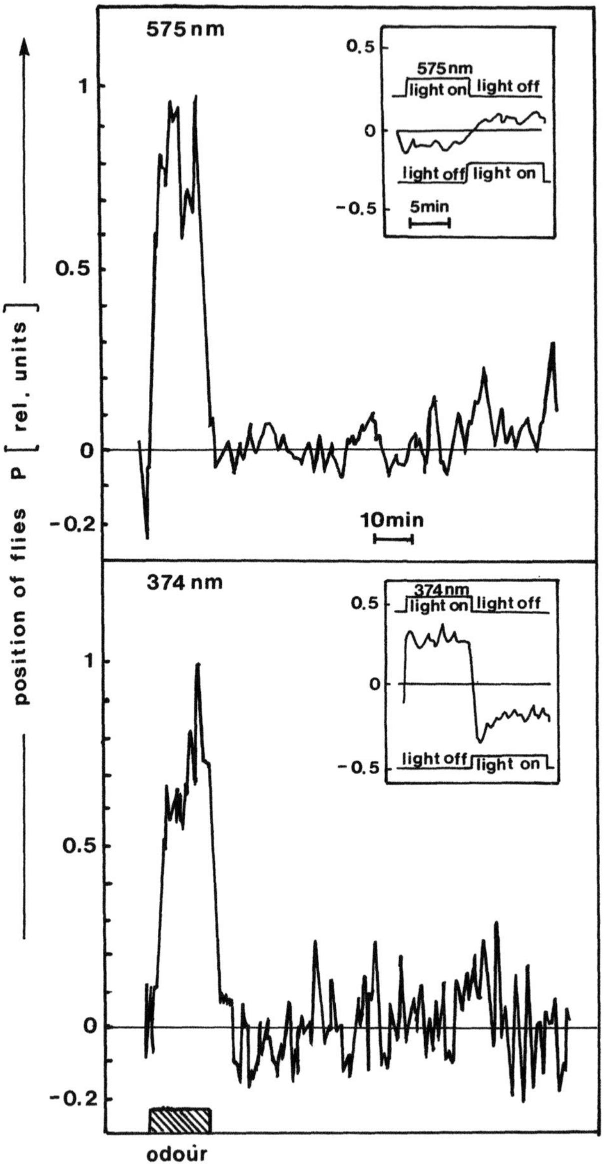

Fig. 111b

different light sources to make them equally attractive for the flies. He now exchanged the well-fed animals for hungry ones and found that the intensity ratio at the point of equilibrium was different.

At the torque compensator the inter-individual variability of the position function in *Musca* (Reichardt, 1973), its variability in long-term experiments with *Drosophila* (Bülthoff, 1980; see also p. 131 ff) are further indications of context dependence. It is difficult to study orientation with *Drosophila* by measuring spontaneous preferences for simple visual stimuli. Only a few will turn out to elicit stable context invariant behavior. The literature on phototaxis of insects is a good example of this difficulty (e.g., Kühn, 1919; Fraenkel and Gunn, 1961; Schöne, 1980).

Context dependence is less dominant in situations in which the significance of the stimulus is well defined, as with certain mostly complex natural stimuli or in conditioning experiments when a positive or negative significance (e.g., food, electroshock) has been attached to some stimulus.

Visual Components of Courtship Behavior

An example of orientation behavior with a complex natural stimulus of obvious significance is courtship tracking (for large flies see Chap. 10). It has long been known that in *Drosophila melanogaster* vision is important although not decisive for courtship success (Connolly et al., 1969). In the dark, sound and chemical cues are sufficient to allow for copulation which, then, however, is preceded by only a rudimentary courtship ritual. Without visual control the male has to follow the female at a smaller distance and loses contact more often.

Tracking performance in the light has been analyzed in detail (Cook, 1979, 1980, 1981 a, b), using an apparatus in which the female to be courted is glued to a thin wire and is kept stationary inside a revolving chamber which forces the tracking fly to run on the moving floor if it is to stay at a fixed distance from the "leading" fly. The positions of the two flies are monitored by an image-sensing integrated circuit. Alternatively, the fly on the wire is moved around in a circle, the courtship is filmed and analyzed frame by frame. The data suggest that at least two visual control systems are operating during courtship tracking. In wild-type flies the distance to the leading fly linearly determines the translatory velocity of the tracking fly. When the leading fly is moved with a sinusoidally modulated velocity ($f = 1.2$ Hz), the velocity of the tracking male follows the modulation with a 148 ms delay and a 54% modulation amplitude (Cook, 1979).

Secondly, the error angle ψ^* between the longitudinal body axes of the tracking and the leading fly is positively correlated with the angular velocity w_f of the tracking fly turning toward the rear end of the leading fly (Fig. 112). This result is remarkable in several respects. First of all, strictly speaking this is not angular tracking, since the tracking male itself generates the error angle (ψ^*), whereas the leading fly by its artificial circular course adds only a slight constant angular bias. Thus for orientation toward the target the tracking fly is in nearly undisturbed closed loop. For fixation under undisturbed closed loop conditions at the torque meter no such correlation be-

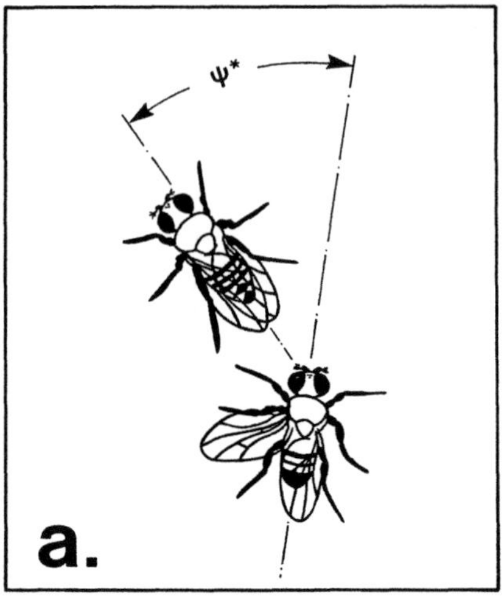

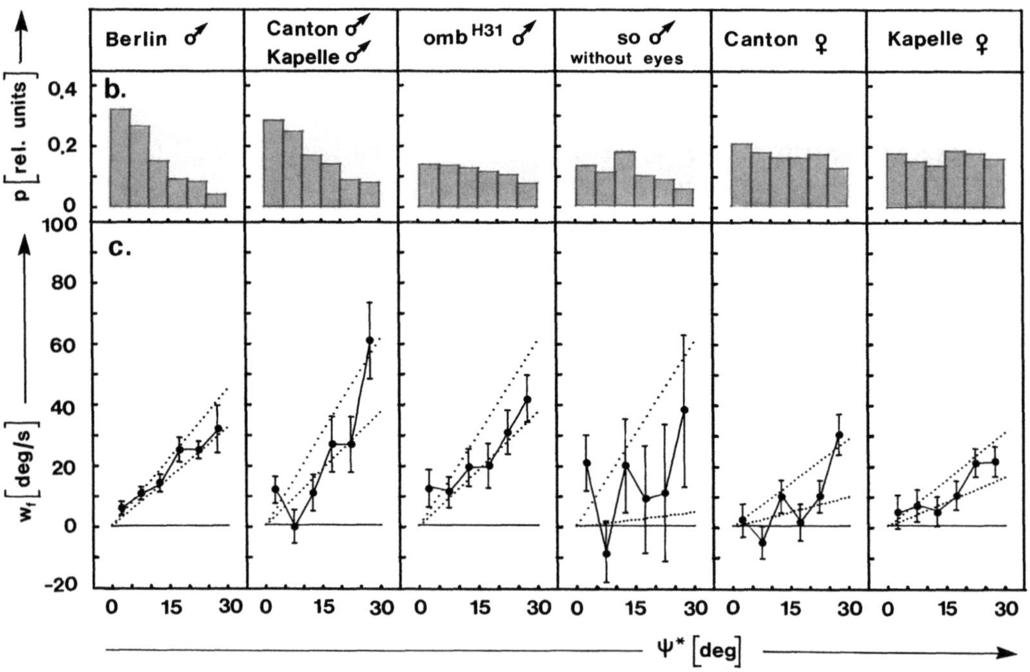

Fig. 112. Angular position control function of male and female *Drosophila* during pursuit of female flies on the ground. With increasing excentricity (ψ^*) of the leading fly in the pursuing fly's visual field angular velocity of pursuing fly increases. [As shown in (a) ψ^* refers to the tip of the abdomen of the leading fly.] (b) Probability to find the leading fly at certain angular positions ψ^*; only with WT males is probability high for having the target at $\psi^* = 0°$. (c) Angular velocity w_f of the pursuing fly as function of angular position (ψ^*) of the leading fly (for interpretation see text). *so*: mutant *sine oculis*, individual flies without any eyes (After Cook, 1980, 1981 a)

tween ψ^* and w_f is found. On the locomotion recorder, as we have mentioned before, not even under open loop conditions has a satisfactory position function been found with a rotating black square (Bülthoff, 1980; see Fig. 72). It seems that, as in other flies (Land and Collett, 1974; Collett and Land, 1975 a; Wehrhahn, 1979), the special significance of the courtship interaction involves a special control system (angular position control) which does not necessarily operate under different circumstances.

As in the case of the aerial pursuits of large flies, tracking flies on the ground is not only a male behavior. Cook (1975, 1981 a, b) observed *Drosophila* females following other females at small distance. By motion picture analysis he demonstrated that, just as in males, the angular position of the leading fly controls the angular velocity of the following one. The concomitant angular position control function,[10] however, is less steep than in males. Moreover, it has not been conclusively demonstrated yet whether female tracking is indeed visually controlled, since blind male flies during courtship tracking generate angular position control functions with positive slopes similar to visually competent females (Cook, 1980, 1981 a, b).

The role of vision in courtship and particularly in male courtship tracking is well documented by visual mutants (Cook, 1980; Tompkins et al., 1982; Markow and Manning, 1980). Interestingly, the mutant *omb*^{H31} exhibits a normal angular position control function in courtship tracking (Fig. 112; Cook, 1980). Thus, the giant fibers of the lobula plate are not crucial for this performance. Even so courtship in *omb* is severely impaired in the light, while being normal in darkness (Tompkins et al., 1982). In the mutants *white*, *ebony* and *sevenless* (Rendel, 1951; Willmund and Ewing, 1982; Tompkins et al., 1982) residual visual processes even seem to impair courtship, since in the light this is more suppressed than in darkness. Mutants finally reveal that visual stimuli related to the behavior or to the presence of the male influence the receptivity of the female (Willmund and Ewing, 1982; Spieß and Schwer, 1978).

This short excursion into the visual processes of *Drosophila melanogaster* courtship behavior illustrates that it may be useful and even necessary to study visual orientation during episodes during which the motivational state of the animal and the significance of the object of orientation are well defined.

Conditioning

Alternatively, the experimenter may resort to manipulating the motivational state of the animal. The only method of manipulation explored so far

[10] "Angular position control function" is used here as distinct from "position function" to indicate the positive and roughly linear dependence of angular velocity upon error angle during undisturbed closed loop

is conditioning (for olfactory conditioning see Quinn et al., 1974; for visual conditioning: Spatz et al., 1974). In such experiments the otherwise ambiguous significance of a stimulus becomes positive or negative, depending on whether the stimulus occurs together with a second one of explicitly positive or negative significance. So far in choice experiments certain colors and intensities have been made aversive by associating them with shaking (Spatz et al., 1974; Folkers and Spatz, 1981), electroshock or quinine sulfate (Quinn et al., 1974; Fischbach, unpublished). Some of these experiments served to investigate the fly's discrimination ability for light intensity (Bicker and Reichert, 1978) or spectral wavelengths (Hernandez de Salomon and Spatz, 1983). To what extent these artificial significances of certain intensities or colors are context-invariant remains to be tested. Pattern discrimination learning and landmark memory (Collett and Land, 1975 b) have yet to be demonstrated in *Drosophila*. In any case, manipulating the fly's motivations is a most powerful strategy for visual system analysis,in particular if this can be achieved with stationarily walking or flying flies.

As stated above, we have not even touched the main problems one faces if one tries to assess an animal's value system. In doing so one would first have to study the animal's natural living conditions. How could one account for them? Would the fly's needs match the fly's goals? The next step would be to investigate the concert of mental states and, finally, to describe their neural representations. Such questions are perhaps avoidable as long as one is interested only in the peripheral steps of visual processing; but we have to face them if we want to understand the visual system as a whole.

Visual orientation involves the concept of "object". In the natural history of animals objects have developed from the "objective", the matter of significance. Visual attributes of objects may be most important for flies and humans, but they are not constitutive properties. What we call objects are guesses about significances. The large degree of certainty humans eventually gain about some of these guesses should not let us forget this basic fact.

Synopsis

The Three Levels of Orientation

Our account of vision in *Drosophila* has focused on visual orientation behavior. It has been useful to distinguish three levels, which in the behavior of freely moving flies are intimately interwoven. At the first level the fly stabilizes its orientation against internal and external perturbations, at the second level it chooses a certain orientation with respect to landmarks and thirdly, it adapts its orientation to the pursuit of an objective.

Experiments with stationarily flying animals provide a partial separation of these levels. Textures, i.e., large patterns of homogeneously distributed and coherently moving elements coarse enough to be resolved by the eye but small enough not to be taken as individual landmarks can serve for course stabilization without eliciting object responses. On the other hand, striped drums and in particular single stripes contribute to both levels. The flies use such patterns for course stabilization and at the same time single out pattern elements as landmarks. Stripes in front of a textured background serve mainly as directional cues, not for course stabilization.

The distinction between the second and third level so far is a conceptual one. It is feasible to remove or present landmarks, but it is difficult to manipulate goals. With the fly at the torque meter we do not even know them. Only under fortunate circumstances, for instance when the fly tries to fly 360°-loops and when it readjusts its visuo-motor coordination under inverted feedback conditions, does it tell us about its one imperative goal: to acquire orientedness. The experience of a fly at the torque meter must be most weird. Unless the experimenter provides visual feedback artificially, the fly cannot do much to improve its orientedness. The normal motivational structure must be strongly impaired under these conditions. Do the goals vanish during flight at the torque meter? Or does the goal to find a landing site get more and more imperative? Torque measurements alone will not solve these problems. We would learn much more if we could manipulate the fly's motivational structure during stationary flight. So far the experiments have taken advantage of the fact that large field course control is not permanently switched off at the torque compensator and that also the object response can be recorded under open loop conditions for a long time and under diverse circumstances. It seems that the motivational structure is sufficiently uniform from fly to fly and stable enough to allow the experimenter to disregard it and to treat orientation and course control as if goals did not exist. Our understanding of vision in flies is accordingly small.

Parallel Processing

The study of vision in flies is based upon the concept of independent behavioral subunits such as large field course control, the object response and visually released start and landing. Most anatomical and physiological detail available to date refers to this concept. One of the more general results which has been slowly emerging during the last decade is the extensive parallel organization in the visual system. It is not new that the array of visual elements feeds into an array of parallel channels which preprocess the local visual information. However, what neurogenetics has revealed is that different behavioral subunits use separate neural machinery. Brain structure reflects not only functional requirements and developmental processes, but also evolutionary history. In evolution behavior is a primary object of selection. During the phylogeny of a species behavioral subunits are optimzed separately. This process, based upon the modification of genes, requires some mutual independence of those subunits (Fischbach, 1983 b). What makes neurogenetics such a useful tool is that the modification of a single gene often causes the modification of a certain behavioral function and leaves many other functions unaltered. Parallel processing, documented by the existence of mutants with specific defects in visual perception becomes apparent already in the retina where different types of photoreceptors mediate different aspects of behavior. At the neuronal level different elementary movement detectors (presumably in the medulla) are found for large field course control, the object response and for the directional motion sensitivity of the landing response. At the level of spatial integration of local movement information (presumably in the lobula complex) large field course control, object response and landing response apparently rely on different sets of neurons. Yet all three may be elicited by the same front-to-back motion stimuli, and large field course control and object response both lead to the generation of yaw torque in flight.

On purely anatomical grounds the differently stratified branching patterns of columnar neurons in the medulla (Fischbach, 1983 b) and the roughly 20 sets of isomorphic small field columnar neurons linking the lobula complex to different foci in the central brain (Strausfeld, 1976; Fischbach, 1983 b) must be interpreted as part of the structural substrate of parallel processing.

A most telling example for parallel organization is the sex-specific trakking system of male *Musca* and *Calliphora* flies mentioned in Chapter 10 (see p. 181). The structural correlate of this behavior on the sensory side consists of specialized photoreceptors in a certain part of the eye, and of specialized neurons in the lobula and presumably central brain. It happens to be identifiable due to its absence in females. Different matters of concern – in this case finding a female – may, as a general rule be represented by separate neural subsystems operating in parallel. To analyze the different kinds of preprocessing in the visual ganglia will be one of the chief problems lying ahead.

At the level of motor programs neurogenetics is only just starting to provide some data. In the mutants *cacophony* (Schilcher, 1976) and *bendless* (see p. 76) certain motor programs are specifically modified. The structural correlates of the defects appear to be localized in the ventral ganglion (Schilcher and Hall, 1979; Thomas and Wyman, 1982). We would like to assume that extensive parallel processing will be discovered in the motor system as well.

Actions and Responses

Vision in flies is more than a collection of behavioral subunits, but the general structure of the visual system can only be deduced from their properties. As an important feature of visual systems we have emphasized throughout the book the interplay between activity and responsiveness. Large field course control, for instance, has originally been regarded as purely responsive. This view is supported by the fact that optomotor responses can be elicited from nearly all healthy animals under the most extreme physiological conditions. But under closed loop conditions even for course control the activity of the fly seems to play a major role. The experiments on optomotor balance and on back-to-front movement of single stripes have shown that the closed loop behavior is different from what would be anticipated knowing the fly's responses under open loop conditions. What is missing under open loop conditions are the movements of the pattern which are due to the fly's own torque fluctuations. It remains to be investigated in what detail the reafferent visual signal is required for proper closed loop behavior. In addition, the experiments at the polyamid thread show that flies can switch off large field course control if they receive contradictory information from other sensory modalities.

The object response again has many characteristics of a true response. After all, presumably it is this response which causes *Musca* invariably to fixate a vertical stripe. A close look at this behavior in *Drosophila*, however, has shown that it is not a stereotyped reflex. After a front-to-back displacement of a stripe in closed loop it occurs in only about 80 % of the cases and with most variable latency and vigor. In open loop with a steadily oscillating object it may occur even before the onset of front-to-back motion, or the response may be omitted and occasionally it even occurs with back-to-front motion. The input–output relations apparently are complex (Chap. 7). We think that this by no means reflects the sloppiness of the neural apparatus, but would rather believe that the latency and vigor of the response inform us about the attitude the fly takes toward the stimulating object. Thus, like active behavior, this response tells us something about the disposition of the fly. A better documentation of the landing response beyond counting how often it occurs would presumably show the same. Many responses may consist of more active behavior than the crude correlation between input and output suggests.

The other behavior we encounter at this level is the torque spike or saccade. We have made the point that saccades are actively generated by the fly rather than being stimulated. It is now time to explain the concept of initiation as announced in the Introduction (see also Heisenberg, 1983). We deliberately confine the discussion to the initiation of complex motor patterns and the problem of voluntariness.

Voluntary Behavior

First of all, let us make sure that the reader does not conceive of voluntary behavior as being elicited by hidden stimuli, external or inner ones. Nor would "automatic" behavior caused by rhythmical events in the body or by a combination of stimuli and rhythmical changes qualify as voluntary. This is to say that voluntariness is more than just a term for our ignorance.

On the other hand, voluntariness does not reflect the "ground noise" of the brain, the "error rate" intrinsic to all neural processing. It would be very dangerous for an animal if behavior were released without a good reason as an accident of the brain. Also, it seems most unlikely and is not supported by any data that the brain should not be able to do better. Randomly generated behavior in humans or animals is strikingly pathological. We maintain that voluntary behavior is neither automatic nor elicited by (releasing) stimuli and that it is nevertheless meaningful, i.e., not an undesirable by-product of neural processing.

Fortunately, from what has been described in this book, we can be fairly certain that torque spikes (saccades) fulfill all these criteria. With the tethered fly we can monitor to a high degree of precision the stimuli reaching the animal from its environment. We have convinced ourselves that the fly generates torque spikes without any events of significance triggering them from the outside, although, of course, the probability of their occurrence depends upon the situation.

The behaviorist objecting to this statement would have argued that even in the open loop situation not all sensory variables can be fully controlled. This argument neglects a general point concerning the relation between stimulus and response. It is now generally accepted that responses can consist of complex behavior patterns elicited by complex perceptions like, for instance, the behavior of young chicken running for shelter when the "releaser", the silouette of a bird in the sky, appears. Certain features of the silouette are critical. If these are changed the response rate sharply drops. At the level of fixed action patterns releasing stimulus and response belong to a common biological context. This context limits the search for hidden variables. Fixed action patterns are not elicited by insignificant stimuli. As pointed out in Chapter 6, saccades can be regarded as such fixed action patterns. The context in which they occur is that of object-related orientation. The above consideration makes it very unlikely that they are triggered by some still uncontrolled event outside this context. We consider this possibility sufficiently remote to be disregarded.

"Inner stimuli" as causes for saccades we also reject. Events in the animal qualifying as releasing stimuli would imply that the saccades were a relevant answer to these events. Again such a possibility seems most unlikely. In our opinion ethologists created the notion of "inner stimuli" as an explanation of spontaneous behavior; but mental states such as moods, drives, and dispositions merely change the probability for a certain behavior. They are no releasing stimuli in themselves. Initiation is needed in addition. With the concept of initiation "inner stimuli" are unnecessary and in fact no evidence supports their existence. The alternative to "releasing stimuli" are "modulating stimuli". They again do not generate behavior but change the animal's disposition. They serve to shape orientedness.

So far we have argued that saccades are not responses to stimuli. We have seen in Chapters 6, 9 and 12 that they are nevertheless employed in a meaningful way. They are used to generate 360°-loops and to stabilize objects in a certain part of the visual field. The fly applies them in the pursuit of goals. Thus, they fulfill all criteria for voluntary behavior.

The ultimate hope of the behaviorist had been that, if we could look at behavior in the greatest possible detail, the puzzling problem of voluntary behavior would dissolve and we would be left with a purely deterministic scheme of how behavior is generated. Orientation research seemed a perfect case in favor of this idea. With tethered *Drosophila* we are in the position to test it. One can record orientation behavior with the precision of the individual wing beat (Götz, 1983) and one is unlimited in specifying stimuli. The result is the opposite of what the behaviorist had been hoping for. The phenomenon of voluntary behavior does not disappear with a better specification of stimuli and responses. On the contrary, pure responses entirely determined by stimuli seem to be a rare exception in the behavior of animals.

If voluntary behavior is real it is interesting to speculate about its nature. As we have pointed out in the first pages of this book, one of the most important properties of brains is initiation. We suggest that voluntary behavior is based on initiation. We conceive initiation as a class of chance events which are needed in evolutionary processes taking place in the brain. As mentioned in the introduction, these processes consist of variation and selection or, to put it in the words of Popper (1976), of trial and the elimination of error. Initiating acts in the nervous system are equivalent to mutations in the evolution of species. As chance events, mutations and initiating acts have in common that the causes generating them often are biologically irrelevant, yet they may become most important for the organism. The main point with initiation is not the existence of stochastic processes in the brain, but the existence of a type of stochastic process which is at the basis of behavioral adaptation in a manner quite analogous to Darwinian evolution. Thus we expect initiation to be a highly organized affair.

Initiation in the central nervous system occurs at all levels: in establishing perceptual hypotheses, in directing selective attention, in search, play, thought, the choice of behavioral subroutines, the positioning of a limb etc. In voluntary behavior initiation generates behavioral options (i.e., mental representations of behavior). As a chance event, initiation is the alternative to releasing stimuli, since, as stated above, a behavior is said to be the response to a stimulus if cause and effect are tied into a biological context, whereas our definition for chance is precisely the absence of this context.

We propose that the whole behavioral repertoire of an animal is represented in a "field of probabilities", indicating for each behavioral option the likelihood by which it will be brought up in the next moment. The probabilities may be different for each situation and may depend upon the genetic endowment and life history of the animal. Under extreme circumstances, for instance facing a predator, only "flight" and "hiding" would have high probabilities, whereas without urgent needs for the animal the field would contain about equal probabilities for a large variety of potential behaviors.

Selection limits and channels variation (preselection) by all the processes shaping the probability field; but selection as the elimination of error (post-selection) is equally important. The simplest post-selective influence may lie in the length of time a behavioral state is maintained. But even small brains may have already the ability to store one or several behavioral options, to compare them with the situation, with neural representations of goals and to discard them if necessary before they are executed.

The main benefit of voluntary behavior has been mentioned in the introduction. As, for instance in politics, negotiations with other countries on a purely reactive basis will soon lead to undesirable situations, also the individual organism would have too little information about the future to survive long with purely reactive strategies. The success of "trial and the elimination of error", particularly in connection with goals, does not need to be specially emphasized. It helps to start something useful on one's own and it helps to manage new situations. That this behavior is found with *Drosophila* at the torque meter, though, comes as a surprise. One may now be in the position to investigate comparatively simple forms of internalized selection and of neurally represented goals as the first chapter of a natural history of thinking.

We have mentioned in Chapter 12 that the ability of *Drosophila* to reorganize its visuo-motor coordination at the torque meter may be closely related to classical operant conditioning. In the painstaking discussion of operant conditioning by the behaviorist school (e.g., Skinner, 1938) the fact has been underrated that the animal is solving a problem which had never been solved before by this animal or any of its ancestors. When the rat happens to press the lever delivering a food pellet, it does not mistake the lever for a root under which to dig, and it does not press it by accident, e.g., falling off the side of the cage. The nature of the rat is to look for unconventional solutions to new problems. It is initiation which helps the rat to survive in a Skinner-box as well as in a modern industrial city.

Our model of voluntary behavior calls for experimental investigations. Central brain mutants and animals with localized brain lesions should show behavioral deficits in some of these functions. In decapitated flies behavioral subroutines like grooming, walking, and flight no longer occur spontaneously but can still be specifically stimulated. Huber (1960) has shown that small lesions in the region of the central body of crickets often lead to a severe suppression of active behavior. Thus we are confident that evidence will accumulate in favor of or against this specific proposal.

If voluntary behavior of animals is no illusion, free will of humans should be real as well. Kant (1783; cited edition 1910) postulated that free will, while having empirical consequences, could not have an empirical cause, since as soon as we discovered a cause for what we were doing we would not call it free. In contrast, we propose to regard free will as a behavioral category based on initiation and developed from voluntary behavior of our animal ancestors. In our view an act of free will can be thought of as the product of an evolutionary process, of variation and selection, and thus as a starting point for something new. The causes leading to this behavior are often biologically irrelevant and do not need to be known introspectively. Its consequences, however, may be of utmost importance. What matters for the subjective feeling and objective state of freedom is the absence of releasing stimuli, of influences manipulating the organism. If freedom is not merely a keyword of 18th century idealism but part of reality, and especially of most successful strategies for the survival of ambulatory organisms, we should make use of it.

Orientedness Revisited

What is orientedness? We introduced this term at the beginning of the book and have used it on various occasions in the interpretation of data. Is this concept necessary and what is the experimental evidence for it?

Orientedness is closely related to voluntary behavior. Animals initiating behavior on their own need information about the situation. In our discussion of voluntariness we have made the distinction between releasing and modulating stimuli (p. 221). We have argued that releasing mechanisms require very specific sensory constellations to operate whereas, the bulk of the everyday sensory data is not tied into stimulus – response chains. We propose that they serve to up-date and to improve the representation of the world in the brain. How is the world represented in the fly brain? This is a most intriguing question hardly discussed at all in this book. We have pointed out that in the fly brain phylogenetic information about the world is stored abundantly. As an example we have mentioned the visuo-motor interface of large field course control (p. 63 ff). In this case the behavior itself reflects the properties of the world (of which the fly is a part). Extrapolating from this one result it seems reasonable to conceive of the phylogenetic representation of the world in the fly brain as being composed of behavioral options.

Do the modulating sensory data, however, provide a three-dimensional model of the momentary surroundings – a model in which the fly can test in advance the behavioral options and their consequences? It is too early to speculate. We have mentioned that the columnar neuropil of the optic lobes of the fly must be interpreted as a stack of "maps" of the visual sphere; but these maps to all our knowledge consist of the different stages in the momentary data flux. Even if series of "snapshots" of the visual world could be stored there like the frames of a motion picture on a film, this would not constitute the essential feature of the neural correlate of orientedness. By interacting with the world the fly gathers information about its relation to the world, and thus about its behavioral options.The concept of orientedness does not require a snapshot-like spatial map; a modulation of the above-mentioned probabilities of the behavioral options (p. 222) would suffice.

In any case, our evidence in favor of orientedness is not based upon any special nature of the representation. It is based upon the observation that in many of our experiments the fly tries to structure the situation, i.e., to make as much sense out of it as possible. Consider for instance the experiment in which the experimenter inverts the visuo-motor coupling of orientation at the torque compensator (Chap. 12). As long as the fly fails to maneuvre the stripe to the frontal position, it keeps trying. Once it has achieved its goal and has learned how to do it, the fly no longer tries to fixate the stripe permanently. It must be a powerful incentive for the fly to be able to direct the stripe "at will". Possessing orientedness means to be on top of the situation. It is just this that the fly is struggling for. The fly tries to keep the number of behavioral options large and to keep the incoming information as consistent as possible. We propose basing orientedness upon behavioral strategies and considering the momentary representation of the world as an ontogenetic elaboration of the phylogenetic one.

The concept of voluntary behavior and orientedness in *Drosophila* emphasizes a certain quality of vision well known introspectively. For ourselves, our eyes are little windows through which we look out into the world, and our actions are means by which we interact with it. A whole private world of mental activity is interposed between sensations and actions. We hope that our attempts to formulate this aspect of behavior in *Drosophila* will stimulate the investigation of central brain function in insects. Naturally, our general description cannot replace a detailed analysis in operational terms, which this book tries to encourage. Considering the Babylonian confusion of languages in the discussion of human brain function it may be useful, however, to develop a biologically founded language for small brains, in particular with neurogenetics as a guide to its natural design.

Appendix 1

List of Neurological Mutants

The following list contains, in alphabetical order, the neurological mutants of *Drosophila melanogaster* mentioned in this text. (Eye color mutants are omitted since they are well documented in the standard list of Lindsley and Grell, 1968). Most of the mutants can be found in an exhaustive list of neurological mutants published as Appendix to the recent review *Genetics of the Nervous System of Drosophila* by J. Hall (1982). Regarding details of nomenclature and accuracy of map positions the reader is refered to this review. For convenience we list under "page reference" the page number(s) where the respective mutant is mentioned in the text.

Name: *abnormal pattern orientation*^{S129} (*apo*^{S129})
Chromosome/map position: X/?
References: Heisenberg (1979); Bülthoff (1982 a)
Page reference: 133, 143
Phenotype: Vertical dark stripes have low attractiveness for flies walking freely in a brighty illuminated arena. Low turning tendency towards dark spots in multiple – Y maze. ERG normal.

Name: *abnormal pattern orientation B*^{S71} (*apo B*^{S71})
Chromosome/map position: X/?
References: –
Page reference: 133
Phenotype: Similar to *apo*^{S129}

Name: *abnormal pattern orientation C*^{S131} (*apo C*^{S131})
Chromosome/map position: X/?
References: –
Page reference: 133
Phenotype: Similar to *apo*^{S129}

Name: *bendless* (*ben*)
Chromosome/map position: X/35
References: Thomas and Wyman (1982)
Page reference: 76, 219
Phenotype: Abnormal axon terminal of the giant descending neuron (GDN). The fiber which in the mesothoracic neuromer of the wild type bends laterally to synapse with the tergal depressor motor neuron of the trochanter (jump muscle) ends befor the bend. Mutant flies do not jump. Electrical stimulation of mutant giant fiber leads to abnormal response of trochanter.

Name: *cacophony* (*cac*)
Chromosome/map position: X/34.6
References: Schilcher (1976)
Page reference: 219, 240
Phenotype: Altered courtship song: multiple pulses of increased amplitude (instead of a single pulse) during "pulse song". Reduced mating success of wingless mutant males as compared to wingless wildtype males. Females normal.

Name: *dunce* (*duc*)
Chromosome/map position: X/4.6; 6 mutant alleles including viable deletion
References: Dudai et al. (1976); Byers et al. (1981)
Page reference: 203, 233, 234
Phenotype: Impaired learning or retention in various tasks of associative and operant conditioning. Slow habituation. Low sensitivity of landing response to front-to-back motion of visual patterns. Increased rate of heart beat. Females sterile. Increased cAMP levels due to reduction or elimination of one form of cAMP phosphodiesterase.

Name: *ebony* (*e*)
Chromosome/map position: 3/70.7; many mutant alleles
References: Lindsley and Grell (1968); Kyriacou (1981)
Page reference: 58, 215, 240
Phenotype: Body color dark; Light-on and -off transients of ERG missing; reduced optomotor turning response at low light intensity; male courtship behavior aberrant; elevated levels of β-alanine and dopamine. Alleles differ with respect to behavioral impairment.

Name: *extra lamina fiber*^{H37} (*elf*^{H37})
Chromosome/map position: X/?
References: Heisenberg and E. Buchner (1977)
Page reference: 59
Phenotype: Tangential crossections of lamina cartridge shows additional fiber profile inside the ring of retinula cell fibers R1–6. Light-off transient of ERG is weak. Threshold of optomotor turning response at low light intensity is shifted to higher intensity. Increased UV-preference in spectral wavelength-specific phototaxis. (According to I. Meinertzhagen (personal communication) the structural phenotype is not apparent in the present stock.)

Name: *lobulaplate - less*^{N684} (lop^{N684})
Chromosome/map position: 2/?
References: Heisenberg (1984); Fischbach (1983 b)
Page reference: 65 ff, 239
Phenotype: Cellular cortex and columnar neuropil of lobula plate are reduced to about 20%. Rudiment is grafted to the caudal surface of the lobula. Medulla neuropil retains columnar organisation but is somewhat reduced in volume as well. Distal cell cortex of medulla contains histologically abnormal cell bodies. Giant fibers of lobulaplate are present but partially misrouted. Optomotor roll and pitch responses are reduced yaw torque is close to normal. Normal ERG. Olfactory discrimination learning normal.

Name: *minibrain* (*mnb*); (originally the symbol *min* was used)
Chromosome/map position: X/58.2; part of the Sh-complex
References: Heidenreich (1982); Heisenberg (1984); Fischbach (1983 b)
Page reference: 207, 236
Phenotype: Volume of brain (including medulla and lobula complex but not lamina) reduced to about 50%, but overall architecture normal; elevated level of Drosophilin in the adult (K. Scheller, personal communication); abnormal pattern fixation of freely walking flies; odor discrimination learning impaired.

Name: *no object fixation A*^{S100} (*nof A*^{S100})
Chromosome/map position: X/?
References: Heisenberg and Wolf (1979); Bülthoff (1982 a); Götz (1980)
Page reference: 133 ff, 140, 143
Phenotype: Low attractiveness of vertical dark stripe in walking flies and during stationary flight at the torque meter under closed loop conditions. Pattern fixation can be improved by continuously disturbing the closed loop (noise). Mean frequency of body saccades (abrupt turns) increased. ERG normal.

Name: *no object fixation* D^{B11} (*nof* D^{B11})
Chromosome/map position: X/?
References: Bülthoff (1982a, b)
Page reference: 136
Phenotype: Abnormal orientation to patterns; abnormal optomotor turning responses; normal ERG.

Name: *no object fixation* E^{B12} (*nof* E^{B12})
Allele of *sol*; should be renamed sol^{B12}
References: Bülthoff (1982a, b)
Page reference: 133, 136
Phenotype: Similar to *nof* D^{B11}

Name: *no on-transient* A^{H2} (*non* A^{H2})
Chromosome/map position: X/52.3; several mutant alleles
References: Heisenberg (1972); Heisenberg and E. Buchner (1977)
Page reference: 124
Phenotype: Light-on and -off transients of ERG missing; optomotor response suppressed, remaining response shows reduced spatial resolution; special impairment of optomotor response to front-to-back motion; UV-preference in spectral wavelength-specific phototaxis suppressed.

Name: *no on-transient* C^{P37} (*non* C^{P37})
Chromosome/map position: X/?; 2 mutant alleles
References: Heisenberg and Götz (1975); Heisenberg and E. Buchner (1977)
Page reference: 59
Phenotype: Light-on and -off transients of ERG missing; low spatial resolution expressed in optomotor turning response. Poor pattern fixation; UV-preference in spectral wavelength-specific phototaxis suppressed.

Name: *optomotor-blind*^{H31} (*omb*^{H31})
Chromosome/map position: X/7.5; mutant phenotype associated with a large inversion on the X chromosome (from 4 C 3–7 to 12 D2–E1) which effectively prevents recovery of recombinants; *omb* gene must reside close to distal breakpoint. If this area is uncovered by a deletion all *omb* phenotypes are expressed.
References: Heisenberg et al. (1978); Blondeau and Heisenberg (1982); Cook (1980); Tompkins et al. (1982)
Page reference: 65ff, 77, 80, 126ff, 139, 169ff, 195, 206f, 215, 232, 237
Phenotype: Reduced optomotor torque and turning responses; other movement dependent responses normal; visual component of male courtship disturbed although courtship tracking function appears normal. Normal ERG. Giant fibers of lobula plate are absent or greatly reduced. Neurons appear not to be formed. Brain structure otherwise normal; olfactory discrimination learning normal.

Name: *outer rhabdomeres absent*^{JK84} (*ora*^{JK84})
Chromosome/map position: 3/65.3; 8 mutant alleles; *ora*^{JK84} does not complement *nina E*, a tightly linked mutant which might be the structural gene for opsin component of R1–6 xanthopsin. *nina E* however forms close to normal sized rhabdomeres (4 mutant alleles). The 7 other alleles of ora form normal amounts of xanthopsin and (presumably) normal sized rhabdomeres as well. Absence of rhabdomeres thus appears as a synergistic effect of two closely linked genes both affected in *ora*^{JK84}.
References: Harris et al. (1976); Heisenberg and E. Buchner (1977); Koenig and Merriam (1977); Scavarda et al. (1983)
Page reference: 21f, 53ff
Phenotype: Large rhabdomeres of outer retinula cells R1–6 in each ommatidium form only as small rudiments. Visual motion-dependent behavior is much reduced or absent. Central retinula cells develop normally and mediate phototaxis. Spectral wavelength-specific phototaxis suppressed. ERG reduced to that of central retinula cells R7/8.

Name: *retinal degeneration* B^{KS222} (*rdg* B^{KS222})
Chromosome/map position: X/42.7; many mutant alleles
References: Harris and Stark (1977); Heisenberg and E. Buchner (1977); Homyk et al. (1981)
Page reference: 21 f, 53 ff
Phenotype: Light-induced degeneration of most outer retinula cells R1–6. Prolonged depolarizing afterpotential (PDA) decays faster than in wildtype and can not be reversed by orange light as in wild type. Central cells R7 and 8 remain functional and mediate phototaxis but no responses to visual motion.

Name: *sevenless*^{LY3} (*sev*^{LY3})
Chromosome/map position: X/33.4; several alleles
References: Harris et al. (1976); Heisenberg and E. Buchner (1977); Wolf et al. (1980); Campos-Ortega et al. (1979)
Page reference: 21 f, 53 ff, 204, 215, 233
Phenotype: Retinula cell R7 absent or greatly reduced; rhabdomere no 7 missing in all ommatidia. Rhabdomere of retinula cell R8 ends below focal plane of cornea lens but more distally than in wild type. About 35% of R8 cells are found between R1 and R6 (instead of R1 and R2). Motion dependent visual behavior normal. Spectral wavelength-specific phototaxis suppressed. Color and light intensity discrimination learning impaired. Other visual and olfactory learning tasks normal.

Name: *sine oculis*^{1} (*so*^{1})
Chromosome/map position: 2/57.1; 2 mutant alleles
References: Götz (1980); Fischbach (1983 a); Fischbach and Technau (1984)
Page reference: 79, 214, 234
Phenotyp: Eyes rudimentary or absent. Massive cell degeneration in optic lobes during early pupal development due to mutant genotype of eye. Laminae of adult flies without eyes are missing; other visual neuropil areas are reduced in volume but retain gross topological relationship among each other as in wildtype. Rudimentary eyes with few ommatidia still mediate optomotor responses.

Name: *small optic lobes*^{KS58} (*sol*^{KS58})
Chromosome/map position: X/68; 16 mutant alleles
References: Fischbach and Heisenberg (1981); Fischbach and Technau (1984)
Page reference: 67, 77, 80, 163, 165, 197, 199, 204, 206 ff, 228, 234
Phenotype: Optic lobes with the exception of lamina reduced in volumne and cell number by about 50%. Cells die during early pupal development. Flies show normal optomotor turning behavior and other basic visual responses including color vision. But many properties of visual behavior are altered. Habituation of the landing response is slow. Flies have severe difficulties to adjust to inverted visuo-motor coupling. Attractiveness of vertical dark stripes is low for freely walking flies. Certain directions of motion which are ineffective in the wildtype elicit landing responses etc. Color and odor discrimination learning are normal.

Name: *tan* (*t*)
Chromosome/map position: X/27.5; 30 mutant alleles
References: Lindsley and Grell (1968); Heisenberg and Götz (1975); Heisenberg and E. Buchner (1977)
Page reference: 58
Phenotype: Body pigmentation in many alleles is lighter than normal. Light-on and -off transients of ERG are missing due to genetic constitution of eye (not lamina). Different alleles show different degrees of reduced light sensitivity in phototaxis. Spatial resolution in tests of optomotor responses is reduced and UV-preference in spectral wavelength-specific phototaxis suppressed.

Appendix 2

Symbols, Dimensions, Abbreviations

α — Inclination of direction of pattern movement. $-180° < \alpha < 180°$. For upward (downward) motion α is positive (negative). $\alpha = 0$ for front-to-back motion along the h-axis of lattice of visual elements (Fig. 20) or along ψ-axis (Fig. 27).

$A_{p\text{-}p}$ — Peak-to-peak amplitude of an oscillation

abs — Absolute value

A.C. — Alternating current

c — Coefficient relating yaw torque T to angular velocity w_f of fly in free flight. $w_f(t) = cT(t)$. ($c \approx 1/k$)

cd m^{-2} — Candela per square meter; $1\ \text{cd m}^{-2} = 10^{-4}$ sb (stilb) $= \pi$ asb (apostilb)

D — Diameter

$D(\psi)$ — Position function. Yaw torque in stationary flight toward the side of a landmark at the position ψ. A response to motion from front to back (object response), responses to flickering light and, in *Drosophila*, a response to stationary patterns are recorded in the position function.

d — Distance

D.C. — Direct current

deg — Degree

$\Delta\varrho$ — Half-width of angular sensitivity distribution of photoreceptor or visual element.

δ — Stripe width measured in degree as seen by the fly in the center of the arena. Stripe width is given for the horizon.

EMD — Elementary movement detector

ERG — Electroretinogram

F — Focal length of cornea lens

f — Frequency

G — Closed loop gain; relation between rotatory bias w_b and angular velocity of fly w_f; $G = w_f/w_b$

h — Horizontal axis of the lattice of visual elements as shown in Figs. 3b and 21a.

H — Vertical extent of a pattern in degree as seen by the fly in the center of the arena.

Hz — Hertz: cycles per second.

I — Light intensity or pattern luminance; $\bar{I}$: spatial average; ΔI: spatial modulation of pattern luminance.

k — Coefficient indicating the yaw torque required to overcome aerodynamic friction. $T(t) = k\dot{\psi}(t)$; $k \approx 1/c$.

L — Latency of object response in closed loop (see Fig. 87 for details).

$\lambda_{h\nu}$ — Spectral wavelength of light

λ_{sp} — Spatial wavelength (pattern wavelength)

m — Pattern contrast

m_{eff} — Actual pattern contrast is reduced by the contrast transfer function of the eye.

mV — Millivolt

μ — Micron, micrometer

n — No of flies, experiments,...

N(t) — Torque "noise" generated by the fly independent of visual responses

Nm — Newton meter; $10^{-10}\ \text{Nm} = 1\ \text{mdyn cm}$

nm — Nanometer

ωt — Phase angle of an oscillation

p_e — Degree of polarization (of linearly polarized light) is defined as $(I_{\parallel} - I_{\perp})/(I_{\parallel} + I_{\perp})$ with $I_{\parallel}$ being the light intensity of the ray oscillating in parallel to the structure of the polarizing filter, $I_\perp$ the intensity of the ray oscillating perpendicular to it.

P Average positon of flies (recorded as center of gravity by a balance; see Fig. 111)

p Probability

PDA Prolonged depolarizing afterpotential

Φ Phase difference between two oscillations (e.g., intensity modulations in adjacent visual elements).

φ Angular coordinate of eye;
$\Delta\varphi$: divergence angle between visual elements (interommatidial angle);
$\Delta\varphi_\alpha^*$: divergence angle between adjacent rows of synchronously stimulated visual elements with the grating moving into the direction given by α.

ψ Horizontal angular coordinate of the fly's visual sphere (Fig. 3a). $\psi = 0°$ is in the direction of the fly's longitudinal body axis. To the fly's right (left) $\psi > 0$ ($\psi < 0$). ψ refers to stationary flight; $\dot{\psi}$ indicating angular motion of panorama. In experiments with freely rotating flies ψ^* is used. Thus $\dot{\psi}^*$ indicates angular velocity of fly ($\dot{\psi}^* = w_f$).
$\Delta\psi_{sep}$: angular distance between landmarks;
$\Delta\psi_{disp}$: displacement angle, when a pattern is shifted from one position to another.
ψ_{tr}: deviation from $\psi = 0°$ of the peak in a stripe position histogram if a rotatory bias (w_b) is added to the closed loop in experiments with *Musca domestica* (see Fig. 60).

R(t) Yaw torque responses to visual stimuli.

$r(\psi)\dot{\psi}$ Directionally selective torque response to visual motion at the position ψ; $r(\psi)\dot{\psi} = r^*(\psi)$

r Radius

$rev._R/rev._F$ Turning response on locomotion recorder. Average curvature of trajectory is expressed as the rotations of the styrofoam ball around its vertical axis divided by the rotations around its transverse axis.

S(t) External disturbance of closed loop situation; $S(t) = w_b(t)$

S_T Slope of rising phase of object response (for details see Fig. 87).

T(t) Torque;
T: time average of torque;
$\bar{T} = 1/t \int T(t)\,dt$

$\theta(\dot{\psi})$ Angular momentum of fly

ϑ Vertical angular coordinate of the fly's visual sphere (Fig. 3a)

τ Transmittance of arena wall

$U(\psi)$ Running integral of $D(\psi)$

V_{air} Velocity of air current

V_f Forward velocity of fly

v Vertical axis of lattice of visual elements as shown in Figs. 2 and 3b

w Angular velocity;
w_f: angular velocity of fly;
w_b: rotatory bias caused by a disturbance;
w_r: retinal slip speed.

w/λ_{sp} Contrast frequency; frequency of intensity modulation for visual elements when fly is stimulated with a moving periodic grating.

x Axis of lattice of visual elements as shown in Figs. 2 and 3b

x_d Axis of two-dimensional photosensitive diode (Fig. 46)

y Axis of lattice of visual elements as shown in Figs. 2 and 3b

y_c Output of elementary movement detector

y_d Axis of two-dimensional photosensitive diode (Fig. 46)

References

Arnett DW (1972) Spatial and temporal integration properties of units in first optic ganglion of dipterans. J Neurophysiol 35:429–444

Autrum H, Wiedemann I (1962) Versuche über den Strahlengang im Insektenauge. Z Naturforsch 17b:480–482

Bausenwein B (1984) Eigenschaften der Objektreaktion von *Drosophila melanogaster*. Dipl Thesis, Univ Würzburg

Bennet-Clark HC, Ewing AW (1970) The love song of the fruit fly. Sci Am 223:84–92

Benzer S (1967) Behavioral mutants of *Drosophila* isolated by countercurrent distribution. Proc Natl Acad Sci USA 58:1112–1119

Benzer S (1973) Genetic dissection of behavior. Sci Am 229:24–37

Bicker G, Reichert H (1978) Visual learning in a photoreceptor degeneration mutant of *Drosophila melanogaster*. J Comp Physiol 127:29–38

Blest AD (1961) Some modifications of Holme's silver method for insect central nervous system. Q J Microsc Sci 102:413–417

Blondeau J (1977) Electrically evoked motor activity in the fly. (*Calliphora erythrocephala*). Doct Thesis, Univ Tübingen

Blondeau J (1981) Electrically evoked course control in the fly *Calliphora erythrocephala*. J Exp Biol 92:143–153

Blondeau J, Heisenberg M (1982) The 3-dimensional optomotor torque system of *Drosophila melanogaster*. Studies on wild-type and the mutant *optomotor-blind*^{H31}. J Comp Physiol 145:321–329

Booker R, Quinn WG (1981) Conditioning of leg position in normal and mutant *Drosophila*. Proc Natl Acad Sci USA 78:3940–3944

Boschek CB (1971) On the fine structure of the peripheral retina and the lamina of the fly *Musca domestica*. Z Zellforsch 110:336–349

Braitenberg V (1967) Patterns of projections in the visual system of the fly. I. Retina lamina projections. Exp Brain Res 3:271–298

Braitenberg V (1970) Ordnung und Orientierung der Elemente im Sehsystem der Fliege. Kybernetik 7:235–242

Braitenberg V (1972) Periodic structures and structural gradients in the visual ganglia of the fly. In: Wehner R (ed) Information processing in the visual system of arthropods. Springer, Berlin Heidelberg New York, pp 3–15

Braitenberg V (1973) Gehirngespinste. Neuroanatomie für kybernetisch Interessierte. Springer, Berlin Heidelberg New York

Braitenberg V, Debbage P (1974) A regular net of reciprocal synapses in the visual system of the fly *Musca domestica*. J Comp Physiol 90:25–31

Braitenberg V, Hauser-Holschuh (1972) Patterns of projection in the visual system of the fly. II. Quantitative aspects of second order neurons in relation to models of movement perception. Exp Brain Res 16:184–209

Braitenberg V, Strausfeld NJ (1973) Principles of the mosaic organization in the visual system's neuropil of *Musca domestica*. In: Jung R (ed) Handbook of sensory physiology, vol VII/3a. Springer, Berlin Heidelberg New York, pp 631–659

Braitenberg V, Taddei-Ferretti C (1966) Landing reaction of *Musca domestica*. Naturwissenschaften 53:155–156

Buchner E (1971) Dunkelanregung des stationären Flugs der Fruchtfliege *Drosophila*. Dipl Thesis, Univ Tübingen

Buchner E (1974) Bewegungsperzeption in einem visuellen System mit gerastertem Eingang. Doct Thesis, Univ Tübingen

Buchner E (1976) Elementary movement detectors in an insect visual system. Biol Cybern 24:85–101

Buchner E (1984) Behavioural analysis of spatial vision in insects. In: Ali MA (ed) Photoreception and vision in invertebrates. Plenum Press, New York

Buchner E, Buchner S (1980) Mapping stimulus-induced nervous activity in small brains by (^{3}H)2-deoxy-D-glucose. Cell Tissue Res 211:51–64

Buchner E, Buchner S (1984) Neuroanatomical mapping of visually induced nervous activity in insects by ^{3}H-deoxyglucose. In: Ali MA (ed) Photoreception and vision in invertebrates. Plenum Press, New York

Buchner E, Götz KG, Straub C (1978) Elementary detectors for vertical movement in the visual system of *Drosophila*. Biol Cybern 31:235–242

Buchner E, Buchner S, Hengstenberg R (1979) 2-Deoxy-D-glucose maps movement-specific nervous activity in the second visual ganglion of *Drosophila*. Science 205:687–688

Buchner S, Buchner E (1982) Functional neuroanatomical mapping in insects by (^{3}H)2-deoxy-D-glucose at electron microscopical resolution. Neurosci Lett 28:235–240

Bülthoff H (1980) Orientation behaviour of *Drosophila melanogaster*. Doct Thesis, Univ Tübingen

Bülthoff H (1981) Figure–ground discrimination in the visual system of *Drosophila melanogaster*. Biol Cybern 41:139–145

Bülthoff H (1982 a) *Drosophila* mutants disturbed in visual orientation. I. Mutants affected in early visual processing. Biol Cybern 45:63–70

Bülthoff H (1982 b) *Drosophila* mutants disturbed in visual orientation. II. Mutants affected in movement and position computation. Biol Cybern 41:71–77

Bülthoff H, Götz KG (1979) Analogous motion illusion in man and fly. Nature (London) 278:636–638

Bülthoff H, Poggio T and Wehrhahn C (1980) 3-D Analysis of the flight trajectories of flies (*Drosophila melanogaster*). Z Naturforsch 35c:811–815

Bülthoff H, Götz KG, Herre M (1982) Recurrent inversion of visual orientation in the walking fly *Drosophila melanogaster*. J Comp Physiol 148:471–481

Bullock TH (1977) Introduction to nervous systems. Freeman, San Francisco

Burkhardt D (1983) Wavelength perception and colour vision. In: Cosens DJ, Vince-Price D (eds) The biology of photoreceptors. Exp Biol Symp XXXVI

Byers D, Davis R, Kiger JA Jr (1981) Defect in cyclic AMP phosphodiesterase due to the *dunce* mutation of learning in *Drosophila melanogaster*. Nature (London) 289:79–81

Campos-Ortega JA, Strausfeld NJ (1972) The columnar organization of the second synaptic region of the visual system of *Musca domestica*.L. I. Receptor terminals in the medulla. Z Zellforsch 124:561–585

Campos-Ortega JA, Jürgens G, Hofbauer A (1979) Cell clones and pattern formation. Studies on *sevenless*, a mutant of *Drosophila melanogaster*. Wilhelm Roux' Arch/Entwickluingsmech Org 186:27–50

Collett TS (1970) Centripetal and centrifugal visual cells in the medulla of the insect optic lobe (*Sphinx ligustri*). J Neurophysiol 33:239–256

Collett TS (1980 a) Some operating rules for the optomotor system of a hoverfly during voluntary flight. J Comp Physiol 138:271–281

Collett TS (1980 b) Angular tracking and the optomotor response. An analysis of visual reflex interaction in a hoverfly. J Comp Physiol 140:145–158

Collett TS, Land MF (1975 a) Visual control of flight behaviour in the hoverfly *Syritta pipiens*. J Comp Physiol 99:1–66

Collett TS, Land MF (1975 b) Visual spatial memory in a hoverfly. J Comp Physiol 100:59–84

Collett TS, Land MF (1978) How hoverflies compute interception courses. J Comp Physiol 125:191–204

Colonnier M (1964) The tangential organisation of the visual cortex. J Anat 98:327–344

Connolly K, Burnet B, Sewell D (1969) Selective mating and eye pigmentation. An analysis of the visual component in the courtship behaviour of *Drosophila melanogaster*. Evolution 23:548–559

Cook R (1975) 'Lesbian' phenotype of *Drosophila melanogaster*? Nature (London) 254:241–242

Cook R (1979) The courtship tracking of *Drosophila melanogaster*. Biol Cybern 34:91–106

Cook R (1980) The extent of visual control in the courtship tracking of *Drosophila melanogaster*. Biol Cybern 37:31–41

Cook R (1981 a) Courtship-like tracking behaviour in wild-type female *Drosophila melanogaster*. Z Naturforsch 36 c:475–483

Cook R (1981 b) Sex-specific tracking identified in single mosaic *Drosophila*. Naturwissenschaften 68:267–268

Cosens DJ, Briscoe D (1972) A switch phenomenon in the compound eye of the white-eyed mutant of *Drosophila melanogaster*. J Insect Physiol 18:627–632

David CT (1978) The relationship between body angle and flight speed in free-flying *Drosophila*. Physiol Entomol 3:191–195

David CT (1982) Compensation for height in the control of groundspeed by *Drosophila* in a new 'barber's pole' wind tunnel. J Comp Physiol 147:485–493

Del Pezzo EM, Hoffman HSM (1980) Attentional factors in the inhibition of a reflex by a visual stimulus. Science 210:673–674

DeVoe RD (1980) Movement sensitivities of cells in the fly's medulla. J Comp Physiol 138:93–119

Dietrich W (1909) Die Facettenaugen der Dipteren. Z Wiss Zool 92:465–539

Dudai Y, Jan YN, Byers D, Quinn WG, Benzer S (1976) *dunce*, a mutant of *Drosophila* deficient in learning. Proc Natl Acad Sci USA 73:1684–1688

Dvorak DR, Bishop LG, Eckert HE (1975) On the identification of movement detectors in the fly optic lobe. J Comp Physiol 100:5–23

Eckert H (1981) The horizontal cells in the lobula plate of the blowfly *Phaenicia sericata*. J Comp Physiol 143:511–526

Eckert H, Bishop LG (1978) Anatomical and physiological properties of the 'vertical cells' in the third optic ganglion of the blowfly *Phaenicia sericata* (calliphoridae; diptera). J Comp Physiol 126:57–86

Eckert H, Hamdorf K (1981) The contrast frequency-dependence. A criterion for judging the non-participation of neurones in the control of behavioural responses. J Comp Physiol 145:241–247

Eckert H, Fligge B, Hamdorf K (1979) Excitation and inhibition in the activation of the landing response of the blowfly *Calliphora*. Naturwissenschaften 66:368

Fischbach KF (1979 a) Simultaneous and successive colour contrast expressed in "slow" phototactic behaviour of walking *Drosophila melanogaster*. J Comp Physiol 130:161–171

Fischbach KF (1979 b) Influence of homogeneously applied olfactory stimuli on phototaxis of *Drosophila melanogaster*. In: Giles R (ed) Animals and environmental fitness. Abstr 2nd Escpb Congr, Pergamon Press, Oxford New York, pp 91–92

Fischbach KF (1981) Habituation and sensitization of the landing response of *Drosophila melanogaster*. Naturwissenschaften 68:332

Fischbach KF (1983 a) Neural cell types surviving congenital sensory deprivation in the optic lobes of *Drosophila melanogaster*. Dev Biol 95:1–18

Fischbach KF (1983 b) Neurogenetik am Beispiel des visuellen Systems von *Drosophila melanogaster*. Habilitation, Univ Würzburg

Fischbach KF, Heisenberg M (1981) Structural brain mutant of *Drosophila melanogaster* with reduced cell number in the medulla cortex and with normal optomotor yaw response. Proc Natl Acad Sci USA 78:1105–1109

Fischbach KF, Lyly-Hünerberg I (1983) Genetic dissection of anterior optic tract. Cell Tissue Res 231:551–563

Fischbach KF, Reichert H (1978) Interactions of visual subsystems in *Drosophila melanogaster*. A behavioural genetic analysis. Biol Behav 3:305–317

Fischbach KF, Technau G (1981) Ganglion cell degeneration in normal and mutant (*sol*) optic lobes of *Drosophila melanogaster*. Abstr IX Congr Int Soc Dev Biol, p 216

Fischbach KF, Technau G (1984) Cell degeneration in the development of the *sine oculis* and *small optic lobes* mutants of *Drosophila melanogaster*. Dev Biol 104:219–239

Folkers E, Spatz HCh (1981) Visual learning behaviour in *Drosophila melanogaster* wildtype. J Insect Physiol 27:615–622

Fraenkel GS, Gunn DL (1961) The orientation of animals. Dover Publ, New York

Franceschini N (1972) Pupil and pseudopupil in the compound eye of *Drosophila*. In: Wehner R (ed) Information processing in the visual systems of arthropods. Springer, Berlin Heidelberg New York, pp 75–82

Franceschini N (1975) Sampling of the visual environment by the compound eye of the fly. Fundamentals and applications. In: Snyder AW, Menzel R (eds) Photoreceptor optics. Springer, Berlin Heidelberg New York, pp 97–125

Franceschini N, Kirschfeld K (1971) Etude optique in vivo des éléments photorécepteurs dans l'oeil composé de *Drosophila*. Kybernetik 8:1–13

Franceschini N, Kirschfeld K (1976) Le controle automatique du flux lumineux dans l'oeil composé des diptères. Biol Cybern 21:181–203

Franceschini N, Hardie R, Ribi W, Kirschfeld K (1981) Sexual dimorphism in a photoreceptor. Nature (London) 291:241–244

Geiger G (1975) 'Short-term learning' in flies. Naturwissenschaften 62:539

Geiger G (1981) Is there a motion-independent position computation of an object in the visual system of the housefly? Biol Cybern 40:71–75

Geiger G, Nässel DR (1981) Visual orientation behaviour of flies after selective laser beam ablation of interneurones. Nature (London) 293:398–399

Geiger G, Nässel DR (1982) Visual processing of moving single objects and wide-field patterns in flies. Behavioural analysis after laser-surgical removal of interneurons. Biol Cybern 44:141–150

Geiger G, Poggio T (1975) The orientation of flies towards visual patterns. On the search for the underlying functional interactions. Biol Cybern 19:39–54

Gibson JJ (1979) The ecological approach to visual perception. Houghton Mifflin, Boston

Götz KG (1964) Optomotorische Untersuchung des visuellen Systems einiger Augenmutanten der Fruchtfliege *Drosophila*. Kybernetik 2:77–92

Götz KG (1965) Die optischen Übertragungseigenschaften der Komplexaugen von *Drosophila*. Kybernetik 2:215–221

Götz KG (1968) Flight control in *Drosophila* by visual perception of motion. Kybernetik 4:199–208

Götz KG (1969) Movement discrimination in insects. In: Reichardt W (ed) Processing of optical data by organism and by machines. Rendiconti SIF, Course XL III. Academic Press, London New York, pp 494–509

Götz KG (1971) Spontaneous preferences of visual objects in *Drosophila*. Dros Inf Serv 46:62

Götz KG (1972 a) Processing of clues from the moving environment in the *Drosophila* navigation system. In: Wehner R (ed) Information processing in the visual systems of arthropods. Springer, Berlin Heidelberg New York, pp 255–263

Götz KG (1972 b) Principles of optomotor reactions in insects. In: Cerebral control of eye movements and perception of motion. Bibl Ophtalmol 83:411–417

Götz KG (1975 a) The optomotor equilibrium of the *Drosophila* navigation system. J Comp Physiol 99:187–210

Götz KG (1975 b) Hirnforschung am Navigationssystem der Fliege. Naturwissenschaften 62:468–475

Götz KG (1980) Visual guidance in *Drosophila*. In: Siddiqi O, Babu P, Hall LM, Hall JC (eds) Development and neurobiology of Drosophila. Plenum Press, New York London, pp 391–407

Götz KG (1983) Bewegungssehen und Flugsteuerung bei der Fliege *Drosophila*. In: Nachtigall W (ed) Biona report 2. Fischer, Stuttgart, pp 21–34

Götz KG, Buchner E (1978) Evidence for one-way movement detection in the visual system of *Drosophila*. Biol Cybern 31:243–248

Götz KG, Wenking H (1973) Visual control of locomotion in the walking fruitfly *Drosophila*. J Comp Physiol 85:235–266

Götz KG, Hengstenberg B, Biesinger R (1979) Optomotor control of wing beat and body posture in *Drosophila*. Biol Cybern 35:101–112

Goodman LJ (1960) The landing responses of insects. The landing response of the fly *Lucilia sericata* and other calliphorinae. J Exp Biol 7:854–878

Hall JC (1982) Genetics of the nervous system in *Drosophila*. Q Rev Biophys 15:223–479

Hardie RC (1977) Electrophysiological properties of R7 and R8 in dipteran retina. Z Naturforsch 32 c: 887–889
Hardie RC (1979) Electrophysiological analysis of fly retina. I. Comparative properties of R1–6 and R7 and 8. J Comp Physiol 129: 19–33
Hardie RC, Franceschini N, McIntyre PD (1979) Electrophysiological analysis of fly retina. II. Spectral and polarisation sensitivity in R7 and R8. J Comp Physiol 133: 23–39
Hardie RC, Franceschini N, Ribi W, Kirschfeld K (1981) Distribution and properties of sex-specific photoreceptors in the fly *Musca domestica*. J Comp Physiol 145: 139–152
Harris WA, Stark WS (1977) Hereditary retinal degeneration in *Drosophila melanogaster*. A mutant defect associated with the phototransduction process. J Gen Physiol 69: 261–291
Harris WA, Stark WS, Walker JA (1976) Genetic dissection of the photoreceptor system in the compound eye of *Drosophila melanogaster*. J Physiol (London) 256: 415–439
Hassenstein B (1951) Ommatidienraster und afferente Bewegungsintegration. (Versuche an dem Rüsselkäfer *Chlorophanus viridis*). Z Vergl Physiol 33: 301–326
Hassenstein B (1959) Optokinetische Wirksamkeit bewegter periodischer Muster nach Messungen am Rüsselkäfer *Chlorophanus viridis*. Z Naturforsch 14 b: 659–674
Hassenstein B, Reichardt W (1956) Systemtheoretische Analyse der Zeit-, Reihenfolgen- und Vorzeichenauswertung bei der Bewegungsperzeption des Rüsselkäfers *Chlorophanus*. Z Naturforsch 11 b: 513–524
Hausen K (1981) Monocular and binocular computation of motion in the lobula plate of the fly. Verh Dtsch Zool Ges 1981: 49–70
Hausen K (1982) Motion sensitive interneurons in the optomotor system of the fly. I. The horizontal cells. Structure and signals. Biol Cybern 45: 143–156
Hausen K, Strausfeld NJ (1980) Sexually dimorphic interneuron arrangements in the fly visual system. Proc R Soc London Ser B 208: 57–71
Hausen K, Wehrhahn C (1983) Microsurgical lesion of horizontal cells changes optomotor yaw responses in the blowfly. Proc R Soc London Ser B 219 (1215): 211–216
Hauser-Holschuh H (1975) Vergleichend quantitative Untersuchungen an den Sehganglien der Fliegen *Musca domestica* und *Drosophila melanogaster*. Doct Thesis, Univ Tübingen
Heide G (1971 a) Die Funktion der nicht-fibrillären Flugmuskeln von *Calliphora*. I. Lage Insertionsstellen und Innervierungsmuster der Muskeln. Zool Jahrb Abt Allg Zool Physiol Tiere 76: 87–98
Heide G (1971 b) Die Funktion der nicht-fibrillären Flugmuskeln von *Calliphora*. II. Muskuläre Mechanismen der Flugsteuerung und ihre nervöse Kontrolle. Zool Jahrb Abt Allg Zool Physiol Tiere 76: 99–137
Heide G (1978) Proprioceptorische Beeinflussung der Impulsmusterbildung im neuromotorischen System fliegender Dipteren. Verh Dtsch Zool Ges 1978: 256
Heide G (1983) Neural mechanisms of flight control. In: Nachtigall W (ed) Biona report 2. Fischer, Stuttgart, pp 35–52
Heidenreich D (1982) Die Genetik der Mutante *minibrain* (*min*) von *Drosophila melanogaster*. Dipl Thesis, Univ Würzburg
Heimburger L, Poggio T, Reichardt W (1976) A special class of nonlinear interactions in the visual system of the fly. Biol Cybern 21: 103–105
Heisenberg M (1971 a) Separation of receptor and lamina potentials in the electroretinogram of normal and mutant *Drosophila*. J Comp Physiol 80: 119–136
Heisenberg M (1971 b) Isolation of mutants lacking the optomotor response. Dros Inf Serv 46: 68
Heisenberg M (1972) Comparative studies on two visual mutants of *Drosophila*. J Comp Physiol 80: 119–136
Heisenberg M (1979) Genetic approach to a visual system. In: Autrum H (ed) Handbook of sensory physiology. vol VII/6a. Springer, Berlin Heidelberg New York, pp 665–679
Heisenberg M (1983) Initiale Aktivität und Willkürverhalten bei Tieren. Naturwissenschaften 70: 70–78
Heisenberg M (1984) Genetically induced changes of brain structure in *Drosophila melanogaster* and their behavioral correlates. 15 Int Congr Genet 1983, Oxford and Ibh Publ Co, New Dehli

Heisenberg M, Boehl K (1979) Isolation of anatomical brain mutants of *Drosophila* by histological means. Z Naturforsch 34:143–147

Heisenberg M, Buchner E (1977) The role of retinula cell types in visual behavior of *Drosophila melanogaster*. J Comp Physiol 117:127–162

Heisenberg M, Götz KG (1975) The use of mutations for the partial degradation of vision in *Drosophila melanogaster*. J Comp Physiol 98:217–241

Heisenberg M, Wolf R (1979) On the fine structure of yaw torque in visual flight orientation of *Drosophila melanogaster*. J Comp Physiol 130:113–130

Heisenberg M, Wonneberger R, Wolf R (1978) *optomotor-blind*^{H31} – a *Drosophila* mutant of the lobula plate giant neurons. J Comp Physiol 124:287–296

Heisenberg M, Blondeau J, Fischbach K-F, Wolf R (1983) Structural brain mutants and the visual system of *Drosophila melanogaster*. A group report. In: Hotta Y (ed) Genetic dissection of *Drosophila* behavior. Taniguchi Symp 1981, Tokyo Univ

Helmholz VH (1910) Handbuch der physiologischen Optik. vol III, 3rd edn, chap 31. Voss, Hamburg Leipzig, p 373

Hengstenberg R (1971) Das Augenmuskelsystem der Stubenfliege *Musca domestica*. I. Analyse der 'Clock spikes' und ihrer Quellen. Kybernetik 9:56–77

Hengstenberg R (1973) The effect of pattern movement on the impulse activity of the cervical connective of *Drosophila melanogaster*. Z Naturforsch 28:593–596

Hengstenberg R (1981) Visuelle Drehreaktionen von Vertikalzellen in der Lobula Platte von *Calliphora*. Verh Dtsch Zool Ges 1981:180

Hengstenberg R (1984) Roll stabilization during flight of the blowfly's head and body by mechanical and visual cues. In: Varju D, Schnitzler U (eds) Localization and orientation in biology and engineering. Springer, Berlin Heidelberg, pp 121–134

Hengstenberg R, Götz KG (1967) Der Einfluss des Schirmpigmentgehalts auf die Helligkeits- und Kontrastwahrnehmung bei *Drosophila*-Augenmutanten. Kybernetik 3:276–285

Hengstenberg R, Hengstenberg B (1980) Intracellular staining of insect neurons with procion yellow. In: Miller TA, Strausfeld NJ (eds) Experimental entomology. Neuroanatomical techniques. Springer, Berlin Heidelberg New York, pp 307–324

Hengstenberg R, Hausen K, Hengstenberg B (1982) The number and structure of giant vertical cells (VS) in the lobula plate of the blowfly *Calliphora erythrocephala*. J Comp Physiol 149:163–177

Hernandez de Salomon C, Spatz HCh (1983) Colour vision in *Drosophila melanogaster*. Wavelength discrimination. J Comp Physiol 150:31–33

Holst E von, Mittelstaedt H (1950) Das Reafferenzprinzip. Wechselwirkungen zwischen Zentralnervensystem und Peripherie. Naturwissenschaften 37:464–476

Homyk T Jr, Pye Q, Pak WL (1981) Visual defective mutants in *Drosophila*. Genetics 97: Suppl 50

Horn E (1978) The mechanism of object fixation and its relation to spontaneous pattern preferences in *Drosophila melanogaster*. Biol Cybern 31:145–158

Horn E, Mittag J (1980) Body movements and retinal pattern displacements while approaching a stationary object in the walking fly *Calliphora erythrocephala*. Biol Cybern 39:67–77

Horn E, Wehner R (1975) The mechanism of visual pattern fixation in the walking fly *Drosophila melanogaster*. J Comp Physiol 101:39–56

Horridge GA (1962) Learning of leg position in headless insects. Nature (London) 193:696–698

Hotta Y, Benzer S (1969) Abnormal electroretinograms in visual mutants of *Drosophila*. Nature (London) 222:354–356

Hoyle G (1980) Learning using natural reinforcements in insect preparations that permit cellular neuronal analysis. J Neurobiol 11:323–354

Hu KG, Stark WS (1980) The roles of *Drosophila* ocelli and compound eyes in phototaxis. J Comp Physiol 135:85–95

Hu KG, Reichert H, Stark WS (1978) Electrophysiological characterization of *Drosophila* ocelli. J Comp Physiol 126:15–24

Huber F (1955) Sitz und Bedeutung nervöser Zentren für Instinkthandlungen beim Männchen von *Gryllus campestris* L.. Z Tierphysiol 12:12–48

Huber F (1960) Untersuchungen über die Funktion des Zentralnervensystems und insbesondere des Gehirns bei der Fortbewegung und der Lauterzeugung der Grille. Z vergl Physiol 44:60–132

Jacob KG, Willmund R, Folkers E, Fischbach KF, Spatz HCh (1977) T-maze phototaxis of *Drosophila melanogaster* and several mutants in the visual systems. J Comp Physiol 116:209–225

Jaervilehto M, Moring G (1976) Spectral and polarization sensitivity of identified retinal cells of the fly. In: Zettler F, Weiler R (eds) Neural principles in vision. Springer, Berlin Heidelberg New York, pp 214–226

Kaiser W (1975) The relationship between visual movement detection and colour vision in insects. In: Horridge GA (ed) The compound eye and vision of insects. Clarendon Press, Oxford, pp 359–377

Kant I (1910) Prolegomena zu einer jeden künftigen Metaphysik, die als Wissenschaft wird auftreten können (1783). Akademie-Ausgabe, vol IV. Akademie Verlag, Berlin, p 33

Kaplan WD, Trout WE III (1974) Genetic manipulation of an abnormal jump response in *Drosophila.* Genetics 77:721–739

King DG, Wyman RJ (1980) Anatomy of the giant fibre system in *Drosophila*. I. Three thoracic components of the pathway. J Neurocytol 9:753–770

Kirschfeld K (1967) Die Projektion der optischen Umwelt auf das Raster der Rhabdomere im Komplexauge von *Musca.* Exp Brain Res 3:248–270

Kirschfeld K (1972) The visual system of *Musca.* Studies on optics, structure and function. In: Wehner R (ed) Information processing in the visual system of arthropods. Springer, Berlin Heidelberg New York, pp 61–74

Kirschfeld K (1973) Das neurale Superpositionsauge. Fortschr Zool 21:229–257

Kirschfeld K (1979) The visual system of the fly. Physiological optics and functional anatomy as related to behaviour. In: Schmitt FO (ed) The neurosciences. Fourth study program. MIT Press, Cambridge Mass, pp 297–310

Kirschfeld K, Franceschini N (1968) Optische Eigenschaften der Ommatidien im Komplexauge von *Musca.* Kybernetik 5:47–52

Kirschfeld K, Vogt K (1980) Calcium ions and pigment migration in fly photoreceptors. Naturwissenschaften 67:516

Kirschfeld K, Franceschini N, Minke B (1977) Evidence for a sensitizing pigment in fly photoreceptors. Nature (London) 269:386–390

Kirschfeld K, Feiler R, Franceschini N (1978) A photostable pigment within the rhabdomere of fly photoreceptors no.7. J Comp Physiol 125:275–284

Koenig J, Merriam J (1977) Autosomal ERG mutants. Dros Inf Serv 52:50–51

Kohler I (1956) Brillenversuche in der Wahrnehmungspsychologie mit Bemerkungen zur Lehre von der Adaptation. Z Exp Angew Psychol 3:381–417

Kühn A (1919) Die Orientierung der Tiere im Raum. Fischer, Jena

Kyriacou CP (1981) The relationship between locomotor activity and sexual behavior in *ebony* strains of *Drosophila melanogaster*. Anim Behav 29:462–471

Labhart T (1977) Electrophysiological recordings from the lateral ocelli of *Drosophila*. Naturwissenschaften 64:99–100

Land MF (1973) Head movements of flies during visually guided flight. Nature (London) 243:299–300

Land MF (1975) Head movements and fly vision. In: Horridge GA (ed) The compound eye and vision of insects. Clarendon Press, Oxford, pp 469–489

Land MF, Collett TS (1974) Chasing behavior of houseflies (*Fannia cannicularis*). J Comp Physiol 89:331–357

Laughlin S (1981) Neural principles in the peripheral visual systems of invertebrates. In: Antrum HJ (ed) Handbook of sensory physiology. vol VII/6b. Springer, Berlin Heidelberg New York, pp 133–280

Lea JY, Mueller CG (1977) Saccadic head movements in mantids. J Comp Physiol 114:115–128

Lindsley DL, Grell EH (1968) Genetic variation of *Drosophila melanogaster*. Carnegie Inst Washington Publ 627

Liske E (1977) The influence of head position on the flight behaviour of the fly *Calliphora erythrocephala.* J Insect Physiol 23:375–379

Lorenz K (1965) Über tierisches und menschliches Verhalten. vol II. Piper, München (e.g. pp 210–211)

Manning A (1959) The sexual behavior of two sibling *Drosophila* species. Behavior 15:123–145

Markow TA, Manning M (1980) Mating success of photoreceptor mutants of *Drosophila melanogaster*. Behav Neur Biol 29:276–280

McCann GD, MacGinitie GF (1965) Optomotor response studies of insect vision. Proc R Soc London Ser B 163:369–401

Menne D, Spatz HCh (1977) Color vision in *Drosophila melanogaster*. J Comp Physiol 114:301–312

Mimura K (1972) Neural mechanism subserving directional selectivity of movement in the optic lobe of the fly. J Comp Physiol 80:409–437

Mimura K (1982) Discrimination of some visual patterns in *Drosophila melanogaster*. J Comp Physiol 146:229–233

Minke B, Wu CF, Pak WL (1975) Isolation of light-induced response of the central retinula cells from the electroretinogram of *Drosophila*. J Comp Physiol 98:345–355

Mittelstaedt H (1951) Zur Analyse physiologischer Regelungssysteme. In: Verh Dtsch Zool Ges. Zool Anz Suppl 150

Nakashima-Tanaka E, Matsubara K (1979) An anomalous response (jumping behaviour) to light in *Drosophila melanogaster*. Jpn J Genet 54:345–357

Pak WL (1975) Mutants affecting the vision of *Drosophila melanogaster*. In: King RC (ed) Handbook of genetics. Plenum Press, New York London, pp 703–733

Pak WL (1979) Study of photoreceptor function using *Drosophila* mutants. In: Breakfield X (ed) Neurogenetics. Genetic approaches to the nervous system. Elsevier North Holland, Amsterdam New York, pp 67–99

Pak WL, Grossfield J, White NV (1969) Nonphototactic mutants in a study of vision of *Drosophila*. Nature (London) 222:351–354

Pak WL, Conrad SK, Kremer NE, Larrivee DC, Schinz RH, Wong F (1980) Photoreceptor function. In: Siddiqi O, Babu P, Hall LM, Hall JC (eds) Development and neurobiology of *Drosophila*. Plenum Press, New York London, pp 331–346

Paschma R (1982) Strukturelle und funktionelle Defekte der Drosophilamutante *lobulaplateless*^{N684}. Dipl Thesis, Univ Würzburg

Pick B (1974) Visual flicker induces orientation behaviour in the fly *Musca*. Z Naturforsch 29 a:310–312

Pick B (1976) Visual pattern discrimination as an element of the fly's orientation behavior. Biol Cybern 23:171–180

Pick B, Buchner E (1979) Visual movement detection under light- and dark- adaptation in the fly *Musca domestica*. J Comp Physiol 134:45–54

Pierantoni R (1973) Su un tratto nervoso nel cervello della mosca. In: Atti della prima riuniore scientifica plenaria. Soc Ital Biofis Pura Appl Camogli, pp 231–249

Poggio T, Reichardt W (1973 a) A theory of the pattern induced flight orientation of the fly *Musca domestica*. Kybernetik 12:185–203

Poggio T, Reichardt W (1973 b) Considerations on models of movement detection. Kybernetik 13:223–227

Poggio T, Reichardt W (1976 a) Nonlinear interactions underlying visual orientation behavior of the fly. Cold Spring Harbor Symp Quant Biol 40:635–645

Poggio T, Reichardt W (1976 b) Visual control of orientation behaviour in the fly. Part II. Towards the underlying neural interactions. Q Rev Biophys 9:377–438

Poggio T, Reichardt W (1981 a) Characterization of nonlinear interactions in the fly's visual system. In: Reichardt W, Poggio T (eds) Theoretical approaches in neurobiology. MIT Press, Cambridge Mass, pp 64–84

Poggio T, Reichardt W (1981 b) Visual fixation and tracking in flies. Mathematical properties of simple control systems. Biol Cybern 40:101–112

Poggio T, Reichardt W, Hausen K (1981) A neuronal circuitry for relative movement discrimination by the visual system of the fly. Naturwissenschaften 68:443–446

Popper K (1976) Unended quest. An intellectual autobiography. Collins, Glasgow Fontana

Posner MI, Snyder CRR, Davidson BJ (1980) Attention and detection of signals. J Exp Psychol Gen 109:160–174

Power ME (1943 a) The effect of reduction in numbers of ommatidia upon the brain of *Drosophila melanogaster*. J Exp Zool 94:33–71

Power ME (1943 b) The brain of *Drosophila melanogaster*. J Morphol 72:517–559

Preiss R, Kramer E (1984) Control of flight speed by minimization of the apparent ground pattern movement. In: Varju D, Schnitzler U (eds) Localization and orientation in biology and engineering. Springer, Berlin Heidelberg New York, pp 140–142
Preißler M (1974) Struktur des inneren Chiasmas im Sehsystem der Fliege *Musca domestica*. Dipl Thesis, Univ Tübingen
Quinn WG, Harris WA, Benzer S (1974) Conditioned behavior in *Drosophila melanogaster*. Proc Natl Acad Sci USA 71:708–712
Reichardt W (1957) Autokorrelations-Auswertung als Funktionsprinzip des Zentralnervensystems. Z Naturforsch 12 b:448–457
Reichardt W (1970) The insect eye as a model for analysis of uptake transduction and processing of optical data in the nervous system. In: Schmitt FO (ed) The neurosciences. Second study program. Rockefeller Univ Press, New York, pp 494–511
Reichardt W (1973) Musterinduzierte Flugorientierung der Fliege *Musca domestica*. Naturwissenschaften 60:122–138
Reichardt W, Poggio T (1975) A theory of the pattern induced flight orientation of the fly *Musca domestica*. II. Biol Cybern 18:69–80
Reichardt W, Poggio T (1976) Visual control of orientation behaviour in the fly. Part I. A quantitative analysis. Q Rev Biophys 9:311–375
Reichardt W, Poggio T (1979) Figure-ground discrimination by relative movement in the visual system of the fly. Part I. Experimental results. Biol Cybern 35:81–100
Reichardt W, Poggio T (1981) Visual control of flight in flies. In: Reichardt W, Poggio T (eds) Theoretical approaches in neurobiology. MIT Press, Cambridge Mass, pp 135–150
Reichardt W, Varju D (1959) Übertragungseigenschaften im Auswertesystem für das Bewegungssehen. (Folgerungen aus Experimenten an dem Rüsselkäfer *Chlorophanus viridis*.) Z Naturforsch 14 b:674–689
Reichardt W, Wenking H (1969) Optical detection and fixation of objects by fixed flying flies. Naturwissenschaften 56:674–689
Reichardt W, Poggio T, Hausen K (1983) Figure-ground discrimination by relative movement in the visual system of the fly. Part II. Towards the neural circuitry. Biol Cybern (Suppl) 46:1–30
Reichert H, Bicker G (1979) A visual learning study of brightness perception in two mutants of *Drosophila melanogaster*. J Comp Physiol 133:283–290
Rendel J (1951) Mating of *ebony*, *vestigial* and wild-type *Drosophila melanogaster* in light and dark. Evolution 5:226–230
Ridder M de (1974) Plastizität des Fühlerputzens bei *Gryllus bimaculatus* Degeer und mögliche neurale Grundlagen. Dipl Thesis, Univ Köln
Riehle A, Franceschini N (1984) Motion detection in flies: Parametric control over on-off pathways. Exp Brain Res 54:390–394
Roeder K (1968) Neurale Grundlagen des Verhaltens. Huber, Bern
Rossel S (1980) Foveal fixation and tracking in the praying mantis. J Comp Physiol 139:307–331
Sandeman DC, Erber J, Kien J (1975) Optokinetic eye movements in the crab *Carcinus maenas*. I. Eye torque. J Comp Physiol 101:243–258
Scavarda NJ, O'Tonsa J, Pak WL (1983) *Drosophila* locus with gene dosage effects on rhodopsin. Proc Natl Acad Sci USA 80:4441–4445
Schilcher F von (1976) The behavior of *cacophony* a courtship song mutant in *Drosophila melanogaster*. Behav Biol 17:187–196
Schilcher F von, Hall J (1979) Neural topography of courtship song in sex mosaics of *Drosophila melanogaster*. J Comp Physiol 129:85–95
Schmidt B (1975) Die neuronale Organisation in den Ocellen von *Drosophila*. Doct Thesis, Univ Freiburg
Schöne H (1980) Orientierung im Raum. Stuttgart, Wiss Verlagsges
Scholes J (1969) The electrical responses of the retinal receptors and the lamina in the visual system of the fly *Musca*. Kybernetik 6:149–162
Schümperli RA (1973) Evidence for colour vision in *Drosophila melanogaster* through spontaneous phototactic choice behavior. J Comp Physiol 86:77–94
Skinner BF (1938) The behavior of organisms. Appleton-Century-Krofts, New York

Smola U, Gemperlein R (1972) Übertragungseigenschaften der Sehzellen der Schmeissfliege *Calliphora*. 2. Die Abhänigkeit vom Ableitort: Retina-Lamina ganglionaris. J Comp Physiol 79:363–392

Sokoloff L (1978) Mapping cerebral functional activity with radioactive deoxyglucose. Trends Neuro Sci 1:75–79

Spatz HCh, Emanns A, Reichert H (1974) Associative learning of *Drosophila melanogaster*. Nature (London) 248:359–361

Spiess EB, Schwer WA (1978) Minority mating advantage of certain eye color mutants of *Drosophila melanogaster*. I. Multiple-choice and single-female tests. Behav Genet 8:155–168

Spüler M, Heide G (1978) Simultaneous recordings of torque, thrust and muscle spikes from the fly *Musca domestica* during the optomotor responses. Z Naturforsch 33 c:455–457

Srinivasan MV, Dvorak DR (1980) Spatial processing of visual information in the movement-detecting pathway of the fly. J Comp Physiol 140:1–23

Stark WS (1975) Spectral selectivity of visual response alterations mediated by interconversions of native and intermediate photopigments in *Drosophila*. J Comp Physiol 96:343–356

Stratton GM (1897) Vision without inversion of the retinal image. Psychol Rev 4:341–360, 463–481

Strausfeld NJ (1970) Golgi studies on insects. Part II. The optic lobes of diptera. Philos Trans R Soc London 258:135–223

Strausfeld NJ (1976) Atlas of an insect brain. Springer, Berlin Heidelberg New York

Strausfeld NJ (1979) The representation of a receptor map within retinotopic neuropil of the fly. Verh Dtsch Zool Ges 1979:167–179

Strausfeld NJ (1980) Male and female visual neurons in dipterous insects. Nature (London) 283:381–383

Strausfeld NJ, Bacon JP (1983) Multimodal convergence in central nervous system of insects. In: Horn E (ed) Multimodal convergence in sensory systems. Fortschr Zool 28. Fischer, Stuttgart pp 47–76

Strausfeld NJ, Bassemir UK (1983) Cobalt-coupled neurons of a giant fiber system in dipterans. J Neurocytol 12:971–991

Strausfeld NJ, Hausen K (1977) The resolution of neuronal assemblies after cobalt injection into neuropil. Proc R Soc London ser B 199:463–476

Strausfeld NJ, Nässel DR (1980) Neuroarchitecture of brain regions that subserve the compound eyes of crustacea and insects. In: Autrum H (ed) Handbook of sensory physiology. vol VII/6b. Springer, Berlin Heidelberg New York pp 1–132

Strausfeld NJ, Obermayer ML (1976) Resolution of intra-neuronal and trans-synaptic migrations of cobalt in the insect visual and nervous system. J Comp Physiol 110:1–12

Strausfeld NJ, Singh RN (1980) Peripheral and central nervous system projections. In: Siddiqi O, Babu P, Hall LM, Hall JC (eds) Development and neurobiology of *Drosophila*. Plenum Press, New York London, pp 267–291

Tanouye MA, Wyman RJ (1979) Mutations affecting the giant fiber system of *Drosophila melanogaster*. Neuro Sci Abstr 5:263

Tanouye MA, Wyman RJ (1980) Motor outputs of giant nerve fiber in *Drosophila*. J Neurophysiol 44:405–421

Thomas JB (1980) Mutations affecting the giant fiber system of *Drosophila*. Neursci Abstr 6:742

Thomas JB, Wyman RJ (1982) A mutation in *Drosophila* alters normal connectivity between two identified neurones. Nature (London) 298:650–651

Thorson J (1964) Dynamics of motion perception in desert locust. Science 145:69–71

Thorson J (1966 a) Small-signals analysis of a visual reflex in locust. I. Input parameters. Kybernetik 3:41–53

Thorson J (1966 b) Small-signals analysis of a visual reflex in locust. II. Frequency dependence. Kybernetik 3:53–66

Timm F (1958) Zur Histochemie der Schwermetalle. Das Sulfid-Silberverfahren. Dtsch Z Gerichtl Med 46:706–711

Tompkins L, Gross AC, Hall JC, Gailey DA, Siegel RW (1982) The role of female movement in the sexual behavior of *Drosophila melanogaster*. Behav Genet 12:295–307

Trujillo-Cenoz O (1965) Some aspects of the structural organization of the intermediate retina of dipterans. J Ultrastruct Res 13:1–33

Varju D (1959) Optomotorische Reaktion auf die Bewegung periodischer Helligkeitsmuster. Z Naturforsch 14 b: 724–735

Varju D (1975) Stationary and dynamic responses during visual edge fixation by walking insects. Nature (London) 255: 330–332

Varju D (1976) Visual edge fixation and negative phototaxis in the mealworm beetle *Tenebrio molitor*. Biol Cybern 25: 17–26

Varju D, Reichardt W (1967) Übertragungseigenschaften im Auswertesystem für das Bewegungssehen. II. Z Naturforsch 22 b: 1343–1351

Vigier P (1907) Mechanisme de la synthèse des impressions lumineuses recueillies par les yeux composés des diptères. C R Acad Sci 63: 122–124

Virsik R, Reichardt W (1974) Tracking of moving objects by the fly *Musca domestica*. Naturwissenschaften 61: 132–133

Virsik R, Reichardt W (1976) Detection and tracking of moving objects by the fly *Musca domestica*. Biol Cybern 23: 83–98

Vogt K (1983) Is the fly visual pigment a rhodopsin? Z Naturforsch 38 c: 329–333

Wagner H (1982) Flow-field variables trigger landing in flies. Nature (London) 297: 147–148

Wehner R (1972) Spontaneous pattern preferences of *Drosophila melanogaster* to black areas in various parts of the visual field. J Insect Physiol 18: 1250–1268

Wehner R (1979) Mustererkennung bei Insekten. Lokalisation und Identifikation visueller Objekte. Verh Zool Ges 72: 19–41

Wehner R (1981) Spatial vision in arthropods. In: Autrum H (ed) Handbook of sensory physiology. vol VII/6c. Springer, Berlin Heidelberg New York pp 287–616

Wehner R, Horn E (1975) The effect of object distance on pattern preferences in the walking fly *Drosophila melanogaster*. Experientia 31: 641–643

Wehner R, Wehner-von Segesser S (1973) Calculation of visual receptor spacing in *Drosophila melanogaster* by pattern recognition experiments. J Comp Physiol 82: 165–177

Wehrhahn C (1979) Sex specific differences in the orientation behaviour of houseflies. Biol Cybern 32: 239–241

Wehrhahn C (1981) Fast and slow flight torque responses in flies and their possible role in visual orientation behaviour. Biol Cybern 40: 213–221

Wehrhahn C, Hausen K (1980) How is tracking and fixation accomplished in the nervous system of the fly. A behavioural analysis based on short time stimulation. Biol Cybern 38: 179–186

Wehrhahn C, Hausen K, Zanker J (1981) Is the landing response of the housefly (*Musca*) driven by motion of a flow field? Biol Cybern 41: 91–99

Wehrhahn C, Poggio T, Bülthoff H (1982) Tracking and chasing in houseflies (*Musca*). Biol Cybern 45: 123–130

Willmund R (1979) Light induced modification of phototactic behaviour of *Drosophila melanogaster*. II. Physiological aspects. J Comp Physiol 129: 35–41

Willmund R, Ewing A (1982) Visual signals in the courtship of *Drosophila melanogaster*. Anim Behav 30: 209–215

Willmund R, Fischbach KF (1977) Light induced modification of phototactic behaviour of *Drosophila melanogaster* wildtype and some mutants in the visual system. J Comp Physiol 118: 261–271

Wolf R, Heisenberg M (1980) On the fine structure of yaw torque in visual flight orientation of *Drosophila melanogaster*. II. A temporally and spatially variable weighting function for the visual field ('visual attention'). J Comp Physiol 140: 69–80

Wolf R, Gebhardt B, Gademann R, Heisenberg M (1980) Polarization sensitivity of course control in *Drosophila melanogaster*. J Comp Physiol 139: 177–191

Wu C-F, Pak W (1975) Quantal basis of photoreceptor spectral sensitivity of *Drosophila melanogaster*. J Gen Physiol 66: 149–168

Zeil J (1979) A new kind of neural superposition eye. The compound eye of male Bibionidae. Nature (London) 278: 249–250

Zettler F, Jaervilehto M (1972) Lateral inhibition in an insect eye. Z Vergl Physiol 76: 233–244

Zimmermann G (1973) Der Einfluß stehender und bewegter Musteranteile auf die optomotorische Reaktion der Fliege *Drosophila*. Doct Thesis, Univ Tübingen

Subject Index

abnormal pattern orientation^{S129} (*apo*^{S129}) 133, 143, 226
abnormal pattern orientation B^{S71} (*apo* B^{S71}) 133, 226
abnormal pattern orientation C^{S131} (*apo* C^{S131}) 133, 226
acceptance angle 17, 18
action 3, 158, 219, 225
action potential 64, 73, 74
active behavior 8, 104, 191, 209, 220, 223
actively controlled movement 166
active process 3, 182, 192, 209
activity 1, 3, 95, 110, 192, 219
acuity 58
adaptation 19, 25, 58
adaptation, dark 19, 21, 58ff., 239
aerodynamic friction 114, 230
aerodynamics of fly 89
afference 100
air current 46, 47, 82, 146, 191, 231
air friction 83
amacrine cell (α) 28, 31, 61
angular acceleration 83, 96, 98
angular bias 213
angular motion 96, 231
angular position 84, 91f., 110, 114, 143, 215
angular sensitivity 17, 18, 38, 53, 149, 152, 230
angular slip speed 88
angular velocity 38f., 42, 44, 48, 64, 78, 83, 87, 89, 93ff., 110f., 114, 119ff., 126, 128, 153, 156, 165f., 169, 172, 176, 180f., 184ff., 191f., 196f., 203, 206, 213ff., 230f.
antennae 74, 76, 82, 191, 194
antennal nerve 76
antifixation 200, 206f., 210
apo B^{S71} see *abnormal pattern orientation* B^{S71}
apo C^{S131} see *abnormal pattern orientation* C^{S131}
apo^{S129} see *abnormal pattern orientation*^{S129}
attention see selective attention
attention, direction of 192
attractiveness 141, 160, 209ff., 226ff.
autoradiograph 62f.

background 108, 158ff., 188, 190, 195, 198, 206ff.
back-to-front displacement 152, 157
back-to-front motion 47, 49f., 62ff., 107, 116, 120, 122, 124, 127ff., 135, 148, 153ff., 171, 179, 187f., 219
back-to-front movement see back-to-front motion
back-to-front stimulation 123
bascet cell 28
base line of torque 122, 130, 134, 188
basement membrane 18
behavioral subroutine 7, 182, 209, 222f.
bendless (*ben*) 76, 219, 226
ben see *bendless*
Bibionidae 181
binocular overlap 11, 71, 129
biphasic response 202
body movement 237
body posture 71, 82, 235
body posture, optomotor control of 235
body saccade 95f., 100ff., 146, 180f., 227
brown 74
burst of torque 97, 102, 155, 169, 184, 200, 202

cac see *cacophony*
cacophony (*cac*) 219, 226, 240
Calliphora 31, 61, 63, 67, 69f., 98, 129, 162, 181, 218, 232ff., 241
cartridge 13, 27f., 60f.
Cataglyphis bicolor 194
cell degeneration 64, 66, 229, 234
cellular cortex 227
central brain 25ff., 31, 63, 73, 102, 209, 218, 223, 225
central nervous system 80, 192, 194, 222, 232, 241
central process 190
cervical connective 63, 74, 194, 237
chasing 180ff., 238, 242
chiasm 31, 60, 64, 66
Chlorophanus 37, 236, 240
chromatic adaptation 107
cinnabar 19

closed loop 43, 82ff., 94ff., 115f., 119, 122, 131ff., 144f., 148, 150, 152ff., 165ff., 195ff., 213, 215, 219, 230f.
closed loop coupling 94, 96, 105, 112, 147, 155
closed loop gain 87, 91f., 94, 182, 194f., 230
closed loop, inverted 202
CNS see central nervous system
cockroach 203
coherence of motion 175
coherent motion 64, 94, 152
color contrast 53, 234
color discrimination 7
color discrimination learning 204, 229
color vision 7, 53, 56, 229, 233, 237ff.
columnar neuron 25ff., 218
columnar neuron, small field 63, 79, 218
columnar neuropil 227
compound eye 10, 183, 234ff.
conditioning 56, 203, 215f.
contralateral sensitization 80
contrast frequency 36, 43f., 50f., 54, 74, 77, 93, 165, 231, 234
contrast perception 195
control of flight speed 240
control of groundspeed 234
control of head posture 48
copulation 180, 213
cornea 14, 17f.
cornea lens 14, 18, 21, 229f.
correction response 44, 101, 196ff.
corrective flight torque 167
correlation model 36, 38, 152
coupling coefficient 83, 107, 154
coupling conditions 96, 174
coupling conditions, natural 98
coupling conditions, reversed 203
coupling constant 97f., 140
course control 7, 37, 46ff., 70, 77, 82, 119, 174, 177, 179, 206, 218f., 232, 242
course control, large field 37, 59, 65, 67, 125ff., 139, 152, 166f., 172, 174ff., 195, 217ff., 224
course control, motion dependent 91
course control, nonvisual mechanism 93
course control, rotatory 64, 70, 73
course control, visual 33, 43f., 49, 52, 71, 73, 76, 87
course correction response 42
course stabilization 174, 217
courtship 33, 213f., 227, 233f., 240, 242
courtship tracking 182, 213f., 228, 233f.
cricket 223
cross-habituation 80
cross-interaction 161f.
cruising flight 47, 107, 192

dark adaptation 19, 21, 58ff., 239
Darwinian evolution 222
Darwinian paradigm 8
deep pseudopupil 14, 19
degree of polarization 230
depolarization 17, 20, 64
descending neuron 27, 63, 73f.
deutocerebrum 76
developmental period 66
developmental process 69, 71, 73, 218
D(ψ)-function 114f.
dichroic absorption 22f.
dichroism 23
difference function 114f., 121, 129, 132, 136f.
direction insensitive response to motion 162
direction of attention 192
direction of displacement 156
direction of flight 93, 97, 107, 116, 177, 182, 192
direction of torque spike 108
dishabituation of landing response 77
displacement angle 231
displacement, back-to-front 152, 157
displacement, direction of 156
displacement experiment 184, 188, 191
displacement, front-to-back 156, 219
divergence angle 12, 14, 40, 52, 123, 147
double arena 164, 174, 176, 195
downward motion 47, 63f., 69, 230
duc see *dunce*
dunce (*duc*) 203, 227, 233f.
dynamics of torque 104

e see *ebony*
ebony (*e*) 58, 215, 227, 240
edge fixation 87, 89, 91, 242
efference copy 100, 102, 146, 196
eigen-behavior 148
electroretinogram (ERG) 17, 20ff., 53, 56, 58, 226ff., 236ff.
electrostatic force 104
elementary movement detector (EMD) 37ff., 51, 59ff., 69ff., 87, 89, 122, 125, 127, 146, 152, 160, 162, 218, 230ff.
elf^{H37} see *extra lamina fiber*^{H37}
EMD see elementary movement detector
equator 16
ERG see electroretinogram
Eristalis 100, 181
error angle 116, 119
escape 49, 193, 210
E-vector 22, 24f., 56, 125, 194
evoced potential 20
exafference 100f., 179
excitatory input 163

extra lamina fiber^{H37} (*elf*^{H37}) 59, 227
eye movement 194, 235, 240
eye pigmentation 233

Fannia cannicularis 166, 180f., 238
fast phototaxis 55f.
feedback, mechanosensory 100
feedback loop 88, 100, 147, 153
feedback, negative 87, 172, 174, 195, 197, 199, 203
feedback, visual 4, 83, 89, 100ff., 105, 108, 139,
147, 154, 157f., 217
figure-ground algorithm 162
figure-ground discrimination 158ff., 233, 240
figure-ground experiment 165, 168ff., 174, 179
fine structure of yaw torque 95, 237
fixation 87, 91ff., 119, 131, 134f., 137, 140ff., 157, 160, 166, 171, 179, 183, 206f., 210, 213, 237, 239, 240, 242
fixation, edge 87, 89, 91, 242
fixation, forced 137, 139f., 145, 157, 179
fixation, voluntary 137, 139, 145f., 196
fixed action pattern 2, 49, 98, 179, 194, 221
flicker 112f., 116, 123, 154, 160, 239
flicker, local detector 122
flight, direction of 93, 97, 107, 116, 177, 182, 192
flight, force of 45, 91, 95, 192, 203
flight, forward 43, 45, 50, 176, 191
flight muscle 74, 76
flight speed 48, 234, 240
flight speed, control of 240
flight speed, over-ground control of 48
flight stability 195
flight torque, corrective 167
flight velocity 177
flow field 34, 50, 73, 192, 242
forced fixation 137, 139f., 145, 157, 179
force of flight 45, 91, 95, 192, 203
foreground-background experiment 152, 158f.
foreground 163f., 168, 172, 175f., 201f., 206
foreleg extension 50f.
forward flight 43, 45, 50, 176, 191
forward motion 146, 158
forward speed 146, 163
forward velocity 48, 107, 144, 176f., 180, 182, 231
fourth-order interaction 162
fovea 13, 180
free flight 82ff., 87, 89, 92, 94ff., 98, 104, 111, 114f., 119, 146, 154, 166, 179ff., 191, 230
free flight tracking 181
free flight saccade 96
free will 223
frequency of torque spikes 108, 134
front-to-back displacement 156, 219
front-to-back motion 45, 47, 50f., 60ff., 69, 73f., 77, 80, 116, 120ff., 127, 129f., 135, 148, 155ff., 166, 171, 176, 178, 187f., 204, 218f., 227ff.
front-to-back movement see front-to-back motion

gain control 162ff.
geometrical interference 39, 52, 87, 174
giant descending fiber 80
giant descending neuron 74, 76, 226
giant fibers of lobula plate 126, 129, 215, 227, 228
giant fiber system 70, 238, 241
giant neuron, horizontal 31
giant neuron, lobula plate 63, 67, 171, 237
giant vertical cells 237
Gomphocerus rufus 194
grand commissure intrinsic neuron 74
grasshopper 194, 203
gravitational field 45
gravitational vector 44
gravity 44f., 74, 146, 211, 231
groundspeed, control of 234
guidance of visual attention 190

habituating units 78, 80
habituation 77, 209, 227f., 234
habitation, cross- 80
habituation of landing response 77, 229, 234
haltere 82, 94, 97
HAS see high acuity system
^{3}H-deoxy-glucose 61ff., 69, 80, 233
headless flie 203
head movement 98, 100, 166, 238
head movement, saccadic 238
head posture 48
head saccade 98, 100, 166
heat 190
high acuity system (HAS) 58f.
high sensitivity system (HSS) 58f.
horizontal cell see HS-cell
horizontal giant neuron 31
horizontal system (HS) 63
housefly 181, 183, 235, 238, 242
hover flight 47, 71, 95, 107, 111, 146
hoverfly 48, 50, 94, 146, 166, 180f., 191, 233
HS-cell 63ff., 77, 126, 129, 234, 236
HS-neuron 70, 74, 128, 162
HSS see high sensitivity system
hyperpolarization 64

initiating activity 2 ff., 8, 104
initiating behavior 223
initiation 1 ff., 7, 50, 179, 182, 192, 220 ff.
inner stimuli 2, 221
input channels 38, 55 ff., 161 f.
input-output analysis 3
input-output process 33
input-output relation 3, 33, 182, 219
In(1)sc^{8}*sc*^{8}*w*^{a} 47 f.
intensity discrimination learning 204
intentionality 182
internal state of fly 179
interneuron 28
interommatidial angle 231
intracellular recording 22 f., 80
inverted closed loop 202
inverted coupling 174, 197 ff., 217, 229
inverted phasic torque response 202
inverted position function 133, 136

jet of air 190
jumping 76, 239
jump reflex 74
jump muscle 226
jump response 238

lamina 17, 21, 25 ff., 53, 56, 58 ff., 67, 209, 227, 229, 232, 240
lamina cartridge 27 ff., 56, 227
lamina ganglionaris 13, 20, 27, 241
lamina neuron 17, 28
lamina potential 17, 20, 58, 236
landing 7, 49 f., 76, 77, 242
landing behavior 49 f., 76 f.
landing maneuvre 73
landing reaction see landing response
landing response 51, 65, 74, 77 f., 218, 220, 227, 229, 232, 234 f., 242
landing response, dishabituation of 77
landing response, habituation of 77, 229, 234
landing response, sensitization of 78, 234
landing site 50, 176, 183, 217
landmark 7, 83 ff., 110, 125, 127, 144, 154, 170, 174, 182, 192, 216 f., 230 f.
large commissure 76
large field course control 37, 59, 65, 67, 125 ff., 139, 152, 165 f., 172, 174 ff., 195, 217 ff., 224
large field neuron 25 f., 162 f.
large field tangential neuron 61 ff., 73, 181
laser beam ablation 235
laser microbeam 69, 129
latency of object response 156
lateral inhibition 60, 61, 125, 242
leading fly 181, 213 ff.
leg extension 50 f., 76 ff.

Lepidoptera 61
lift 46, 49
lift control 46, 71, 74, 146
lift response 48, 65, 73
light-induced response 239
light intensity 49, 54, 59, 141
light intensity discrimination learning 229
linearly polarized light 230
lobula 25 f., 30 f., 61 f., 66 f., 74, 76, 79 f., 181, 227
lobula complex 32, 163, 218, 227
lobula plate 25 f., 30 f., 59 ff., 128, 162, 227, 234, 236, 237
lobula plate giant fiber 63 ff., 126, 129, 215, 227, 228
lobula plate giant neuron 63, 67, 171, 237
lobulaplate-less^{N684} (*lop*^{N684}) 65 ff., 227, 239
local flicker detector 122
local turbulence 104
locomotion recorder 34, 37, 40, 129, 133, 136, 141, 164, 215, 231
locomotor activity 37, 238
locust 152, 203
lop^{N684} see *lobulaplate-less*^{N684}
Lymantria dispar 48

mantids 166, 238
Mantis 240
mechanosensory feedback 100
mechanosensory fiber 76
medulla 25 ff., 53, 60 ff., 163, 218, 227, 233, 234
medulla cortex 234
menotaxis 111, 135, 137, 143, 147 ff., 165 ff., 174, 207
metaxanthopsin 20
microvillus 22 f., 56
min see *minibrain*
minibrain (*mnb*) 207, 209, 227, 236
mnb see *minibrain*
monocularly blinded fly 195
monopolar cell (L4) 60 f.
mood 221
motion after effect 120, 127, 131, 133, 136
motion-dependent course control 91
motion parallax 145 f., 176 f., 180
motion perception 34, 36, 38, 50, 241
motion sensitive interneuron 236
motion sensitivity 33, 49, 63, 218
motivational state 2, 182, 191, 195, 209, 215
motor program 2, 27, 73, 98, 104, 107, 219
movement, actively controlled 166
movement detection 36, 63, 87, 89, 148, 160, 162, 235, 239
movement detector 37, 60, 70, 91, 116, 123, 160 ff., 234

movement discrimination 235, 239
Musca 31 f., 50, 58, 61, 80, 82 f., 100 ff., 129, 131, 137, 140, 144, 158, 160 ff., 178 ff., 204, 209, 212, 218 f., 231 f., 236 ff.

natural coupling conditions 98
navigation system 235
negative feedback 87, 172, 174, 195, 197, 199, 203
negative phototaxis 211
neural crystal 25
neural superposition 13 f., 27 f., 53, 242
neurocrystallinity 69
neurogenetics 6 f., 218 f., 225
neuro-ommatidium 27
nina E 228
nof A^{S100} see *no object fixation* A^{S100}
nof D^{B11} see *no object fixation* D^{B11}
nof E^{B12} see *no object fixation* E^{B12}
noise 104 f, 137, 178
non A^{H2} see *no on-transient* A^{H2}
non C^{P37} see *no on-transient* C^{P37}
nonfixation 111
nonvisual course control mechanism 93
no object fixation A^{S100} (*nof* A^{S100}) 133 ff., 140, 143, 227
no object fixation D^{B11} (*nof* D^{B11}) 228
no object fixation E^{B12} (*nof* E^{B12}) 133, 136, 228
no on-transient A^{H2} (*non* A^{H2}) 124, 228
no on-transient C^{P37} (*non* C^{P37}) 59, 228
nystagmus 5, 107

object response 65, 119, 125 ff., 134, 137, 146, 154 ff., 165 ff., 182, 184, 186 f., 197, 202, 217 ff., 230 f.
object response, latency of 156
object response, steepness of 156
ocellus 23, 237 f.
odor 104, 190 f., 211
olfaction 7
olfactory conditioning 216
olfactory learning 227 ff.
olfactory signal 192
olfactory stimulus 234
omb^{H31} see *optomotor-blind*^{H31}
ommatidium 10, 13 f, 18, 22, 25, 27, 228 f., 239
ommochrome 19
open loop 3 f., 33, 43 f., 82 f., 85, 88, 91, 100, 107, 113, 119 f., 128, 132 f., 135, 146, 148, 150 ff., 169, 178 f., 184, 188, 191, 194, 197, 215, 217, 219, 220
open loop gain 34, 44, 116
operant conditioning 203, 223, 227
operant learning 203
optical axis 10 ff., 19, 21, 27, 82
optic anlage 69
optic foci 73
optic lobe 7, 25 ff., 32, 62, 67, 76, 197, 207, 224, 229, 234, 239, 241
optomotor balance 83, 87 ff., 99, 111, 134, 140, 146, 152 ff., 157, 166 ff., 175, 177, 179, 186, 191 ff., 206, 219
optomotor blindness 58
optomotor control of body posture 235
optomotor control of wing beat 235
optomotor equilibrium 43 f., 82, 91, 192, 235
optomotor pathway 61, 73, 165
optomotor pitch response 44, 47, 66, 227
optomotor response 5, 36, 38 ff., 51 ff.,56, 58, 60, 63 ff., 83, 88, 91, 95, 126, 131, 137, 171, 178, 204, 219, 228 f., 233, 236, 239, 241
optomotor roll response 44, 66, 227
optomotor torque response 45, 52, 59, 61, 64, 71, 91, 127 f., 228
optomotor yaw response 44, 66, 234, 236
optomotor yaw torque 34, 39, 41, 43, 48, 51 ff., 66, 69, 74, 76 f., 87, 129
optomotor-blind^{H31} (*omb*^{H31}) 65 ff., 77, 80, 126 ff., 139, 169 ff., 195, 206 f., 215, 228, 232, 237
ora^{JK84} see *outer rhabdomeres absent*^{JK84}
orientation 5, 33, 42, 44 f., 82, 206, 217
orientation behavior 110, 119, 133, 141, 221, 233, 239 ff.
orientation histogram 110, 112, 116, 131, 134, 139 ff., 167, 196, 199, 207, 209
orientedness 5, 6, 33, 82 f., 95, 110, 158, 174 f., 191, 203, 217, 221, 223 f.
oscillating mode 105, 108 f., 137, 139
outer rhabdomeres absent^{JK84} (*ora*^{JK84}) 21 f., 53 ff., 228
over-ground control of flight speed 48

panorama 84 ff., 91, 93 ff., 102, 107, 114, 135, 141, 146, 154 f., 166, 172, 188, 231
pattern preference 141, 210, 237, 242
PDA see prolonged depolarizing afterpotential
periodic grating 38 f., 69, 88, 91, 126, 129, 151, 174, 231
Phaenicia 60, 63, 234
phasic burst of torque 169
photon absorption 20
photopigment 241
photoreceptor 10, 13, 18, 87, 152, 218, 230, 232, 235 ff., 242
photoreceptor, sex-specific 236
phototactic behavior 55 f., 234, 242

phototaxis 23, 55, 58, 210ff., 227ff., 234, 237f., 242
phototaxis, fast 55f.
phototaxis, negativ 211
phylogenetic information 6, 73, 224
pigment cell 18
pigment granule 18f.
pigment migragtion 238
pitch 36, 45, 65ff.
pitch cell 70
pitch response 44f.
pitch, optomotor response 44, 47, 66, 227
pitch torque 34, 49
plasticity 7, 80, 131, 194
polarity of torque spike 107, 129, 187, 196, 203
polarization 24
polarization, degree of 230
polarization sensitivity 22f., 56, 58, 61, 236, 238, 242
polarized light 29, 56, 194, 230
pool neuron 163
position function 69, 114ff., 119ff., 129ff., 146, 165, 212, 215, 230
position function, inverted 133, 136
posterior slope 63, 66, 74
potential 115
potential profile 116
potential trough 119, 137, 139
Principle of Reafference 100, 102
prolonged depolarizing afterpotential (PDA) 17, 20ff., 53ff., 229, 231
proprioreceptor 98, 102
prosternal organ 48, 100
pseudo-fixation 137
pseudopupil 235
pupal period 66

quantum bumps 20
quick start behavior 194
quiet mode 105, 108, 140, 143, 146, 148

R1-6 17ff., 27ff., 53ff., 181, 227ff.
R7/8 16ff., 28f., 53, 55, 58, 141, 181, 228f.,236
rdg B^{KS222} see *retinal degeneration* B^{KS222}
reafference 100ff., 155, 157, 166, 179
receptive field 12, 60, 69, 78, 80, 87
receptor potential 20, 236
red eye color 18
releasing stimulus 2, 8, 104, 182, 220ff.
responsiveness 219
retina 20, 27, 180, 182, 232, 236, 241
retinal degeneration B^{KS222} (*rdg* B^{KS222}) 21f., 53ff., 229
retinal slip speed 86ff., 93, 168, 170ff., 231
retinal subsystems 52f., 56
retina potential 20, 58
retinotopic map 25f., 31, 69
retinotopic projection 64, 70
retinotopy 27
retinula cell 13f., 19ff., 27, 53ff., 141, 181, 227ff., 239
retinula cell axon 27
reversed coupling conditions 203
rhabdomere 13ff., 53, 55ff., 181, 228, 238
rhodopsin 20, 240, 242
rhythmical change 220
right-left inverting goggles 195
rising phase of torque spike 99
roll 36, 45, 65, 69, 71ff.
roll optomotor response 44, 66, 227
roll response 44
roll stabilization 237
roll torque 34, 45, 49
rotational bias 195
rotational inertia 98, 114
rotatory bias 84ff., 94, 116, 119, 156, 168, 170ff., 194, 230f.
rotatory course control 64, 70, 73
rotatory response 40, 43f., 54f., 59f., 65, 67, 164
rotatory self-motion 43, 192

S100 see *no object fixation* A^{S100}
saccade 95ff., 101, 104, 182, 220f.
saccade, body 95f., 100ff., 146, 180f., 227
saccade generator 102
saccade, head 98, 100, 166
saccade in free flight 96, 104
saccadic head movement 238
saccadic program 100
saccadic suppression 102
saccadic tracking 166ff., 174, 180, 182
scarlet 19, 74
screening pigment 18ff., 23, 48f.
search behavior 112, 178
searching maneuvre 105
selective attention 129, 182ff., 203, 222, 242
self-interaction 161
sensitization of landing response 78, 234
sensory hair 100
sev ^{LY3} see *sevenless* ^{LY3}
sevenless ^{LY3} (*sev* ^{LY3}) 21f., 53ff., 204, 215, 229, 233
sev ora 55
sex-specific photoreceptor 236
sex-specific tracking 218, 234
sexual behavior 238, 241
sexual dimorphism 181, 235
short-term learning 235
sideways flight 192

sideways force 192
sideways velocity 180, 181, 182
sine oculis[1] (*so*[1]) 79, 214, 229, 234
Skinner-box 223
slef motion 73
small angle fluctuations 147, 150ff.
small field columnar neuron 63, 79, 218
small field neuron 26, 76, 80
small optic lobes^{KS58} (*sol*^{KS58}) 67, 77, 80, 163, 165, 197, 199, 204ff., 229, 234
so[1] see *sine oculis*[1]
sol^{KS58} see *small optic lobes*^{KS58}
spatial frequency 40, 59f., 210
spatial resolution 27, 53
spatial sensitivity 38
spatial vision 242
spatial wavelength 38ff., 51f., 60, 93
spectral sensitivity 20ff., 29, 53, 236, 238, 242
spike detector 104
spontaneity 2
spontaneous activity 2
start behavior 194
start reflex 74, 76
stationary flight 34, 39, 42, 45ff., 84, 93, 96, 98, 101, 111, 139, 146, 153, 163f., 177, 180f., 184, 192, 210, 217, 227, 230f.
stationary object 113, 158, 176, 180, 182, 237
steepness of object response 156
stochastic process 2, 3, 113, 222
straight flight 102, 107
streaming velocity of air 146
styrofoam ball 34, 37, 231
subject-oriented walking 194
superposition 14, 53
superposition, neural 13f., 27f., 53, 242
suppression of torque spike 137
Syritta 48, 94, 180ff., 191ff., 205, 233

tan (*t*) 58, 229
temporal frequency 36
temporal resolution 25
Tenebrio 87, 242, 209
tergotrochanteral muscle 74, 76
thrust 45, 48f.
thrust control 46, 71, 74, 146
thrust response 48, 65, 73
topophobia 207
torque 94, 120, 123
torque base line 102, 108, 122, 123, 130, 134, 166, 175f., 187f.
torque, burst of 97, 102, 155, 169, 184, 200, 202
torque compensator 34f., 43, 85ff., 91ff., 98ff., 108, 110ff., 115, 119, 135, 137, 140f., 145ff., 154, 158f., 165, 176, 178ff., 195, 197, 204, 206, 209, 212, 217, 224
torque, dynamics of 104
torque fine structur 188
torque fluctuations 105, 108, 113, 115, 137, 145, 148, 152, 154, 178f., 219
torque meter 34, 36, 45, 51, 101, 104, 177, 182, 205, 209, 213, 217, 222, 227
torque mode 111
torque modulation 108, 162, 188
torque noise 137, 230
torque response, inverted phasic 202
torque response, optomotor 45, 52, 59, 61, 64, 71, 91, 127f., 228
torque spike 95ff., 111, 129ff., 134, 146, 166ff., 172, 175, 177ff., 187f., 195ff., 200ff., 220
torque spike, direction of 108
torque spike, polarity of 107, 129, 187, 196, 203
torque spike, rising phase of 99
torque spike, suppression of 137
torque spike frequency 107f., 134, 143, 169
torque spike rule 104, 107, 129, 130
tracking 163, 165, 166, 169, 176, 180, 181, 182, 213, 215, 233, 239, 240, 242
tracking, courtship 182, 213f., 228, 233f.
tracking, free flight 181
tracking, saccadic 238
tracking, sex-specific 218, 234
translatory bias 71
translatory optomotor responses 45
translatory self-motion 43, 192
translatory sideways motion 180
translatory velocity 107, 213
tread compensator 34, 120, 210
trial and error elimination 154, 203, 221f.
turning behavior 195, 229
turning command 98, 100
turning response 41ff., 52ff., 58ff., 64ff., 120, 125, 129, 163, 227f., 231
turning tendency 34, 129, 210, 226

upward motion 47f., 69, 73, 80, 230
UV-light 210f., 227ff.
UV-preverence 227ff.

ventral ganglion 27, 63, 73, 219
vermilion 18
vertical cell see VS-cell
vestigial 240
vibration 104
visual attention see selective attention
visual cortex 233
visual course control 33, 43f., 49, 52, 71, 73, 76, 87
visual element 11ff., 21ff., 32, 34, 38ff., 52ff., 58, 60, 69f., 73, 78, 80, 82, 87, 110, 125, 127, 147, 149, 152, 160, 218, 230, 231

visual feedback 4, 83, 89, 100ff., 105, 108, 139, 147, 154, 157f., 217
visual field 31ff., 39ff., 50ff., 61, 64ff., 69ff., 78ff., 85, 102, 110f., 115, 120, 123, 125, 127, 129ff., 137, 139, 145, 147, 152, 154, 156, 160, 165, 169, 171, 174ff., 196f., 200, 205f., 214, 221, 242
visual foci 26, 27
visual ganglia 24, 64, 76, 192, 219, 232
visual guidance 182, 195, 235
visual learning 229, 232, 234, 240
visual neuropil 29, 41, 69, 229
visual pigment 18f., 242
visual space 10ff., 52, 183
visual sphere 10, 19, 25ff., 34, 38, 224, 231
visual system 25, 34, 43, 74, 81, 93f., 100, 102, 109, 113, 154, 160, 191f., 204ff., 216, 218, 232ff.
visually naive fly 71, 204
visuo-motor coordination 7, 194ff., 200ff., 217, 222
Volucella 181
voluntariness 220, 223
voluntary behavior 1ff., 49, 104, 182, 220ff.
voluntary fixation 137, 139, 145f., 196
VS-cell 63ff, 77, 126, 129, 234
VS-cell, giant 237
VS-neuron 71, 74

w see *white*
w^a 48
walking, subject-oriented 194
walking fly 49, 54, 65, 136, 144, 163, 206, 209f., 227, 233, 237, 242
walking mode 129, 144
walking speed 47f.
wavelength discrimination 237
wavelength perception 233
white 18, 21f., 53ff., 107, 140ff., 215
white sevenless 141ff.
wind 74, 190, 192
wind tunnel 182, 234
wing beat 48f., 102, 111, 179, 221, 235
wing beat, optomotor control of 235
wing hitch 95
wings 82, 84, 91f., 95, 98, 135, 141, 143, 191, 194
w ora 21
w rdg B sev 21
w sev 21
w sev ora 21
WT (Berlin) 42, 45, 48, 54, 59, 65, 77f., 87, 121, 124, 126ff., 132, 135f., 139f., 167, 208f.
WT (Canton S) 78, 209
WT (Kapelle) 54, 89, 108, 208, 209
WT (Sevelen) 144

xanthopsin 20, 23, 228

yaw 36, 45, 65, 67, 71ff.
yaw, optomotor response 44, 66, 234, 236
yaw torque 34f., 41f., 47ff., 77, 83ff., 95, 98, 110, 113f., 120, 124ff., 134, 139, 154, 160, 171, 188, 197, 203, 218, 227, 230, 237, 242
yaw torque, fine structure of 95, 237
yaw torque, optomotor 34, 39, 41, 43, 48, 51ff., 66, 69, 74, 76f., 87, 129
yaw torque response 34, 39, 44f., 56, 65, 83, 88, 100, 115f., 122, 127, 150, 162, 178, 184, 231

zero adjustment 83
zero torque 44, 83, 85, 91, 130, 146, 153, 195

GPSR Compliance

The European Union's (EU) General Product Safety Regulation (GPSR) is a set of rules that requires consumer products to be safe and our obligations to ensure this.

If you have any concerns about our products, you can contact us on ProductSafety@springernature.com

In case Publisher is established outside the EU, the EU authorized representative is:

Springer Nature Customer Service Center GmbH
Europaplatz 3
69115 Heidelberg, Germany

Batch number: 10683525

Printed by Printforce, the Netherlands